PHYSICAL PRINCIPLES OF QUANTUM BIOLOGY

PHYSICAL PRINCIPLES OF QUANTUM BIOLOGY

Nathan S Babcock
Brandy N Babcock

World Scientific

NEW JERSEY • LONDON • SINGAPORE • BEIJING • SHANGHAI • HONG KONG • TAIPEI • CHENNAI • TOKYO

Published by

World Scientific Publishing Co. Pte. Ltd.
5 Toh Tuck Link, Singapore 596224
USA office: 27 Warren Street, Suite 401-402, Hackensack, NJ 07601
UK office: 57 Shelton Street, Covent Garden, London WC2H 9HE

Library of Congress Control Number: 2026944612

British Library Cataloguing-in-Publication Data
A catalogue record for this book is available from the British Library.

PHYSICAL PRINCIPLES OF QUANTUM BIOLOGY

ISBN 978-981-98-3141-8 (hardcover)
ISBN 978-981-98-3142-5 (ebook for institutions)
ISBN 978-981-98-3143-2 (ebook for individuals)

For any available supplementary material, please visit
https://www.worldscientific.com/worldscibooks/10.1142/14827#t=suppl

Desk Editor: Muhammad Ihsan Putra

Typeset by Stallion Press
Email: enquiries@stallionpress.com

For
Faithful Enthusiasts
of
Quantum Biology

About the Authors

Dr. Nathan S. Babcock is a distinguished expert in the quantum sciences, with over two decades of experience primarily concentrated in the domain of quantum biology. His academic journey began at two prominent Canadian quantum research institutions: the University of Waterloo in Ontario and the University of Calgary in Alberta. After completing his Ph.D. in Physics, Dr. Babcock further advanced his understanding of the intricate relationship between quantum mechanics and biological processes through postdoctoral research in structural biology at Simon Fraser University in British Columbia. His research then led him to explore spin chemistry, conducting groundbreaking studies on radical electron pair models of avian magnetoreception at the Living Systems Institute at the University of Exeter in the U.K. Dr. Babcock continued to refine his expertise in open quantum systems by investigating the quantum mechanical phenomenon of superradiance in microtubules at the Quantum Biology Laboratory at Howard University in Washington, D.C., where his innovative research on quantum effects in microtubules garnered significant attention worldwide. In addition to research, writing, and lecturing, he offers quantum biology consulting for corporations and academic institutions. His unwavering determination, comprehensive knowledge of the subject, and infectious enthusiasm for the pursuit of physics has given this book its fundamental core. Through his contributions, he aims to ignite curiosity and foster a deeper understanding of the interplay between quantum physics and the living world. You can find out more about his work at www.nathanbabcockphd.com.

Brandy N. Babcock is an enthusiastic advocate for the effective and faithful communication of the natural sciences. Her journey began with the initial pursuit of Astrophysics and Theoretical Science at the University of Pittsburgh. Although she shifted her focus to health and wellness, where she thrived as a small business owner and coach for over 15 years, her passion for science never waned. In addition to learning therapeutic harp, during this time Brandy honed her skills in small-scale project management, developed expertise in educational course creation and print media including publishing and contributing to multiple books, and gained extensive experience coaching others to achieve their personal and professional aspirations. Immersion in the holistic health field made her acutely aware of the prevalence of pseudoscience and the need to properly express the implications of quantum physics for health and medicine with rigor and clarity. Alongside her husband Nathan, she is deeply passionate about educating the public on the true nature and significance of quantum biology, ensuring that accurate and reliable information prevails in the discourse surrounding this emerging field. Brandy's unique blend of visionary thinking, organizational prowess, talent for coaching, and innate ability to communicate complex scientific concepts have made her an invaluable co-author. Her insights and dedication were paramount in composing this work from inception to completion.

Disclaimer

This book is the result of extensive research and collaboration between the authors, and so it is important to clarify the role of artificial intelligence (AI) in its creation. Generative AI was not used in any capacity to write the main content of this book. Nathan utilized AI-enhanced search engines critically in his research, which ironically often asked him to confirm that *he* was not a robot. Brandy employed AI assistance for the development of supplementary materials including the Sample Syllabus and the initial Glossary of Terms (prior to Nathan's revisions).

Preface

Nathan's interest in quantum biology dates back to 2004, when he was studying quantum computing as an undergraduate in Waterloo, Canada. He wrote a short essay on the impact of quantum information science, extrapolating Moore's law of electronics and finding that the scale of transistors would soon reach the size of molecules such as proteins and DNA. He concluded that for computers to reach the scale of molecules, they would soon need to be made out of biological materials which are also cheap, plentiful, durable, efficient, fast, and functional at room temperature. That interest in quantum biology led to an opportunity to attend the *First Workshop on Quantum Technology in Biological Systems* at the National University of Singapore in January 2009. After focusing the ensuing years of his career on this exciting discipline, he spoke at the first *Gordon Research Conference on Emerging Methodologies to Investigate Quantum Effects in Biology* held in Galveston, Texas in March 2023. This monograph was especially motivated by discussions at the event, where graduate students remarked that the field lacked cohesion as a whole. This presented challenges to those wanting to enter into research in quantum biology. It was clear to the authors that quantum biology must be unified by principles of quantum mechanics first and foremost, and these principles must be shared.

A diverse body of literature has emerged debating the importance of quantum mechanics to biology and the question of which quantum effects enable biological function. This lively debate has stimulated a great deal of scientific activity, both experimental and theoretical, by intensifying interest in individual topics as well as in the field as a whole. As stimulating as it has been, this productivity has come at the cost of clarity — presenting quantum biology as a patchwork of disjoint topics and piecemeal definitions, rather than a definite field of inquiry motivated by a unified theoretical framework and a consistent body of rigorously-established experimental results. It is our hope to address this by providing a comprehensive compendium of quantum effects in biology, presented from a fundamental quantum physical perspective.

It was important that this monograph and the citations therein be accessible to those entering into the field, therefore it does not include any complex mathematics, numerical examples, or physical derivations to obscure the essential presentation of the concepts. In this sense, it is written colloquially as an essential primer with a corresponding syllabus included at the back of the book, and a series of online lectures presented by the first author on his YouTube channel https://www.youtube.com/@drbabcock to provide an expository overview of the essential material. A separate, forthcoming book is being created to present an advanced mathematical formulation of quantum biology for those interested in seriously pursuing the field.

Nathan is grateful to the many colleagues and professors who have inspired, encouraged, supported, and challenged his work on quantum biology over the last two decades including Betony Adams, Margaret Ahmad, Clarice Aiello, Janet Anders, Salil Bedkihal, David Beratan, Doug Brash, Hans Briegel, Luca Celardo, Jeremy Choquette, Majed Chergui, Nina Copping, Victor Davidson, Paul Davies, Aurélien de la Lande, Art Du Rea, David Feder, James Freericks, Gilad Gour, Geoffrey Guy, Stuart Hameroff, Lucien Hardy, Stefan Idziak, Daniel Kattnig, Stu Kauffman, Robert Keens, Youngchan Kim, Philip Kurian, Peter Kusalik, Benjamin Lavoie, Tony Leggett, Elliot Martin, Rudy Marcus, Johnjoe McFadden, Alisdair McKenzie, James Murray, Sergei Noskov, Alistair Nunn, Hartwig Peemoeller, Drew Ringsmuth, Chris Rourk, Dennis Salahub, Chris Sampson, Barry Sanders, Marlan Scully, Christoph Simon, David Sivak, Michael Skotiniotis, Sharon Strawbridge, Robert Thompson, Luca Turin, Vlatko Vedral, Nathan Wiebe, and to the many others who have kept this work in their thoughts and prayers.

Contents

About the Authors vii

Disclaimer ix

Preface xi

List of Abbreviations xvii

Introduction 1

Part I Quantum Physics of the Living World 3

Chapter 1 Quantum Theory and the New Observables 5

Chapter 2 Quantum Electrodynamics: Lighting Up Life 13

Chapter 3 Definitions of Non-Triviality for Quantum Biology 23

Chapter 4 Photosynthesis & Open Quantum System Dynamics 31

Chapter 5 Light Receptors, Spin Chemistry, & Cryptochrome 39

Chapter 6 Dynamic Control of DNA Repair in Photolyase 51

Chapter 7 Enzyme Catalysis: Quantum Fundamentals 59

Part II Coherent Quantum Effects in Biology 71

Chapter 8 Ultraweak Photon Emission & Cell Processes 73

Chapter 9 Electromagnetic Oscillations in Biostructures 83

Chapter 10 Functional Chemical Dynamics in Living Cells 91

Chapter 11 Magnetic Biomodulation: Biodynamic Control 103

Chapter 12 Molecular Fluctuations: Solvent Effects & Dispersion 111

Chapter 13 Multiscale Modeling of Biomolecular Systems 121

Chapter 14 Quantum Correlations in Biological Cofactors 129

Part III Nanomedicine & Biotechnology 139

Chapter 15 Photobiomodulation & Electromagnetic Therapies 141

Chapter 16 Photodynamic Therapy & Nanotheranostics 151

Chapter 17 Regenerative Processes: Cells, Tissues, & Organs 161

Chapter 18 Morphogenetic Integration & Immunodynamics 171

Chapter 19 Quantum Biotechnology: Universal Applications 181

Chapter 20 Quantum Biology: Essential Further Research 189

Chapter 21 Conclusion: A Quantum Framework 197

Sample Syllabus 201

Glossary of Terms 245

References 257

List of Abbreviations

3D three-dimensional
ADP adenosine diphosphate
AlkB alkylation B
AMP adenosine monophosphate
ARDS acute respiratory distress syndrome
ATP adenosine triphosphate
CALI chromophore-assisted light-inactivation
CASSCF complete active space self-consistent field
CDT chemodynamic therapy
CI configuration interaction
CICR calcium-induced calcium release
CISS chirality-induced spin selectivity
COPD chronic obstructive pulmonary disease
cQED cavity quantum electrodynamics
CT-QMC continuous-time quantum Monte Carlo
DFT density functional theory
DMFT dynamical mean-field theory
DMRG density matrix renormalization group
DNA deoxyribonucleic acid
ECM extracellular matrix
EDT electrodynamic therapy
ELF extremely low frequency
EPR electron paramagnetic resonance
ET electron transfer
ETC electron transport chain
ETF electron transfer flavoprotein

EZ exclusion zone
FAD flavin adenine dinucleotide
FDA Food and Drug Administration
FeS iron-sulfur
FMO Fenna-Matthews-Olson
GDP guanosine 5′-diphosphate
GGA generalized gradient approximation
GMF geomagnetic field
GTP guanosine 5′-triphosphate
HER2 human epidermal growth factor receptor 2
HF Hartree-Fock
HIF hypoxia inducible factor
HMF hypomagnetic field
HPD hematoporphyrin derivative
ICH intracerebral hemorrhage
IR infrared
ISC intersystem crossing
LDA local density approximation
LED light-emitting diode
LHCII light-harvesting complex II
LHN light-harvesting nanotube
LLLT low level light therapy
MFEs magnetic field effects
ML machine learning
MM molecular mechanical
MO molecular orbital
MR multi-reference
MSC mesenchymal stromal cell
NADH nicotinamide adenine dinucleotide
NADPH nicotinamide adenine dinucleotide phosphate
NIR near-infrared
NISQ noisy intermediate-scale quantum
NMR nuclear magnetic resonance
NO nitric oxide
NP nondeterministic polynomial-time
OXPHOS oxidative phosphorylation

PBM photobiomodulation
PCET proton-coupled electron transfer
PDT photodynamic therapy
PEG polyethylene glycol
PEMF pulsed electromagnetic field
PSII photosystem II
PTT photothermal therapy
QA quantum annealing
QED quantum electrodynamics
QM quantum mechanical
QM/MM quantum mechanics / molecular mechanics
RDT radiodynamic therapy
RIRR ROS-induced ROS release
RNA ribonucleic acid
ROS reactive oxygen species
RPM radical pair mechanism
SCF self-consistent field
SDT sonodynamic therapy
SR stochastic resonance
STM scanning tunneling microscopy
THz terahertz
TRIM tissue resonance interaction method
Trp tryptophan
TTQ tryptophan tryptophylquinone
UPE ultraweak photon emissions
UV ultraviolet
UVR8 ultraviolet resistance locus 8
WMF weak magnetic field

Introduction

As a quantum physicist, my interest in quantum biology was motivated by the hope that quantum mechanics could bring insight to biology. I was convinced that quantum theory would reveal the nature of life itself. In time, I began to see that the reverse was also true, that life is in turn revealing the nature of quantum physics.

Futurists have touted the promises of quantum mechanics, heralding it as the fate of science and technology. Quantum computing is hyped as the destiny of data processing, quantum cryptography is branded as the ultimate unbreakable zenith of communications security, and quantum information is identified as the fabric of reality itself — a modern *æther*.

Lifeless, cold, isolated systems are widely considered to define the essence of quantum theory. Meanwhile we overlook the open-ended aspects of our universe and our native biology, most notably the essential features of life: growth and regeneration, healing and development, acumen and instinct, optimization and synthesis. These are the biological characteristics that hold the spark to illuminating many of humanity's greatest questions.

Quantum biology challenges us to renew our understanding of the foundations of quantum mechanics. This return to established postulates presents the opportunity to formulate a framework for the foundations of quantum biology — a framework that identifies life as the exceptional feature of quantum physics. Rather than inquiring how isolated quantum systems operate independently from the environment, quantum biology asks the quintessential question, *How does life ultimately work?*

Part I

Quantum Physics of the Living World

"One can best feel in dealing with living things how primitive physics still is."

— Albert Einstein, quoted in Refs. [1,2]

Chapter 1

Quantum Theory and the New Observables

To establish quantum biology in its proper context within the history of physics, we are recalled to an evening lecture at the Royal Institution in London on the 27th of April 1900, when Lord Kelvin lamented that two nineteenth century clouds hung over the horizon of science [3]. The scientific observations at the time were incompatible with predictions of the leading theories, presenting a crisis for the dynamical theory of heat and light:

> "The beauty and clearness of the dynamical theory, which asserts heat and light to be modes of motion, is at present obscured by two clouds. I. The first came into existence with the undulatory theory of light, and . . . it involved the question, How could the earth move through an elastic solid, such as essentially is the luminiferous ether? II. The second is the Maxwell-Boltzmann doctrine regarding the partition of energy"

Kelvin's first metaphorical cloud ('Cloud I') was the motion of material bodies with respect to the ether [4]. At that time, a luminiferous ether was believed to enable propagation of electromagnetic waves, such as light or radio waves (analogous to water waves traveling on the surface of the ocean). This ether was believed to exist everywhere throughout all space, leaving the nagging question of how material bodies moved through it without displacing it. Experiments by Michelson and Morley to measure the Earth's movement through the ether had failed, resigning Kelvin glibly to describe Cloud I as "very dense."

'Cloud II' covered the kinetic theory of the time, which could not explain the relationship between temperature and heat. Maxwell and Boltzmann had proposed that, all other things being equal, the energy of a thoroughly mixed substance would be stored equally among its microscopic degrees of freedom. This "equipartition" principle worked well at predicting the amount of heat required to increase the temperature of many materials at high temperatures, but failed scandalously at low temperatures. Moreover, Kelvin felt that Maxwell and Boltzmann's proof of the principle was inconclusive, prompting him to deny the Maxwell–Boltzmann equipartition principle entirely.

The clouds darkening the 19th century theory of heat and light were dispelled with the advent of modern physics. The density of Cloud I was penetrated by the vision of Einstein, who proposed the principle of relativity that requires the speed of light to appear constant to all observers, regardless of their reference frames. As such, the concept of a three-dimensional ether was overtaken by that of four-dimensional spacetime [5]. Spacetime remains the term used today [6], while the idea of a "relativistic ether" [7, 8] has become taboo among physicists [9].

If that were not enough, Einstein dispelled Cloud II by imposing the so-called quantization rules on the molecular motions of atoms in solids [10]. Those motions (i.e., degrees of freedom) become frozen at low temperatures [11, 12]. On the other hand, the spacing between the quantum energy levels becomes negligible at higher temperatures where the classical equipartition theorem was found to work well. Thus, energy quantization rationalized the failure of energy equipartition at low temperatures and in systems where conventional classical assumptions tend to fail.

No believer in the classical theory of heat and light could have foreseen, let alone have accepted, Einstein's solutions to the problems posed by Kelvin's Clouds. Even the pioneers who developed Einstein's work — Fitzgerald and Lorentz on special relativity and Planck on quantization — initially saw their own works as mathematical heuristics without fully appreciating their deep physical implications. The profundity of those implications was, however, appreciated by the chemist Nernst who called an emergency meeting of Europe's most eminent physicists to confront the emerging crisis in the first of the famous Solvay conferences. Nernst's invitation reads [13]:

> "According to all appearances, we are now in the midst of a new development of the principles on which the classical kinetic-molecular theory was based. The systematic development of this theory leads on the one hand, to a radiation formula that disagrees with all experimental results; from this same theory are deduced, on the other hand, assertions on the subject of specific heats ... that are likewise refuted by many measurements. It has been shown, especially by Planck and Einstein, that these contradictions disappear if one sets certain limits on the motions of electrons and atoms oscillating about an equilibrium position (the principle of energy quanta); but this interpretation in turn departs so much from the equations of motion used up to now that its acceptance would necessarily and indisputably entail a vast reform of our current fundamental theories."

That first conference in 1911 focused on the disconnect between classical physics and the new quantum theory — a problem which remains a thorn in the side of fundamental physics today. Over the course of the ensuing Solvay Conference series, the emerging quantum revolution stunned its discoverers foremost because it did away with traditional notions of what observable quantities represented in science — if they represented anything at all [14].

Fully embracing the departure from classical concepts, Heisenberg formulated an abstract theory that admitted only observable quantities by discarding all empirical connections to unobserved quantities [15, 16]. Dirac recognized Heisenberg's essential departure from the old concepts and descriptions of physical reality and formalized it into a new system of algebra [17]. Unlike conventional algebra, Dirac's operator algebra was found to be equivalent to an advanced form of matrix algebra that had been developed by Born and Jordan [18], which was later discovered to reproduce the wave theory developed by Schrödinger [19, 20]. So, it was that quantum mechanics was born.

Kelvin himself had previously proclaimed that one's knowledge is meagre, unsatisfactory, and unscientific unless one can express it in numbers [21], heralding the need for a theory to numerate observable and unobservable quantities alike. Wildly successful as a measurement calculus [22, 23], quantum mechanics' failure as an objective theory of reality is well known [24–26]. Despite its enormous success, quantum mechanics is increasingly recognized as a source of consternation, rather than an authoritative guide that definitively clarifies the nature of reality [27]. This failure was felt early in the effort to unify quantum mechanics and relativity, as Dirac observed [16]:

> "To have a description of Nature is philosophically satisfying, though not logically necessary, and it is somewhat strange that the attempt to get such a description should meet with a partial success, namely, in the non-relativistic domain, but yet should fail completely in the later development. It seems to suggest that the present mathematical methods are not final. Any improvement in them would have to be of a very drastic character..."

The fundamental incompleteness of quantum theory in the relativistic limit was eventually addressed by the advent of quantum field theory in an overhaul of quantum mechanics that brought problems of its own [28, 29]. Those problems are exacerbated by the fact that quantum mechanics are based on a set of abstract mathematical axioms that are postulated *ad hoc* [30] rather than inferred directly from fundamental principles of Nature. Unlike Newton's laws (which govern classical physics) and Einstein's principles of relativity (such as the finite speed of light), there are no universal laws of quantum theory to preclude discoveries of whole new phenomena [31–34].[1] Heisenberg's original text on *The Physical Principles of the Quantum Theory* [36] is now rarely taught or cited.

Perhaps the closest thing that quantum mechanics has to a physical law is an arcane mathematical property known as 'Hermitian symmetry' [20, 37] which preserves all relevant physical characteristics of a quantum system by requiring it to be bounded and closed

[1]To the contrary, even though quantum theory was derived by Heisenberg from the physical principle that the theory should not contain "relationships between quantities that are ... unobservable in principle" [35], this point is rarely, if ever, taught to students as the founding assumption of quantum mechanics or included in standard textbook treatments on the subject.

in the sense of being completely isolated from its surroundings. Although this symmetry is widely considered fundamental to quantum theory [38, 39], it is rarely satisfied for long in systems that are open to the environment. This demanding condition leaves an enormous gap between the limits of conventional microscopic quantum mechanics and the everyday macroscopic phenomena of the classical world [40].

The gap separating the quantum realm from the scale of the classical world is usually so considerable as to leave no ambiguity between the scale of quantum mechanical particles and that of everyday classical objects. There are exceptions to the rule, some as simple as the ordinary bar magnet which holds its magnetic field by virtue of the quantum "spin" interactions hidden inside [41]. Yet the quantum-classical divide is bridged nowhere more vividly than in living systems — from microtubules and mitochondria to night-migrating songbirds and entire ecosystems — which rely on inherently quantum-dynamical effects to survive. To that effect, Schrödinger argued that the physics of his time was unable to account for living processes, posing the question of how living organisms made of atoms could overcome the tendency for all matter to degrade from order into disorder in his 1943 public lecture series, *What is Life?* [42]. He proposed that classical physics would be unable to explain molecular biology even in principle, contending that a quantum theory of biology would lead to the discovery of new physical laws simply because the construction of biological systems is so different from anything tested previously in the laboratory. In his own emphatic words [42], "The classical physicist's expectation, far from being trivial, is wrong."

By the 1960s, the most vital questions in quantum biology were already well known, although it still was not clear how to relate the abstract mathematics of quantum mechanics directly to its role in biology [43]. That subtle issue reflected long-standing concerns about the deficiencies of conventional quantum theory, dating back to the first Solvay Conference, and in turn recognized by Einstein [44], Schrödinger [45], Dirac [16], and later Deutsch [46], Bell [47],

Leggett [48], Gell-Mann [49], and Weinberg [50]. The essence of the issue is that quantum mechanics only describes closed quantum systems which are theoretically observed by classical measuring devices [51, 52].

Ironically, "orthodox" quantum theory implies that the principles of quantum systems do not apply to measuring devices, which are intrinsically governed by classical probabilities [47]. The fundamental distinction between quantum systems and classical measuring devices results from essential differences between open and closed systems, insofar as classical probabilities do not apply to closed systems [53]. This raises questions about whether quantum theory is either consistent [54] or exact [55] in a debate which dates back to original disagreements between Dirac and Heisenberg as to whether quantum mechanics as a theory is "open" or "closed" [56].

That apparent paradox presents no problem to conventional quantum mechanics, where there is a definite — if informal — divide separating the observer and the observed (known as the "Heisenberg cut" [57, 58]). For a quantum system that is open to its environment, the problem is conveniently addressed by re-envisioning it as a closed quantum system that is observed by its classical environment. The convention begins to break down, however, for quantum systems that are open to interact with themselves through the back-action of the surrounding quantum continuum [59]. Heralded by Dirac, the failure of quantum theory to account for that "self-interaction" has long posed an obstacle to forming a complete and coherent interpretation of quantum theory [60, 61]. In the same vein, feedback through self-measurement and self-organization are hallmarks of biological systems [62–65], posing a fundamental challenge to quantum theory to provide a consistent description of self-referential systems [66, 67].

In this respect, biology presents fertile ground to investigate a new class of physical systems with properties quite literally at the boundaries of conventional quantum mechanics. In addition to established examples of quantum physics in biology such as the electronic properties of liquid water [68, 69], biological electron transfer [70, 71], nuclear quantum phenomena [72], enzyme-mediated reactions [73], collective spontaneous emission [74], biological (protein-protein) recognition

[75], and the ultrafast (fs) dynamics of countless photochemical and phonon-mediated biological processes [71, 76–78], there is growing interest in the role of an expanding number of "exotic" non-Hermitian quantum effects that are involved in photosynthesis [79–81], coherent exciton transfer [74, 82], and photobiomodulation [83].

In an echo of the first quantum revolution, the conventional quantum mechanics of last century are now being extended beyond the domain of isolated atoms and molecules to explore a range of open quantum systems [84, 85]. In general, the theory of open quantum systems describes the interaction of a quantum mechanical entity with its external quantum environment, sometimes characterized as a reservoir or bath [86], often with loss or gain [80, 87]. Invariably, living systems rely on the environment to procure energy and expel waste, ensuring that biological order develops robustly [88]. Recently, advances in open quantum systems have inspired renewed interest in foundational questions in relativity and quantum theory alike [89]. As such, quantum biology affords an ideal test bed for studies of open quantum systems and presents fresh opportunities to break through the cloud cover that still obscures a complete view of the fundamental principles of quantum theory. This view, in turn, promises to illuminate a growing list of rich physical effects with no counterparts in conventional quantum theory [90, 91].

Quantum theory is indispensable to the description of many biological processes, including light-harvesting and vibronic charge-separation processes that occur during photosynthesis [92–94], light-sensitive chemical reactions responsible for vision [95–97], clinical light-based therapies [98], biological pigmentation [99,100], ergosterol (vitamin D) production [101, 102], biological hydrogen tunneling [103–105], and a variety of bioluminescent effects such as the chemiluminescent mechanisms involved in firefly flashing [106–109]. There is no greater opportunity in physics today than to explore the emerging field of quantum biology, where we encounter the rich open system dynamics of gain and loss, life and death, growth and decay. Likewise, quantum biology poses unprecedented opportunities to observe, characterize, and ultimately understand many vital noise-driven and symmetry-breaking effects that are distinctive of living systems, such as reproduction, healing, sickness, aging and death. A radical new quantum theory of the 21st century — and the living world — promises to uncover new and extraordinary phenomena as life reveals more about quantum physics than ever before.

Chapter 2

Quantum Electrodynamics: Lighting Up Life

Sunlight provides the energy for life on earth through photosynthesis, yet the relationship between light and life is even more fundamental: the quantum unit of light (or "photon") defines *the* quantity of electron excitation as a bound electron is excited into motion by absorbing a photon of light [110]. In this sense, light is conceptually indispensable from electronic energy [111]. It is not enough to claim that light powers life. Light is inextricable from the electronic energy driving the life force, fixed by the ratio of electronic energy E to the light frequency f. The need to fix E/f to the constant h (known as Planck's constant) is *the* "quantum" of quantum physics [112]. This concept was so stunning to the founders of quantum theory that Planck first thought it to be an artifact or a mathematical "trick" [113]. It took the genius of Einstein to realize that it was no trick, and Planck's formula $E = hf$ was there to stay. This had far reaching consequences, expressed famously by Heisenberg in the "uncertainty principle," which marked the departure from deterministic classical physics to the quantum realm of inherent uncertainty [114, 115].

The quantum nature of the electron's motion was observed by German physicist Heinrich Hertz in 1887 when he first noticed that the ultraviolet (UV) light from a spark produced by an electrified surface (i.e., an electrode) could induce a second spark from another metal surface, unlike light produced by other lower frequency sources [116]. Thus, the "quantum leap" was witnessed (in the observation of a first spark igniting a distant second one) but was not explained until Einstein subsequently published his theory for it in 1905 [117].

Today, this "photoelectric effect" provides an elementary example of quantum theory that continues to be demonstrated in classrooms and laboratories everywhere. The quantum theory of absorption and emission of radiation was later formalized by Dirac [118], before giving rise to the principles of quantum electrodynamics [119]. According to quantum mechanics, a quantum particle such as an electron can pass through a potential energy barrier that would be impassable according to classical physics. Based on this quantum theory of "tunneling" [120], the wave-like particle can slip through a barrier even when it does not have enough energy to overcome it [121]. This is unlike the case in classical mechanics, where a particle must have enough kinetic energy to overcome an intervening barrier in order to pass across it.

Quantum electrodynamics enter biology during electron transfers between proteins [122] where electronic transitions deliver the electric force *via* the quantum "jumps" [71, 123, 124]. In other words, biological electron transfer provides the current of life, driving cell respiration as electrons tunnel quantum mechanically between the proteins that power the metabolism of all known organisms [125], ranging from cells to whole ecosystems [126, 127]. Interest in electron transfer reactions blossomed in the 1960s when those reactions were found to present much greater difficulty and complexity than previously imagined [128]. DeVault and Chance soon realized that quantum tunneling was the simplest way to explain the rate of activationless electron transfer in cytochrome proteins [129]. Although classical physics requires that an object must have sufficient kinetic energy to penetrate (or overcome) a potential barrier, quantum uncertainty allows an electron to spontaneously "tunnel" through an insulating barrier if the tunneling time is short enough [130].

Quantum tunneling is currently the accepted theory of electron transport during biological energy transduction [131]. During decoherent tunneling (sometimes referred to as incoherent tunneling [132]), the long-range coherence of the tunneling particle is destroyed after each quantum jump. Nevertheless, even decoherent tunneling is, strictly speaking, a coherent quantum effect with no counterpart in classical physics [133]. On the other hand, resonant tunneling is widely described as "coherent" because it preserves the quantum mechanical phase of the electron after tunneling [132]. Although the question of whether living systems can exploit resonant tunneling has intrigued scientists for decades, the prevailing view is that most biological electron transfer steps take place over long distances (one to two nanometers) in a series of quantum jumps that occur by decoherent quantum tunneling [131, 134]

Quantum tunneling during respiration is just one example of the crucial role that quantum coherence and entanglement play in enabling the existence of life itself [71]. Coherence (i.e., collective synchronization) is found ubiquitously in living cells [135] where quantum coherence is essential to allow long-range tunneling during electron transfer [136]. It is exactly the coherence of the electronic wave function — delocalized in space — that allows it to tunnel from one site of the electron transfer chain to the next. In other words, if the quantum coherence of the tunneling electron were completely destroyed then no tunneling would occur. This long-range coherence creates an electronic superposition state distributed across two distant locations, constituting an established form of quantum entanglement [137] — another hallmark of quantum mechanics [138].

Respiratory enzymes optimize intervening media to enhance inter-protein electron tunneling dynamics [139], and quantum interferences between optimized tunneling pathways have been identified in biological and biomimetic electron transfer systems [140]. When respiratory enzyme cytochrome *c* [141] binds with its redox partner, the binding mechanism modulates its electron transfer rate by modifying the transfer pathway [142]. Its partner enzyme, cytochrome *c* oxidase (a.k.a. complex IV) uses redox state-dependent solvent organization to control a key proton transfer step. Advances in time-resolved laser spectroscopy have opened the possibility of observing electron tunneling in real time [143]. Ultrafast measurements of electron transfer dynamics in the electron carrier protein flavodoxin have revealed fundamental mechanisms of non-equilibrium dynamics that may be critical to enabling biological functions (including DNA repair) [144], making electron transfer a core topic of biophysical research.

Electron transfer through biomolecules can be surprisingly efficient [145], and experiments have revealed that long-range coherent charge transport is feasible in the chiral helix of deoxyribonucleic acid (DNA) [146]. Vibronic (nuclear-electronic) coupling has arisen as an essential factor determining the efficiency of charge transfer in biological and artificial systems alike [147]. Model simulations have shown how control of inelastic electron tunneling can be used to gate long-range electron transfer using coupling between interfering tunneling pathways [148]. Protein junction experiments have revealed that the shape of the respiratory enzyme azurin can switch conformations along with the type of electron transfer through it [149]. Dynamical methods have also been used to switch between tunneling-mediated and solvent-controlled electron transfer in cytochrome *c* [150]. Resonant tunneling experiments performed on the respiratory protein azurin revealed a role for azurin's Cu(II) metal cofactor in mediating electron transfer, demonstrating coherent quantum mechanical electron transport through biological molecules [151].

Electron transfer reactions were first characterized by Marcus who received the Nobel Prize in Chemistry in 1992 for that work [152, 153]. Biological electron tunneling has been established "indisputably" [122, 154], with a growing number of biologically-inspired technological applications [155]. The electronic excitation that accompanies light absorption during photosynthesis is just one of numerous examples of light harvesting and sensing in biology where light absorption is governed by the principles of quantum electrodynamics [156]. Photosynthesis has also attracted attention for the role that quantum mechanics plays in the efficiency of its energy conversion. Nearly every photon absorbed during photosynthesis generates an excited electron to perform work [157], garnering attention from physicists interested in rationalizing photosynthesis' extraordinary product yields ($> 95\%$ [158, 159]).

Marcus' theory is crucial to understand the very high quantum efficiency of photosynthetic organisms [160]. The predictions of Marcus electron transfer theory differ dramatically from conventional transition state theories of reaction rates, qualitatively and quantitatively [152]. Notably, Marcus' theory predicts the "inverted" effect of an electron transfer reaction rate that decreases with increasing Gibbs free energy of product formation. This is in contrast with conventional reaction rate equations which predict only an increasing reaction rate with an increasing driving force. Marcus' inverted effect arises in the rate equation for a weakly-coupled chemical reaction [161] — derived as a consequence of Dirac's formulation of quantum electrodynamics [162] — where time-dependent perturbation theory is applied to obtain the first-order reaction rate (i.e., using Fermi's "Golden Rule #2").

Electron transfer dynamics are crucial to the efficiency of energy transduction processes in life, which are believed to control the rate limiting step of the metabolic quinol cycle (Q cycle) [163]. The Q cycle is indispensable to the process of aerobic respiration, wherein molecular oxygen (O_2) is reduced to form water (H_2O) in the production of adenosine triphosphate (ATP), the universal energy currency of life [164]. In fact, it has been established that proton uptake controls electron transfer in the enzyme cytochrome *c* oxidase [165], which catalyzes the energy-generating reaction step converting oxygen and hydrogen into water. In addition to the four protons that are used to convert molecular oxygen into water, cytochrome *c* oxidase transports another four protons across the respiratory membrane, increasing the acidity (pH) imbalance across the membrane. The resulting transmembrane protein gradient is exploited in turn by the enzyme ATP synthase, which uses it to drive the synthesis of ATP.

In physical terms, the electronic energy used to drive ATP synthesis is not any different than the energy of an electron excited by the photoelectric effect at frequency f. In this light, one might consider a photon being absorbed by a photosynthetic organism (such as plants, algae, or photosynthetic bacterium) that uses sunlight to generate bioelectric power making sugar for food. While it might seem that electron and proton transfers must occur on different timescales due to the vast difference in inertia between them (the proton being over a thousand times heavier than the electron), simultaneous photo-excited electron–proton transfers have been observed [166] and are, in fact, essential to a plethora of diverse energy conversion processes in both chemistry and biology [167].

Thus, proton and electron transfers are intimately connected in biochemistry, where proton-coupled electron transfer (PCET) is an essential charge transfer mechanism [168], and electron and proton transfer steps may occur consecutively or in concert [169]. In mitochondria, for example, long-range PCET drives cellular respiration in the electron transport chain (ETC) [123, 170], where electron transfer cofactors are arranged into three proton-translating complexes (I, III, IV), along with succinate-dehydrogenating complex II [171] and ATP-synthesizing complex V [172]. The organization of these distinct complexes reveals the subtle quantum engineering principles [173] that govern the precise arrangement of electron and proton transfer sites [174], in turn controlling key physiological effects such as reactive oxygen species (ROS) production as a byproduct of electron transfer [175].

Mitochondrial complex I is the largest and most complicated of the respiratory chain complexes [176]. As an excited electron traverses the ETC, it is first injected into mitochondrial complex I, where it transits a column of iron-sulfur clusters in a proton-translocating mechanism that pumps protons across the mitochondrial membrane into a rich solvent matrix. Complex I couples quinone reduction with conformational changes across its four proton pumps [177]. Simulations have shown how a chain of Fe_4S_4 clusters in complex I enables its tunneling efficiency [178], but a consensus on the mechanism of proton pumping in the membranous part of complex I is lacking [179–181]. Real-time observations have revealed proton-coupled electron transfer via ubiquinone [182] in complex I [183–185], and a blueprint is emerging to explain coupling between electron transfer and proton pumping [179].

Electron bifurcation enables aerobic respiration in the Q cycle where it is crucial to oxidative phosphorylation, the process of synthesizing adenosine triphosphate (ATP) from its precursor adenosine diphosphate (ADP). During respiration, Coenzyme Q_{10} (CoQ_{10}, also known as ubiquinol) shuttles electrons from mitochondrial and bacterial respiratory Complexes I and II to Complex III, where an electron bifurcation reaction enables the simultaneous reduction of cytochromes *b* and *c* by ubiquinol [186, 187]. Cytochrome enzymes have been a topic of research interest for many years due to their central importance to energy transduction in all forms of life [188].

Proton-translocating, membrane-bound complexes I, III and IV interact with each other to form respiratory supercomplexes known as respirasomes [189]. Although the mechanism and purpose of respirasome formation is not entirely clear [190, 191], respirasomes have functional implications for the properties of the respiratory chain. Beyond the immediate kinetic advantage of proximity afforded by supercomplex formation, respiratory complex assemblies help to mitigate the formation of ROS during proton-coupled electron transfer [192]. The respiratory chain is a major source of ROS-generating electrons in the cell [193], and oxygen-based free radicals play a critical role in cell signaling, immunity and growth. However, excessive ROS production can cause a myriad of diseases [194].

Coenzyme Q_{10} (a.k.a. ubiquinone/ubiquinol [195]) is a major source of ROS with concurrent antioxidant properties [196]. It serves as an electron shuttle between complexes I, II and III of the mitochondrial respiratory chain [197, 198], where excess ROS production can result from impaired charge transfer from the electron transfer flavoprotein (ETF) enzyme to ubiquinone during fatty acid metabolism [199]. Ubiquinone deficiency constitutes a rare yet debilitating disorder [200] that is associated with various forms of disease [201]. ETC dysfunction decreases the mitochondrial membrane potential, which compromises oxidative phosphorylation needed for ATP synthesis and impairs redox homeostasis, dysregulating an assortment of metabolic processes implicated in human disease [202].

Today, the main structure and core subunits of cytochrome *c* oxidase (complex IV), its active site, and the transfer paths of electrons, protons, oxygen, and water are all fairly well understood [203]. However, there are still uncertainties in particular aspects of the proton translocation mechanisms, including the exact nature of the proton pump's proton-loading site and the path the proton follows to exit the membrane [203]. Physical models of proton-pumping in respiratory and photosynthetic ETCs are being developed to include aspects of many-particle and self-synchronization effects to fully account for the high quantum yield and thermodynamic efficiency of biological proton pumps [204]. A microscopic description of the Q cycle mechanism based on an open quantum system formalism has been proposed to account for its thermodynamic efficiency under physiological conditions [205].

Beyond its crucial role in respiration, biological electron transfer is essential to metabolic processes including enzymatic reactions controlled by biological photocatalysis [206, 207]. For example, the enzyme cytochrome P450 plays an essential role in many biosynthetic and metabolic reactions [208, 209]. Cytochrome P450 reductase serves as a key electron donor during the oxygen activation by monooxygenase enzymes, but the electron transfer mechanism has been difficult to decipher, requiring quantum mechanical calculations to determine the role played by the local electric field in enhancing biological electron transfer [210]. PCET reactions may appear to occur synchronously (i.e., in one uniform reaction step without forming thermodynamically-stable intermediates), but on ultrafast timescales they entail asynchronous non-equilibrium dynamics [211]. This presents the inherent challenge of modeling PCET systems dynamically using electronically and vibronically non-adiabatic simulations with quantized protons [212]. Biocatalytic reactions often depend on the conditions of the biochemical surroundings including ionic interactions, van der Waals forces, and allosteric effects, making it necessary to incorporate comprehensive information from structural biology, enzyme kinetics, and spectroscopy to devise working models of biological catalysis [213].

Thus, biological electron transfer mechanisms remain key topics of research [214], with renewed interest in mechanisms that control complex effects involving electron spin and molecular chirality [215, 216]. Effects of other quantum mechanical phenomena such as electronic spin are now recognized to have a profound effect on vital processes such as biological electron transfer [217]. Cellular respiration drives homeostasis [218], relying on the proton-coupled electron transfer of the ETC [123] (i.e., for chemiosmosis [219]). In eukaryotes, a chain of respiratory complexes carries out cell respiration on the inner mitochondrial membrane [220]. High-resolution images of the respiratory complexes have now been obtained by cryo-electron microscopy [202, 221, 222], revealing highly-organized arrangements of the electron transfer chain complexes in mitochondria [175]. Life-giving functions such as photosynthesis in plants and nitrogen fixing in soil bacteria are also being increasingly recognized as processes that depend on quantum mechanical correlations due to quantum spin-exchange interactions [223].

Chiral molecules are ubiquitous in biology [224], and chirality-induced spin selectivity (CISS) effects also show promise to utilize the coherent properties of electron spin [225], where the spin polarization induced by this effect occurs dynamically and simultaneously with charge polarization [215]. These effects are biologically relevant, e.g., in proteins containing heme where heme-enhanced spin polarization is predicted [226, 227]. CISS has been observed in various contexts that include photoinduced electron transfer systems, quantum dots, and biological molecules [228]. CISS effects have broad implications for catalysis, enantiomer distinction, long-range electron transfer, and biological recognition [229]. CISS occurs as the chirality of a molecule filters electrons passing through it by their spin-polarization, allowing electrons with one spin (say, clockwise) to pass freely while electrons with opposite spins (say, counterclockwise) are reflected. This presents implications for spin selective transport at physiological temperatures and functional spin control in dynamic environments such as the cellular ETC [228]. Coherent spin control may be essential for ROS regulation in the ETC at the point of ROS formation where magnetic effects govern the relative yields of ROS such as superoxide ($O_2^{\bullet-}$) and hydrogen peroxide (H_2O_2) [230]. CISS effects are typically rationalized by the quantum chemical interaction between electron spin and orbital angular momentum [231] during wave-like electron transfer [232]. As a consequence, these results also carry significant implications for chemical reactions involving radical chemical species more generally [233].

Insights taken from quantum biology have begun to enable applications of quantum coherent processes *in vitro* such as artificial photosynthesis [234], yet fundamental advances in the theory of biologically active electron transport will continue to be needed for the development of present-day and future bioelectronics applications [235]. These findings highlight the essential character of the many quantum electrodynamic processes involved in cell respiration and metabolism, and raise questions about the fundamental physiological mechanisms underlying common therapies intended to target the mitochondrial ETC (including simple, over-the-counter health supplements such as coenzyme Q_{10} [236–238]). The influence of these coherent, coupled electronic and vibronic processes on vital biological electron transfer reactions illustrates the importance of quantum electrodynamics in mediating essential biochemical processes.

Chapter 3

Definitions of Non-Triviality for Quantum Biology

The question of whether non-trivial quantum effects play a role in biology has emerged as a topic of extensive debate, although the meaning of the word "trivial" can vary widely in this context. For example, even though all the electrons in the universe are intrinsically and perpetually "entangled" due to the effect of being identical to and indistinguishable from one another, this entanglement is trivial in the sense that it has no observable consequences for electrons that are far apart unless they have been previously prepared together in a correlated state [239]. When two electrons come close together, their intrinsic entanglement becomes significant with observable consequences (e.g., entanglement is non-trivial in the structure of chemical bonds where the wavefunctions of electrons overlap). Non-trivial entanglement between nearby electrons provides the basis for fundamental research in the field of quantum chemistry [240], where it is considered in the context of density functional theory using the mathematical formalism of exchange-correlation functionals [241, 242]. Nevertheless, the word "trivial" has been used inaptly in quantum biology, where the role of quantum effects in determining biological structure has been overlooked [243].

Whereas discussions of quantum biological triviality have inspired a wide body of topical literature [243], a general lack of consensus concerning the definition of "non-trivial" has been a cause of ambiguity, precluding any rigorous or tangible outcome to the debate. Several vague, incongruous, and inconsistent definitions have obscured the preeminence of quantization, quantum coherence, and entanglement in determining practically every aspect of biological structure and function. The resulting equivocation has obscured the meaning and importance of quantum biology as a contemporary research area [244], with deleterious consequences for progress in the field [245]. It cannot be overemphasized that quantum mechanical effects such as coherence and entanglement determine the structure of biological macromolecules and enable the function of electron tunneling processes during cell respiration [70].

Writings on quantum biology are sometimes premised on the intuition that quantum effects in biology are primarily trivial and that implications for crucial non-trivial effects may be rare [243, 246–248]. However, this perspective is at odds with an extensive body of physics and chemistry literature documenting the challenges and limitations associated with implementing quantum and semiclassical models of condensed biological matter [249–253], not to mention the inherent difficulties associated with predicting parameters for classical biochemical models using quantum mechanical simulations [254]. The structures of biological macromolecules have long posed a challenge to computational chemistry [255]. Likewise, the technical challenges associated with modeling quantum chemical dynamics in biological matter are well known [76]. Whereas those phenomena designate non-trivial problems in quantum chemistry, quantum biology is far more than a subfield of physical chemistry. A growing body of work has increasingly implicated an ever-broadening range of quantum phenomena in living matter, drawn widely from research on quantum optics, photonics, and open quantum systems theory [79–81, 256–258].

In this respect, it is imperative to define exactly what is "trivial" in quantum biology in a meaningful and concrete way. Making contact with the sense of the word as it is meant in quantum theory and designating a set of best practices for its usage in quantum biology are necessary to provide a baseline definition of the word. To develop a rigorous concept of triviality that is suitable for quantum biology, let us employ the concept of a "non-trivial model" [259] as it is defined in quantum field theory [260, 261] and connect it with the quantum chemistry framework already developed for structural biology. The idea of triviality was proposed by Landau and coworkers in the 1950s [262]: A model system of interacting particles is considered trivial if it is equivalent to a "free field" theory with no particle interactions (usually in the asymptotic limit of high energies or short distances) [259]. In other words, a system of interacting particles is considered trivial if its properties can be reduced to those of a system of non-interacting particles, which is in fact equivalent to just one particle [263].

In this construction, a complex many-particle theory becomes trivial in the sense that its predictions are no different than an equivalent one-particle system, or likewise the average of an ensemble of one-particle systems. In fact, such an averaged "mean field" approach provides the most common method for simulating the electronic structure of matter, known as density functional theory (DFT) [264]. DFT is based on the assumption that the many-electron density function describing the electronic structure of a system may be replaced by an equivalent density of non-interacting electrons. This mean field approach to solving a many-body problem — by considering an ensemble of isolated one-body systems — was first adopted as a formal method in quantum chemistry by Hohenberg, Kohn, and Sham [265] in which the structure of a system of interacting electrons is represented as a function of its spatial density only [266]. Given that all the properties of an electronic system may be uniquely determined from its density [267], a system of interacting electrons is re-imagined as a fictitious system of non-interacting ones suspended in a classical force field that reproduces the same electron density as the true many-electron system. This results in a set of one-electron solutions to the many-electron problem, known as the Kohn–Sham equations [268].

In principle, the one-electron Kohn–Sham equations provide an exact solution to the many-electron problem. This would appear to suggest (as Hopfield and others have [243, 269]) that biochemical structure models are "trivial" according to the formal definition of the term provided by quantum field theory. However, this naïve view overlooks the fact that even though the Kohn–Sham equations can provide a semiclassical solution to the exact quantum mechanical many-electron problem in principle, that in practice this method relies on a fictitious potential function which operates as an interaction kernel [270] that precisely reproduces the features of the exact quantum mechanical correlations [268]. The problem of solving a complicated quantum mechanical system of many interacting electrons is replaced by the problem of finding a fictitious classical potential (i.e., the "interaction kernel") to emulate the effect of electron–electron correlations in an effective system of non-interacting electrons. In this respect, an effective physical theory is non-trivial in a rigorous sense if its interaction kernel is non-zero.

Countless efforts have been made to construct effective classical potentials to reproduce these electron exchange and correlation effects in quantum mechanical systems, with varying levels of limited success [271]. Thus, the study of so-called "exchange-correlation" functionals has come to represent a central problem in DFT because there is no general solution to the problem of finding an effective classical potential that reproduces all the consequences of quantum mechanics [272]. In contrast, defining a mean-field approximation to reproduce the exact many-electron density function in quantum chemistry lays a foundational guide post for quantifying quantum mechanical effects in a chemical system by including all of the quantum mechanical interactions in the construction of an effective exchange-correlation functional. To lowest order, quantum mechanical methods used in electronic structure calculations typically begin with mean field theories that do not account for long-range electron correlations, despite the fact that these correlation effects are necessary for realistic simulations of many condensed matter systems [273].

Defining a general approximation to reproduce the exact many-electron density function in quantum chemistry has been shown to be as difficult as solving the hardest verifiable problems in quantum computing [274]. Accurate quantum chemical studies of biological systems have been found intractable because the size and complexity of biological molecules prohibit the elucidation of large-scale biological structures and processes using full-scale quantum-mechanical calculations [255]. As such, efforts to determine viable exchange-correlation functionals came to represent a central problem in molecular structure theory. This fundamental difficulty is well established in quantum chemistry [272], where quantum chemical approximations are ordered in a ladder of increasing complexity from seemingly-trivial classical potentials based on mean-field electron densities to complex functionals incorporating non-local exchange effects [275]. As a quantum mechanical model's complexity increases, the required set of model parameters expands from the local electron density alone to account for such factors as spin, density gradients, kinetic energies, and quantum correlations (often imposed as physically unrealistic empirical parameters) [276]. In practice, the applicability of using semiclassical methods such as the local density approximation (LDA) is extremely limited [277]. Therefore, when assigning degrees of triviality in quantum biology one must consider model complexity as laid out in quantum chemistry and physics, first in terms of increasingly complex static correlations and subsequently in terms of increasingly complex time-dependent quantum dynamics.

Quantum correlations can be static or dynamic. Even in the cases where underlying static quantum correlations are non-trivial, the dynamics of the system may be trivial in the sense that the system is stationary. This provides another proposed definition of triviality in quantum biology that designates the case when the quantum *dynamics* of the system are stationary or negligible in the sense that motion can be ignored [278]. This is convenient, from a classical point of view, because it allows one to devise a static (or slow-moving) picture of a biomolecule using quantum mechanics, which can subsequently be used to model motion using only classical trajectories. In this "quasi-static" approach, one solves for the physical properties of a biomolecular system using quantum mechanics in order to *estimate* the values of the physical quantities that subsequently define parameters of a classical model.

The electrostatic case where quantum dynamics are neglected from the molecular model is known as the Born–Oppenheimer approximation, which is derived using the quantum adiabatic theorem in the limit of the zeroth-order approximation in time [279]. In this framework, stochastic transitions (i.e., quantum jumps) between adiabatic "Born–Oppenheimer" quantum states are the first-order corrections to the zeroth-order limit. Jumps are modeled using a first-order transition rate that predicts the probability of the quantum jump occurring instantly in time [156]. These transitions are not trivial according to any of the definitions of triviality given above. Nevertheless, quantum jumps are not always recognized as coherent quantum effects because the jump rate can be described using classical probability theory once it is known.

These non-trivial quantum transitions include charge transport steps between proteins in the cell respiratory chain [280], where the Marcus theory of electron transfer plays an essential role in accounting for the free energy optimizations and tunneling rates in electron transfer proteins [281]. For example, electron transfer drives proton translocation during respiration, transferring hydrogen nuclei across the membrane to generate a proton gradient [177]. However, biological energy transfer is not limited to this "spontaneous surface hopping" [282], and electron transfer can involve complex correlations between electronic and nuclear motions [283]. For this reason, higher-order quantum electrodynamic effects are more often accepted as non-trivial because they involve explicit time-dependent quantum dynamics [278]. Although the quantum dynamics of a molecule might be considered trivial in the sense that its electronic wavefunction is stationary [269], the static correlations present in that molecule may still be highly non-trivial, as reflected in the complexity of the exchange correlation functional used to represent them in the density functional formulation of quantum mechanics. Therefore, the dynamical concept of non-triviality is inadequate to characterize many (if not most) biomolecular systems because so many electrostatic systems designated as "trivial" actually define notoriously hard problems or even grand challenges of other areas [284].

As a prime example, consider a common biological molecule such as caffeine [285]. The full quantum simulation of the caffeine molecule represents a major outstanding problem at the limit of present-day quantum information processing applications [286]. Quantum simulations of the structure of atmospheric nitrogen-fixing enzyme nitrogenase, while trivial in the same electrostatic sense [269], are far beyond the capabilities of present-day classical and quantum computers to simulate. The enormous challenge of simulating biomolecules has prompted increasing interest in the use of quantum computers to solve many key problems in structural and functional biology [287]. Even though structural questions of quantum biology represent highly non-trivial open problems for other fields, they've often gone unrecognized in defining an essential subfield of quantum biology [243]. Instead, it is crucial to define an ordered set of approximations to describe quantum effects in biology. Ordering phenomena from most-to-least trivial, or rather least-to-most complex, is essential to characterize the complexities of quantum effects when extrapolating beyond the low-order semiclassical approximations that underpin most biochemical models. The so-called "trivial" quantum effects that determine ground-state electronic structures of biomolecules often reflect globally-entangled and highly-correlated (i.e., spatially coherent) quantum states of matter [288].

Although solving for properties of biomolecules may seem trivial in the sense that the characteristics of a quantum mechanical system of interacting electrons can be reproduced exactly by a semiclassical density function, represented as a model of non-interacting electrons under the influence of a "exchange-correlation" potential, the quest to identify such a universal potential remains a central challenge of quantum chemistry [289]. By the same measure, many of the hardest problems in quantum computing would also be considered "trivial" [274], including the complicated task of finding exact solutions to the time-independent Schrödinger equation (i.e., full configuration interaction calculations [290]) using a non-local exchange-correlation potential [276]. The non-local correlations of quantum theory are only "trivialized" in return for solving a highly non-trivial non-local potential function instead.

In closing, the concept of "trivial" as inconsequential, negligible, or stationary does not reflect on the scientific significance or life-sustaining importance of quantum effects in biology [291]. Even though many seemingly trivial effects (such as the existence of stable nuclei, atoms, and molecules) may be defined according to classical electrostatic models, they cannot be predicted by classical theory *a priori*. This classical electrostatic picture of molecular biology is still fundamentally dependent on quantization, coherence, and nonlocality (i.e., single-particle entanglement [137]). Given the limitations of the classical picture, we stress the preeminence of non-trivial quantum effects in all areas of biomolecular structure and function. Static structural effects constitute an indispensable sub-field of quantum biology because they lay the foundation for all other non-trivial and dynamical quantum effects.

Chapter 4

Photosynthesis & Open Quantum System Dynamics

Sunlight is the ultimate source of energy for life as we know it [292]. From elementary phototrophic archaea, cyanobacteria, and phytoplankton to the most complex plants, the process of photosynthesis enables life on Earth by converting solar energy into chemical form [293, 294]. Like cell respiration, photosynthetic processes are described by the principles of quantum electrodynamics [257]. Although the structural components and regulatory mechanisms of photosynthesis are well characterized, physiological consequences of the fundamental processes that control photosynthetic light-matter interactions and electron transport phenomena remain challenging to predict [295, 296].

Photosynthesis provides the oxygen in the air we breathe by splitting water into its component atoms (oxygen and hydrogen). Experiments on green algae in the 1930s showed that many chlorophyll pigment molecules would be required to harvest enough light to produce a single oxygen molecule from water during photosynthesis [292]. This prompted the idea of inter-molecular exciton transfer of photosynthetic quanta [297], later formulated into the contemporary picture of an aggregate of light-absorbing pigments that act collectively as a light-harvesting antenna [298]. The high efficiency of the process inspired a theory that quantum coherence could enable energy transfer between the antenna pigments before funneling it to the photosynthetic reaction center [299–301].

Interest in the role of coherent quantum effects in photosynthetic processes was spurred by findings of wave-like energy transfer in a photosynthetic pigment system known as the Fenna–Matthews–Olson (FMO) complex [302]. The FMO complex is a pigment-rich protein found in light-harvesting antenna systems of green sulfur bacteria [303]. During bacterial photosynthesis, the FMO complex transfers exciton energy from the light-harvesting antenna to the reaction center, making the FMO complex an exemplary system in which to investigate the quantum mechanics of the pigment-protein structures that occupy life's photosynthetic complexes [303, 304].

Coherent oscillations were initially discovered in the FMO complex under cryogenic conditions [305, 306], and were later confirmed to exist at physiological temperatures as well [307]. A series of high-profile studies ensued, bolstering an idea that wave-like energy transfer could enable the efficiency of bacterial photosynthesis [306–312]. Subsequent studies provided estimates of the electronic structure and population dynamics of the complex, confirming the delocalization of the observed excitations across multiple pigments [299], and establishing that the coupling of nuclei to electronic excitations initiated long-lived coherent nuclear vibrations in the FMO complex [313, 314].

This revealed that observations of long-lived coherences were not due to coherent superpositions of excitonic states as had been believed, but were induced by vibrational–excitonic (vibronic) couplings [315,316]. It is now recognized that observations of coherent quantum dynamics longer than 600 fs are far in excess of known estimates of exciton lifetimes (which are no longer than 100 fs) [317,318]. This comparison establishes that long-lived coherences in the FMO complex result from hybrid vibronic states involving coordinated interactions between photo-induced excitons and nuclei [93], thus implicating an unexpected role for hybrid quantum dynamics during photosynthesis [314].

Interest in the role of coupled electronic and nuclear motions in photosynthetic energy transfer [94, 319] grew with acceptance that the coherent oscillations observed in the FMO complex persisted much longer than the lifetimes of superpositions of individual exciton states (i.e., excitonic coherences) [93, 320, 321]. In the FMO complex, interactions between initially-excited electronic states and the surrounding nuclear vibrations have a screening effect, preserving the coherent dynamics from the destructive effect of the surrounding environment and suppressing the rate of decoherence with respect to that of the unscreened excitons [322]. The coherent mixing of excited electronic (exciton) and nuclear vibrational (phonon) motions creates hybrid (polaron) quantum states [322, 323] in a more complex and nuanced picture of the phenomenon [324] than originally proposed in the simple exciton model of these long-lived coherent oscillations [302, 307, 312].

Long-lived oscillations in the FMO complex are now interpreted as vibronic coherences, induced by the initial excitonic impulse through a resonance process that transforms the nuclear and electronic subsystems into an inseparable (i.e., entangled) quantum superposition state [304, 325]. Coupling between the excitonic states and nuclear vibrational modes also enables dissipation that drives energy transport in photosynthetic systems [326, 327]. Rather than suppressing environmental dissipation, photosynthetic systems were shown to exploit environmental interactions to reduce the impact of structural variations on energy transport efficiency [328]. Further studies showed that specifically-tuned vibronic interactions (mediated by cysteine amino acid residues) could funnel potentially-harmful excess energy from photo-activated excitons to vibrational quenching sites in pigments on the periphery of the protein [158,303]. Similar results were found in studies of photosystem II (PSII) [329] — another pigment-protein complex found in cyanobacteria, red and green algae, and plants [294] — where an energy-dependent quenching mechanism governs the photosynthetic response to varying sunlight levels by regulating light-harvesting [300]. This dedicated quenching mechanism mitigates the risk of light-induced damage by judiciously controlling the amount of solar energy that is dissipated as heat during photosynthesis [330, 331].

Just as ultrafast spectroscopy experiments have revealed the presence of long-lived vibronic coherence in the FMO complex, vibronic coupling was also found to facilitate ultrafast energy transport in light-harvesting complex II (LHCII), a key component of photosynthetic light-harvesting systems in green plants [332]. Although models of exciton dynamics may produce reasonable estimates of experimental spectra by including thermal fluctuations, exact simulations of photosynthetic light-harvesting typically require much more detailed models of system-environment coupling to account for the spectral density of a light-harvesting antenna system embedded in its vibrational environment [323]. Photosynthetic electron transfer rates in algae and plants become saturated at about one fourth of the full intensity of sunlight on Earth, due to the large optical cross section of light-harvesting antenna complexes which can capture photons an order of magnitude faster than the rate-limiting electron transfer step [333]. To prevent electronic overloading, the photosynthetic apparatus dissipates excess excitation energy during intense light exposure by a process known as non-photochemical quenching [334]. Although the exact details of this quenching mechanism are not known, it is believed to be controlled by a switching effect triggered by a proton gradient across the photosynthetic membrane [335]. Regardless of the exact mechanistic details, both natural and artificial light-harvesting systems rely on photo-excitation mechanisms to produce a light-activated charge-transfer state which can subsequently decay (non-photochemically) back to its ground state, produce free charge carriers (in photovoltaics), or generate radicals for photochemistry or photosynthesis [336]. Studies of quantum coherence effects in artificial light-harvesting nanotube (LHN) systems have shown how excitonic delocalization and vibronic coherences can be controlled to increase fluorescence quantum yields by up to 30% by manipulating system disorder [296, 335].

Derivations of quantitative macroscopic properties from underlying microscopic processes rely on a variety of methods that range from multi-scale theoretical modeling to machine learning and brute-force trial and error [337]. The failure to apply quantization principles correctly may result in an inappropriate classical description (e.g., by violating Newton's Laws [338]) [339]. However, when performed correctly, studies of quantum dynamics underpinning vibrationally-assisted energy transfer can inform classical models of the photosynthetic energy funnel [320, 340]. For cases where underlying quantum dynamics can be approximated classically (e.g., using molecular mechanics [341]), results obtained from classical approximations can be sensitive to the choice of a classical force field because effective classical models may not uniquely recover the results of quantum mechanics. In-depth knowledge of quantum theory may then be required to derive the appropriate classical limit from the underlying quantum dynamics [342].

Beyond quantum mechanics' broadly established role in enabling vibrationally-assisted exciton transfer [338] and its importance to vibronically-driven excitation processes [343], organisms may also harness quantum effects to gain adaptive advantages in the ecosystem [291]. For example, the quantum efficiency of long-range photosynthetic electron transfer in cyanobacterial photosystem I (PSI) depends critically on long-range tunneling-mediated electron transfer by way of the Marcus 'inverted effect' [160]. The Marcus inverted effect is a quantum mechanism whereby the charge transfer rate counter-intuitively decreases as the force driving transfer grows [344]. As an example of a quantum biological process without any classical counterpart [345], inverted electron transfer can also involve nuclear quantum effects which cannot be ignored [71, 346]. Cyanobacterial PSI relies on the inverted effect to enhance photosynthetic efficiency by preventing the unproductive back-transfer (recombination) of the excited electron to its initial state, unlike many other photosynthetic systems [160]. Results of this kind reveal why coherent vibronic interactions are increasingly recognized as essential to adaptations that enable biological energy transduction [343].

These examples also show how the high efficiency of photosynthetic light-harvesting is not due to the coherence of the exciton transfer itself [328], but rather its efficacy [347]: Nearly 100% of the excitons reach the reaction center rather than escaping or recombining [348]. In practice, near-unit efficiency of exciton capture is achieved only under low light conditions where the photosynthetic reaction center is not saturated by an influx of photo-excitons. Under bright sunlight, light-induced damage is mitigated by nonphotochemical exciton quenching [330, 331, 349], which is achieved through the controlled interchange of light-harvesting and photo-regulation functionalities that occur in concert with structural reorganizations and redox stress signaling in the pigment–protein complexes [350–352]. In this respect, the photosynthetic quantum yield may be better defined as the ratio of the number of successful charge-transfer events to the number of excitons directed to the reaction center by the photosynthetic antenna [353].

Different photosynthetic apparatus may achieve this high efficiency of charge separation at the reaction center in different ways [354], for example by promoting forward transfer vibronically (as in purple bacteria [92]) or by suppressing back transfer by the inverted effect (as in cyanobacterial PSI [160]). Quantum effects are also implicated in exciton migration during light-harvesting in photosynthetic purple bacteria [355, 356]. Electron transfer reactions in purple bacteria exhibit a variety of deviations from Marcus theory, which have been attributed to conceptual and technical problems with predicting the electron transfer factor [357]. Problems with predicting electron transfer factors are compounded by influences of nuclear vibrations and coherent relaxation processes, motivating increasing attention to the advantages that collective coherent quantum dynamics can impart to these processes [358].

Biological light-harvesting systems challenge the present understanding of chemical dynamics with sophisticated arrangements of chromophores that optimize light capture with highly-tuned chromophore site energies and subtle manipulations of electronic excitations [300]. These typical examples of biomolecular networks represent the vast array of astonishingly complex systems of electronically-active molecules found in biology. Respiratory electron transport chains perform similar energy transformations, transferring electrons long distances through intricate molecular machineries that defy a detailed mechanistic description using modern biochemical research [123]. Coherent relaxation processes like those found in the FMO complex [299] also provide an example of a broad class of phenomena found in open quantum systems, as exemplified in the physics of "superradiance" (i.e., collective spontaneous emission) [359]. Superradiance, also known as superfluorescence, has been observed in the photosynthetic light-harvesting systems of purple bacteria [360] and the chlorosomes of green bacteria [361], where collective coupling can improve energy capture efficiency by enhancing the rate of exciton energy transfer to the reaction center [362]. This improves prospects for the initial light capture through non-local pigment-excitation effects in the related process of superabsorption (i.e., collective light absorption [363]) [292, 299]. Superabsorptive light-harvesting was shown to benefit from the presence of environmental noise which can be tuned to enhance energy trapping and storing [363], just as molecular vibrations can enable the distribution and storage of energy during photosynthesis [345].

Photosynthetic processes have taken on an exemplary role in open quantum systems research [79–81], where intrinsic molecular vibrations found in biological light-harvesting systems both influence the optical responses of those systems and drive the exciton transport processes on which they rely [326]. This has motivated research into the optimal balance between coherent and dissipative processes in artificial light-harvesting applications [348, 364] and design principles to protect quantum coherence effects from disorder and environmental noise [365, 366]. Far from being detrimental, noise and losses can be coordinated to amplify spectral intensities, suppress fluctuations, and enhance coherence in resonator systems reminiscent of photosynthetic pathways [367]. Thus, dissipative processes are increasingly becoming recognized as quantum resources that can be engineered to enable key quantum information processing tasks such as quantum state preparation, stabilization, and measurement in quantum computing applications [368, 369]. In other words, the environmental interactions that were once believed to destroy quantum coherence in biological systems have now been identified as non-trivial quantum dissipative effects.

Chapter 5

Light Receptors, Spin Chemistry, & Cryptochrome

The reception of electromagnetic signals is common in the biological world, where photoreception extends far beyond phototropism and light-harvesting (such as photosynthesis) to include a wide range of light-receptive processes including vision, circadian photoentrainment, photobiomodulation, plant photomorphogenesis, and plasmodial phototactic responses [370, 371]. Optical photoreceptors found in the eye carry out the first step in a visual phototransduction process that is fundamentally described by principles of quantum electrodynamics [372, 373].

Rod and cone cells are the two types of photoreceptor cells that allow vision in vertebrates, enabling low light and color vision, respectively [374]. Despite key functional distinctions, rod and cone cells have comparable photo-efficiencies and active lifetimes [375]. Until the year 2000, rod opsins (rhodopsins) and cone opsins (photopsins) were the only two types of opsins known to exist in the mammalian retina. Beginning in 1998 with the discovery of a new type of opsin in the light-sensitive skin cells of African clawed frogs [376], a series of findings led to recognition of a third opsin receptor in the eye. Dubbed melanopsin for its role in modulating skin pigmentation, this blue-light photoreceptor has also now been identified in retinal, brain, blood, and adipose tissues [377].

Opsins are G-protein-coupled light receptors in visual and non-visual light-sensing systems of animals [378], where they typically activate guanine nucleotide-binding proteins (a.k.a. "G proteins") via the photo-isomerization of retinal. More than one thousand opsins have been categorized into seven subfamilies. Rhodopsin, in particular, is considered a prototype opsin for its role in low-light sensing and peripheral vision, enabled by high photo-absorption efficiencies that make it capable of single-photon detection [95, 375]. Rhodopsin gains its light-receptive properties from the prosthetically-bound pigment molecule retinal [379], which initiates phototransduction by the photoabsorption, electronic excitation, and isomerization steps that trigger a neuronal signaling cascade by separating retinal from the surrounding opsin [372]. Visible light absorption is made possible in rhodopsin through the formation of a Schiff base (i.e., a carbon-nitrogen double bond) linkage between the retinal chromophore and a nearby lysine residue of rhodopsin. The typical UV spectral absorbance of the retinal-lysine Schiff base is red-shifted into the visible spectrum by a resonance effect caused by increased electronic delocalization at the Schiff base due to the protonation of the nitrogen atom of the carbon-nitrogen bond there [380]. During the light-activation of rhodopsin, the vitamin A aldehyde 11-*cis*-retinal is rapidly converted into all-*trans*-retinal via *cis-trans* photoisomerization [379]. The subsequent conformational relaxation of the photochemically deformed retinal releases isomerization energy, causing conformational changes to the surrounding protein structure and in turn forming the signaling state of the photoreceptor [379]. As such, retinal photoisomerization in rhodopsin is the topic of fundamental studies that combine computational techniques such as atomistic modeling and hybrid quantum mechanics/molecular mechanics (QM/MM) with ultrafast spectroscopy experiments to infer biological design principles [96].

Rhodopsin has also been suggested as a structural model of olfactory receptors [381]. Although about half of all known olfactory receptors are G protein-coupled receptors (GPCRs), high-resolution structural models of olfactory receptors have been experimentally difficult to obtain. GPCRs are cell surface receptors that detect particles outside the cell, in turn activating cell responses. Each GPCR consists of seven α-helical protein segments that fold back and forth through the cell membrane in a series of six loops, with three intracellular loops interacting with G proteins and three extracellular loops interacting with external ligand molecules. Olfaction is believed to rely on the formation of ligand-receptor hydrogen-bonding networks where the odorants provide ligands that serve as electron donors and/or acceptors in the bonding network [381], thus enacting a mechanism by which to relay external signals into the cell. Like other light-dependent GPCRs, melanopsins are pigments that activate their associated G proteins when exposed to light [382]. Similar to other opsins, melanopsin contains light-sensitive 11-*cis*-retinal which is photoisomerized to form all-*trans*-retinal upon light reception [377].

Unlike the ocular pigments rhodopsin and photopsin, melanopsin is not involved in visual perception. Instead, melanopsin modulates a range of other non-image-forming processes that include circadian regulation, sleep cycles, and pupil dilation [383]. Originating with the Latin expression *circa diem* [377], circadian rhythms are present in most living organisms [384] where they coordinate day-to-day metabolism and physiology [385]. Complex and intrinsically connected to a wide range of biological functions including psychomotor coordination, sleep, digestion, and mood, the circadian clock system has been linked to practically all aspects of health and disease [377]. However, melanopsin does not work in isolation when entraining the circadian clock to daily cycles of light and darkness. Discoveries made in vitamin A deficient mice [386] prompted the hypothesis that an unrelated class of pigments known as "cryptochromes" may also play a central role in circadian light entrainment in mammals [387,388].

Cryptochrome photoreceptors have been found in all known kingdoms of life [389,390], where they regulate growth [391], circadian rhythms [392], morphogenesis, phototaxis [393], DNA transcription [394], and other various physiological responses to blue light [395]. They are widely synthesized in higher plants and animals, and occasionally in prokaryotes and single-celled eukaryotes [396]. Plant recognition of blue light was known in the 19th century [397], and light-activated flavoproteins were anticipated as early as 1950 [398], but the pigments responsible for plant developmental responses to near-UV (350 nm) and blue (470 nm) spectra [399] were not characterized until the 1970s [400]. Around that time, the pigments were dubbed "cryptochromes" for their significance in cryptogamic (i.e., flowerless) plants and their cryptic (i.e., hidden) nature [389, 401]. By the 1980s, researchers discovered the HY4 gene of plant *Arabidopsis thaliana* was essential to its blue light response [402]. In 1993, Ahmad and Cashmore isolated the protein responsible [403], flavin adenine dinucleotide (FAD) was identified as its blue light receptor [404], and the name cryptochrome was appropriated for it [405]. Subsequently, a human homologue of *Drosophila melanogaster* (6-4) photolyase was cloned [406] and identified as another cryptochrome-type blue-light photoreceptor [407].

Cryptochrome research in the late twentieth century focused on its light-dependent role regulating metabolism in plants, as well as circadian rhythms in fruit flies [408, 409] and mammals [410–412]. Having been identified throughout a wide range of organisms across taxa, cryptochromes were put forward as universal circadian photoreceptors [413] needed to entrain organisms' diurnal rhythms with day and night cycles. This regulation serves as an important health indicator because these rhythms regulate active and dormant behavioral patterns (i.e., subject to environmental stimulus) [414]. Disruptions to circadian rhythms are correlated with the onset of neuropsychiatric disorders in humans and other animals [415, 416], as well as metabolic dysfunction [414]. One study suggested that low-intensity electromagnetic fields can help stabilize circadian rhythms [417]. Human rhythms in particular were identified as suitable indicators for evaluating MFEs on living tissue [418]. Metabolic studies demonstrated that electromagnetic fields could also influence sleep/wake cycles in mammals such as mice [419] and rats [420].

Circadian magnetoreception became a topic of interest in the early 1960s after the report of geomagnetic entrainment of circadian rhythms in fiddler crabs and other species [421]. Thus, in addition to their established role in regulating circadian rhythms — including the electromagnetic modulation thereof — cryptochromes are the primary candidate magnetoreceptor in animals [422]. Cryptochromes are mainly identified with circadian rhythm regulation, unlike structurally-similar photolyase enzymes which are primarily known for carrying out DNA repair [423]. This similarity prompted Ritz and colleagues to hypothesize that cryptochromes could act as dedicated magnetoreceptors, just as photolyases rely on blue light-activated radical-pair generation to perform resonant DNA repair [424]. This corroborated findings from bird magnetoreception experiments, which suggested the receptor responsible might be activated under light from the green-blue frequency spectrum [425]. Substantial similarities between cryptochrome and photolyase structures motivated the idea of a light-induced flavin–tryptophan radical pair in cryptochrome, based on established knowledge of radical pair formation in photolyase via a series of three (or four [426]) internal tryptophan cations (Trp^+) [427]. Circadian clock activity of cryptochrome was later confirmed to rely on tryptophan (Trp)-mediated photoreduction and radical-pair generation to enable biological activity [428].

Cryptochrome's central role in circadian regulation inspired the hypothesis that it could act as a magnetoreceptor, as circadian rhythms were known to respond to magnetic field variations [425]. Cryptochrome is the primary candidate magnetoreceptor in animals because cryptochrome and photolyase enzymes are the only vertebrate proteins known to generate the reactive radical pairs needed for the chemical operation of a radical pair compass [429]. Following work of Schulten *et al.* [430, 431], Ritz *et al.* hypothesized that the recombination of photo-generated radical pairs in cryptochrome could enable magnetic field perception in birds [425]. That model of magnetically sensitive radical-pair dynamics became known as the *Radical-Pair Mechanism of Magnetoreception* [432].

That model rose to prominence after it predicted that radio frequencies in the 1–10 MHz range would disrupt the avian compass sense [433–438]. This was exhibited in findings that birds' magnetic orientation with respect to the Earth's magnetic field could be disrupted by radio frequency field noise in the 0.1 to 10 MHz range [434–436, 439–442]. A vast body of scientific work was subsequently carried out to develop and test models of radical pair-mediated magnetoreception [432,443]. Cryptochrome-dependent magnetosensation was demonstrated in fruit flies [421, 444–448], cockroaches [449], and plants [450–454]. Extensive studies of magnetoreception were also performed on birds, employing comprehensive experimental controls to test numerous aspects of avian magnetic faculties [433–436, 438, 442, 455–475]. Further research indicated that weak broadband electromagnetic fields in the 2 kHz to 9 MHz range are more disruptive to the avian magnetic compass than strong narrow-band electromagnetic fields in a similar range [476], whereas broadband 0.1 to 100 kHz noise was found not to disrupt the avian compass sense at all [477].

The radical pair mechanism (RPM) is the most widely accepted theory of the avian magnetic sense [432, 443, 478, 479]. It is based on a theory of magnetically-sensitive radical reaction dynamics which has been known in spin chemistry for many decades [480, 481]. The RPM is a magnetically-sensitive chemical mechanism that is light-driven, insensitive to magnetic field polarity, responsive to a range of field intensities, and vulnerable to radio frequency (rf) field noise [482], consistent with the findings of experiments on the avian magnetic sense. Three decades after the RPM was proposed in 1969 [483, 484], it became associated with the photoreceptor protein cryptochrome because cryptochrome in birds' eyes can generate the radical pairs that theoretically enable the avian visual magnetic sense [425]. Ritz *et al.* invoked the RPM to form the hypothesis that the avian compass sense depends on a delicate balance between singlet and triplet quantum states in the cryptochrome radical pairs. Electromagnetic field noise disrupts this balance by driving transitions between the radical pair quantum states [485–487].

Ritz *et al.*'s theory prompted a feasibility analysis indicating that in principle any detection limit could be satisfied by a sufficiently-sensitive radical-recombination reaction rate [488]. This theory therefore became a popular way to interpret animal magnetoreception because it could rationalize the influence of a magnetic field as weak as the Earth's on a physiological reaction [432, 489]. As a result, magnetic senses in birds and other animals are now widely believed to be mediated by the interconversion between the excited "singlet" and "triplet" electronic spin states of charge-separated radical pairs produced in cryptochrome by photo-excitation [432].

The magnetic field dependency results from differences in the singlet (radical recombination) and triplet (free radical escape) yields of the reaction product. Singlet-triplet interconversion rates are modulated by variations in the external field because the effective strength of each radical's interaction with the field is controlled by its local magnetic environment (e.g., via local anisotropic hyperfine [490, 491], spin-orbit [492], dipolar [482], and/or exchange interactions [493]).

The RPM describes the hyperfine interaction-mediated chemical effect of a nuclear magnetic field on the rate of radical pair recombination for a *geminate* pair of radicals "born together" by photoexcitation in either a singlet or a triplet state. If at least one radical contains a nucleus with spin, the nuclear magnetic field modulates the rate of interconversion between the singlet electronic state (which can recombine) and triplet states (which cannot) [494]. The external influence of a weak magnetic field has a symmetry-breaking effect which splits the energies of the otherwise-degenerate singlet and triplet states [495]. If the radicals are photochemically born as a singlet, then the influence of the weak magnetic field decreases the rate of recombination by transforming some singlets into triplets which can more easily escape as free radicals. For triplet-born radicals, a weak field enables recombination instead [496]. The field-induced energy splitting between these states make them vulnerable to interference by radio frequencies in the MHz range, near the Larmor frequency of electronic spin precession about the Earth's field [497].

The conversion between excited singlet and triplets is known as an intersystem crossing (ISC) reaction. When ISC occurs, the presence of an ambient (i.e., external) weak magnetic field will lift the degeneracy between the three triplet states while in turn modulating the strength of the coupling between singlets and triplets [495]. Modulation of the rate of exchange between singlet and triplet spin states by a weak external magnetic field is the defining feature of the RPM and other similar mechanisms in chemistry [480, 498], where radical pairs are typically formed with the breaking of a chemical bond. If the singlet state of a bound pair of electrons is severed to release a pair of radicals, each containing an open shell spin-half electron, then the two electrons may recombine in a charge transfer step that recovers the original bound singlet configuration. However, the two free electrons cannot directly recombine after transitioning to any of the three triplet states (due to the Pauli exclusion principle [499]).

Despite its success [443], the RPM has not been adequate to interpret many of the magnetic field effects (MFEs) in biology [485]. The effects of dipole-dipole radical and electronic exchange interactions, ignored in most models of the RPM, tend to render the effect of the RPM negligible under realistic conditions by suppressing its anisotropy [482]. However, anisotropic MFEs involving cryptochrome can also be markedly enhanced in the presence of scavenger radicals [500], prompting proposals that the viability of the RPM as a magnetic sensing scheme could be recovered by exploiting these radical scavenger-mediated enhancements [482, 501]. A number of other magnetic effects that operate independently from the RPM have also been suggested as alternative mechanisms, such as magnetite compasses, electromagnetic induction, radical scavenging reactions, or level crossing effects [454, 502–504]. As such, many key properties of the radical-pair compass still remain unknown, exemplifying uncertainties regarding the influence of weak-strength magnetic fields in biology, health, and disease [505].

The role of interference by nuclear spin in the conventional RPM has also prompted the idea of a role for the RPM in the anaesthetic action of noble gases, where the effectiveness of the anaesthetic has been correlated with its nuclear spin [506]. This suggests that nuclear spin coupling may interfere with the anaesthetic effect. Noble gases have large spin orbit couplings [507], indicating that a robust model may need to incorporate multiple radical pair-mediated ISC mechanisms (e.g., hyperfine, spin-orbit coupling, etc.) in order to explain the anaesthetic effects of noble gases. Similar mechanisms have been proposed to explain the distinct pharmacological effects of different isotopes of lithium (i.e., ^{6}Li and ^{7}Li) observed in responses to lithium treatments for hyperactivity in rats [508]. The influence of lithium drugs and magnetic fields on the circadian clock have also be proposed to be due to biophysical mechanisms involving radical pair dynamics [509], although the cellular pathways underlying the neurophysiological effects of lithium isotopes remain unclear [510, 511].

A cryptochrome-dependent magnetic sense has been widely reported in fruit flies [512–514] and flavin-based magnetic sensitivity has been observed in human cells [515]. There is ample evidence to demonstrate magnetically-sensitive photochemical reactions in cryptochrome *in vitro* [516–520], but it remains to be shown that the RPM can constitute the mechanism for a working chemical compass *in vivo* [489]. Difficulties associated with demonstrating the operation of the mechanism *in vivo* are linked to the problem of correlating biophysical models directly to behavioral data [521]. This issue has precluded the determination of the exact mechanism(s) underlying magnetic faculties in birds and other species, where concurrent receptor signaling and amplification provides a topic of ongoing theoretical and experimental interest [475, 522, 523]. In fruit flies, the magnetic sense was found to depend only on the presence of the cryptochrome C terminal tail, and modest effects were observed even without the part of the cryptochrome that produces radical pairs [446]. Likewise, the cryptochrome C-terminal tail is critical to enable neuronal magnetic-field sensitivity [524]. Cryptochrome's intrinsically disordered C-terminal tail is known to play an essential role in the mammalian circadian regulation [525], where it controls circadian timing by regulating cryptochrome's association with transcription factors and master genes [392]. Although the mechanism of cryptochrome's signaling function remains largely unknown [391], it is now recognized that the light-activated opening of cryptochrome's C-terminal tail extension can facilitate its interaction with other circadian clock proteins [526].

Although the two main hypotheses regarding cryptochrome-based magnetoreception (i.e., involving the cryptochrome C-terminal tail versus full-length cryptochrome) are not mutually exclusive, they appear to be at odds [513]. Likewise, a number of competing radical pairs have been proposed in addition to the conventional FAD-Trp pair of the RPM [527], such as a flavin-superoxide radical pair [528–531], a flavin-ascorbate radical pair [532], or a flavin diradical [533]. Thus, the RPM's significance to cryptochrome's magnetosensitive dynamics remains unclear [485,502,534,535]. Many results are contentious, and even the existence of a magnetic sense in fruit flies is hotly debated [512–514]. Moreover, bird magnetoreception was shown to rely on a light-activated process followed by a light-independent magnetosensitive step, ruling out most magnetoreception models based on short-lived photoreduction-generated radical pairs [474]. The lack of a widely accepted theory of biological magnetoreception amidst a plethora of inconclusive results has spawned a proliferation of competing models [503] and a number of alternatives to the conventional RPM have been proposed; those include radical pair processes based on spin-orbit coupling [492, 536], spin-vibronic coupling [537], electron-electron dipole-dipole interactions [482, 538], radical scavenging [503], the quantum Zeno effect [539], and/or combinations thereof [536, 540].

Those complementary models have similar chemical reaction schemes in which the magnetic field influences coherent electron spin dynamics to modulate an observable electron-transfer reaction rate [503]. Thus, each mechanism is initiated by the formation of two or more radicals (i.e., "open-shell" chemical species [541]), where the coherent relaxation dynamics of the radicals are traditionally modeled using the Haberkorn master equation [542], although more advanced methods derived from open quantum systems theory have also been explored [543, 544]. In spite of the many outstanding theoretical and methodological discrepancies, there is a large body of evidence to support the existence of magnetically-sensitive photochemical reactions in cryptochrome *in vitro* in the presence of magnetic fields in the milliTesla (mT) [516–518] and sub-mT range [519, 520]. Magnetic compass effects were also observed in the cryptochrome-related molecule photolyase in the mT [545] and sub-mT range [546], along with demonstrations of magnetic responses to an Earth-strength magnetic field ($\sim 50\,\mu$T) in a synthetic molecular system designed as a prototype chemical compass [547].

It is remarkable that weak magnetic fields on the order of 25 mT [515], 100–500 μT [548], or tens of nT [549] can have a marked effect on cell physiology or animal behavior, because the interaction of the Earth's magnetic field (30–65 μT) with an individual molecule is at least a million times less than typical thermal energies ($\sim k_B T$) at cell temperatures [432]. This makes the energy of the magnetic signal much smaller than typical thermal-fluctuation energies, so that the expected signal-to-noise ratio of the magnetic stimulus becomes negligible [550, 551]. As a consequence, the physics of physiological processes that transcend such seemingly-insurmountable ambient noise have attracted sustained research interest [432], motivating research on nonlinear amplification effects [501, 552].

Cryptochromes have captured scientific attention for their diverse signaling and light-sensing properties [553]. Together, cryptochrome and photolyase enzymes comprise an ancient family of light-absorbing flavoproteins [554, 555]. Cryptochromes act primarily as photoreceptors and transcription regulators, whereas photolyase enzymes absorb UV-visible (300–500 nm) light to repair DNA [556] — although some animal-like cryptochromes carry out dual functions by regulating gene transcription and repairing UV-induced DNA damage [557, 558]. In contrast with speculations that quantum effects are washed out by decoherence in biological systems [559, 560], cryptochromes and photolyases carry out complex multi-stage photocycles [561] involving electron tunneling and superexchange effects that drive complex structural rearrangements in the enzymes [562, 563]. Research has also shown that sizeable MFEs are compatible with fast singlet-triplet dephasing in cryptochrome [564]. Although many questions remain unanswered about the role(s) of radical pair dynamics *in vivo*, fundamental spin chemical mechanisms like the RPM provide the theoretical basis for a range of biological processes influenced by magnetic fields [443, 504, 565–567]. This basis is necessary to understand the physiological effects of magnetic field exposures with broad implications for biomedical and radio-frequency technologies [505, 568, 569].

Chapter 6

Dynamic Control of DNA Repair in Photolyase

Like cryptochromes [396], photolyases are blue-light photoreceptors that are widely synthesized in plants and animals, as well as in prokaryotes and simple eukaryotes [570]. The cryptochrome/photolyase protein family [393] is broadly categorized into two distinct types of cryptochrome photoreceptors (plant and animal), and three distinct types of photolyase enzymes that are categorized by what kind of DNA lesions they repair; either (6-4)-pyrimidine-pyrimidine photoproducts (6-4PPs) or cyclobutane pyrimidine dimers (CPDs) in double-stranded DNA, or CPDs in single-stranded DNA [571]. Cryptochromes generally have disordered C-terminal tail extensions which prevent them from repairing DNA [525]. DNA repair activity was presumed absent from cryptochrome [389, 572] until it was demonstrated in a fungal cryptochrome *in vitro* [573, 574]. As primary sequence homologues, photolyases and cryptochromes share similar structures and molecular weights (50–75 kDa) [390, 404, 556], as well as a common molecular cofactor FAD [575]. Their many similarities allow research on photolyases to inform inferences about cryptochromes [545, 576].

Photolyase enzymes use ambient light to repair carcinogenic damage to DNA by UV light across all domains of life [424, 577] (except placental mammals [578]). DNA repair is crucial to genetic regulation, cellular replication, growth, and healing [579]. Photolyase is an enzymatic photoreceptive protein that absorbs blue light to inject an excited electron into UV-damaged DNA [577], reversing the DNA damage while releasing thermal energy [580]. Direct evidence for magnetosensitivity in these enzymes first emerged when magnetic modulation of light-induced signals was demonstrated in *E. coli* photolyase where photo-activation initiates catalysis [545]. The dynamical evolution of photolyase activity has now been mapped out [581], including the timescales for its multiple catalytic steps [582]; from photo-induced electron transfer, to the creation of a radical pair of electrons, to the recombination of electrons after breaking up DNA lesions. Spectroscopic analyses have shown that the high efficiency of DNA repair by photolyase is due to a synergistic optimization of key steps in the photo-repair process [583].

Photo-induced oxidation is a primary cause of DNA damage, which may be catalyzed by reactions with singlet oxygen ($^1O_2$), removal of a hydrogen atom (forming a free radical), or the loss of an electron from an aromatic base (forming a radical cation) [584]. Photolyase binds to DNA in a light-independent step before catalyzing the repair of DNA lesions upon illumination with 300–600 nm light [585]. Beginning with the photolyase enzyme docking to its DNA substrate, the key steps in DNA repair involve the absorption of a photon by the light-activated pigment FAD. During photo-activation, the FAD pigment couples to a network of Trp amino acid residues to generate a spin-correlated radical pair of electrons [545]. A sophisticated series of mechanisms control the dynamics of the radical electrons as they separate, target the damage, break up the DNA lesion, and ultimately recombine in a complex multi-stage DNA repair reaction that transcends the limits of medical biotechnology today [586].

The first stage of this DNA repair process is photolyase activation. In this initial activation process, the FAD cofactor residing inside the photolyase photoreceptor becomes excited by a photon of blue light which triggers an electron transfer cascade from a chain of three or four Trp residues [426]. The initially-oxidized FAD cofactor ($\mathrm{FAD}^{\mathrm{ox}}$) is excited by blue light to its activated state (FAD*), which is then reduced to the radical $\mathrm{FAD}^{\bullet-}$ by the ensuing electron cascade [545], resulting in the formation of a charge-separated $\mathrm{FAD}^{\bullet-}/\mathrm{Trp}^{\bullet+}$

radical pair. The cascade produces a spatially-separated but still-entangled radical electron pair [508] which can then undergo coherent singlet-triplet oscillations and may become sensitized to the ambient magnetic field by the presence of a third (e.g., nuclear or electron) spin [538]. The outcome of this process is the birth of the $FAD^{\bullet -}/Trp^{\bullet +}$ radical pair which forms the basis for the canonical RPM [443, 587]. In this picture, photo-absorption (first by DNA and then by photolyase) is the initiator of both genetic damage and DNA repair [588–590]. A similar photo-activated electron cascade has also been observed in cryptochrome [591, 592].

To restore the DNA integrity, photolyase relies on a structural scaffold of Trp residues to control light-activated electron tunneling from the FAD cofactor of the repair enzyme into the DNA lesion [207, 593], followed by back-transfer of the electron to photolyase in a complete catalytic cycle [594]. The amplitude and inclination of a weak magnetic field have been shown to influence the rate of DNA lesion repair by using both photolyase and modified cryptochrome enzymes [546]. Advanced time-resolved absorption spectroscopy experiments have shown that photolyase's flavin chromophore can switch between its semi-reduced and fully-reduced forms under physiological conditions [595, 596]. Proton transfer during charge recombination can then follow one of two possible mechanisms which switch near pH 6.5 [597], reminiscent of pH-dependent quantum yields and color changes observed during firefly luminescence [598]). *Ab initio* molecular dynamics simulations have also provided insights into the bond-cleavage steps involved in lesion repair [590].

FAD is a photoactive coenzyme involved in many important metabolic processes that are vital to cell respiration and homeostasis [599]. The biological function of FAD is related to changes in its molecular shape [600], and it is believed that the hairpin-like "U-shape" conformation [601] adopted by the FAD cofactor enables photo-induced electron transfer in both cryptochromes and photolyases [602]. FAD is ideally suited as a magnetosensitive cofactor because it undergoes single electron transfer steps through a stable semiquinone radical intermediate [603], unlike the coenzyme nicotinamide adenine dinucleotide (NADH) which acts primarily as a non-magnetic two-electron carrier in metabolic enzymes [604]. In addition to a common FAD active site, photolyase and cryptochrome enzymes often contain a second photo-active coenzyme that varies between species [555, 605], such as an additional light-harvesting folate or pterin chromophore [606–608].

Tryptophan (Trp) is the primary electron donor in the photoactivation of photolyase [609]. The conservation of the Trp chain motif across the whole cryptochrome/photolyase family is associated with its role in the photo-reduction of the excited FAD cofactor [428] which is enabled by three conserved Trp residues [427, 428]. These Trp residues help to quench the excited FAD singlet state after photo-activation, through a rapid multi-step chemical reduction. In some photolyase and cryptochrome species [426, 591, 610], a fourth Trp extends the Trp chain from the inner FAD cavity to the protein surface. This generates a surface-exposed $Trp^{\bullet +}$ radical upon FAD reduction which may be reduced by the solvent or involved in chemical signaling [611]. The FAD-Trp photo-oxidation cycle drives functional conformational changes in the catalytic cycles of photolyase [207] and cryptochrome [553, 563, 612]. In turn, the enzyme structure controls and stabilizes the radical pair over the course of its lifetime [593].

With the highest photoabsorbance of all the amino acids, Trp has become a common subject of studies of photoinduced processes in biomolecules [613, 614]. As such, Trp chromophores have also been identified for their unique role in enabling UV-initiated dimer monomerization for light sensing in the photoreceptor known as UV resistance locus 8 (UVR8) [615]. Plant photoreceptor UVR8 is the first light receptor known to use a network of Trp amino acids as a funnel to enhance the quantum efficiency of light harvesting [616]. This utilization of a network of intrinsic amino acids for light sensing and harvesting marks a departure from other photoreceptor motifs which rely on a separate cofactor (such as FAD in cryptochrome and photolyase) or pigment (such as retinal in rhodopsin) to enable light detection. The discovery of Trp-mediated photodetection in UVR8 opens up a new scientific horizon to explore collective light-protein interactions in Trp chromophore arrays such as those found in photoreceptor complexes and large biomolecular architectures [74,617] — similar to other natural light-harvesting systems with light capture and transport properties that rely on highly-organized structural scaffolds of chromophores [618–620].

Tryptophan is practically ubiquitous in photosensitive proteins [621], and findings of chains of Trp (as well as photoactive tyrosine) residues in many diverse proteins indicate that these chains form a common link between the internal protein structure and the surrounding biochemical environment [93, 303]. Coherences measurements can provide insight into the excited-state structural dynamics of biological light-harvesting systems, revealing aspects of the electronic structure that are far beyond the level of detail captured by simplified classical models [93]. Deciphering the principles that enable highly efficient energy transfer in biological systems may be expected to facilitate the design of nanotechnological energy transfer mechanisms for biosynthetic systems [93]. The growing understanding of the photo-active properties of Trp networks in biology may have important biomedical implications as Trp metabolism is implicated across a range of disease processes, including cancer and neurodegeneration where it has been identified as a promising diagnostic and therapeutic target [622, 623].

Microtubule Trp networks have been suggested to operate as photochemical energy conduits that mediate cellular signaling and/or photo-protective exciton transport via a coherent collective effect known as "superradiance" [74]. Superradiance is a quantum mechanical effect in which spontaneous light emissions are enhanced by entanglement between light emitters, whereas "subradiance" is the entanglement-mediated inhibition of spontaneous emissions [624]. Beyond research on superradiance in microtubule Trp networks [74, 625], computational studies have revealed the possibility of superradiance effects in Trp networks found in actin filament bundles, amyloid fibrils [617], and photosynthetic light-harvesting complexes [362, 626, 627]. Superradiance has been proposed to play a photo-protective role in cellular physiology by rapidly dissipating of potentially-harmful UV frequency excitations from the cytoskeleton [74], whereas subradiance could enable lossless energy storage or transport [617]. Superradiance has been called one of the most enigmatic phenomena in the history of quantum optics [628]. It has been proposed as a mechanism to speed up quantum computing algorithms based on quantum annealing (QA) techniques [256], where coherent dissipation may be essential to accelerate quantum algorithms that solve hard computational problems using quantum mechanics [629, 630]. Non-Hermitian speed-ups may be most important for specific problems that must be solved rapidly but leave room for significant flexibility [631], as is the case for many biological problems which are solved in real time under variable conditions (e.g., biomolecular recognition [632–634]).

Recently, ultrafast X-ray crystallography was used to observe the coordinated structural changes that stabilize radical pairs and optimize electron dynamics during the electron-transfer cascade that occurs across an arrangement of tunneling electron transfer pathways during photolyase photo-excitation [207]. Distinct electron tunneling pathways are critical to the control of the electronic parameters which determine the catalytic efficiency of DNA repair [635] and optimize its operation with respect to its noisy surroundings [636]. In these cases, the control of electron dynamics is critical to ensure a high DNA-repair quantum yield close to 100% [582]. In other cases, subtle control mechanisms regulate DNA-repair operation and efficiency by exploiting arrangements of multiple electron transfer pathways [637]. Advances in time-resolved imaging methods have enabled studies of DNA repair processes at the atomic level, in order to draw new inspirations for DNA-protective drugs and synthetic DNA repair systems from enzymatic methods that organisms already employ to mitigate DNA damage [638].

Advances in time-resolved spectroscopic and structural imaging methods have enabled profound insight into the ultrafast quantum dynamics that govern the action of light-activated enzymes [639]. In particular, research on the interplay between photochemical and conformational dynamics in photolyase and cryptochrome has motivated efforts to control biomagnetic sensitivity by tuning key features of the protein environment in those enzymes [640]. Photolyase and cryptochrome photocycles are also regulated by environmental factors such as cellular metabolites, oxygen concentrations, and the surrounding temperature, further influencing biological activity [554]. In general, DNA repair enzymes navigate the complex cellular environment to find damaged DNA and repair it rapidly with high efficiency and specificity [641–643], constituting a significant search-optimization and quantum control task.

Fundamental insights into the quantum dynamics of photolyase and related enzymes will continue to open up new avenues of research for biochemistry and new modes of catalysis for novel enzyme systems [644] which are expected to be particularly important for enzyme engineering [645, 646]. The development of novel medical treatments based on DNA-repair enzymes will require a comprehensive understanding of the complex sequence of structural and functional dynamics that control DNA photo-catalysis [647, 648] to enable targeted treatments for cancer and other forms of metabolic disease [649]. The understanding of photo-catalytic enzyme dynamics will cultivate advances in the health and biomedical sciences which hold promise to inform the design of revolutionary therapeutic and bioengineering applications [586, 639].

Chapter 7

Enzyme Catalysis: Quantum Fundamentals

Enzyme catalysis is the basis for biochemistry [650]. All chemical processes — including enzyme-catalyzed ones [651] — involve some form of electronic reorganization, and enzymes can accelerate reaction rates by many orders of magnitude [652]. Most enzymatic rate enhancements are attributed to a decrease in a reaction's free energy of activation as a consequence of the energy invested in the electrostatic pre-organization of the enzyme structure [653–655] that occurs during protein folding [656–660]. However, quantum dynamic effects (e.g., dynamical coupling between the conformational and chemical coordinates) do contribute to the catalytic reactions involving ultrafast processes such as photoisomerization and electron transfer [661]. Thus, enzyme-catalyzed reactions can be adiabatic (involving quasi-static processes) or non-adiabatic (i.e., dynamic). Enzymes and photoreceptors such as photolyase and cryptochrome harness their constituent active sites (such as flavins and other biochemically-active cofactors) to carry out a wide variety of chemical transformations using the quantum effects of delocalization, charge resonance stabilization, and in some cases highly-non-trivial electron spin dynamics. Countless enzymes carry out the various sensory, metabolic, and energy transduction functions of the cell, and the importance of quantum mechanical effects in biocatalytic transformations is well established [662, 663]. These functions include many light reception, charge transfer, and electron bifurcation processes which are exemplified in the physics of flavin coenzymes [664–667].

Flavins play essential roles in countless fundamental biological processes including biological electron transfer, bioluminescence, blue light reception, circadian regulation, vitamin biosynthesis, antioxidant defense, redox sensing, gene expression, and light-driven DNA repair [664, 668]. Flavin's key role in photo-absorption, electron bifurcation, signaling, and catalysis have made it a topic of intensive computational biochemistry investigations [669]. The extraordinary adaptability of flavins to act as chromophores, redox cofactors, and free radicals reveals their pivotal importance to a diverse range of fundamental physiological processes. This adaptability is generally attributed to the highly correlated and delocalized electron structure of flavins' definitive isoalloxazine ring system [668].

Flavins are organic molecules that contain the tricyclic heterocyclic compound isoalloxazine, such as riboflavin (vitamin B_2), flavin mononucleotide (FMN), and flavin adenine dinucleotide (FAD). Flavins are highly versatile, light-sensitive, electron carriers found in many enzymatic systems. The isoalloxazine ring likewise provides the electronic structure needed for the photo-generation of spin correlated radical pairs in flavoproteins such as cryptochrome and DNA photolyase [670]. As such, flavoproteins are prototypical systems in which to study the role of fundamental quantum mechanical effects in diverse areas of molecular biophysics that range broadly from light-activated gene repair to geomagnetic field sensing in plants and animals. Unlike related nicotinamides which are primarily two-electron carriers, flavins can transfer one or two electrons at a time. The capacity to transfer individual electrons imparts flavins with unique spin-chemical properties that aptly stage them for a range of radical reactions. Most notably, photoactivated flavins produce spin-correlated radical pairs, imparting them with magnetic spin-dynamic properties that position them as the leading candidate magnetoreceptor in biology [670].

Its role as a redox sensor, antioxidant, and free radical generator place flavin crucially at the center of cellular immune, energy-transduction, and morphological functions, all of which depend on the redox status of the cell. Cellular redox control is linked to cell cycle control via the initiation of cell replication and apoptotic triggers [671, 672]. This also suggests a central role for flavin and other photo-active redox cofactors in cell signaling and homeostasis, where quantum effects can have a subtle dependence on the surrounding electrostatic environment. In general, enzymes accelerate reaction rates by many orders of magnitude by providing a chemical environment with substantially different properties than those of the bulk solvent — in effect lowering the activation energy (i.e., the free energy barrier) needed to access the reaction's transition state while progressing from the reactant to the product state [651]. This catalytic environment is understood largely in terms of the electrostatic pre-organization of the enzyme complex [653, 654] with the *caveats* that catalytic rates may also be limited by the timescales of stochastic enzyme–solvent fluctuations [657, 673], and in some cases also the dynamics of ultrafast photochemical or electron transfer processes [661]. This is especially true for flavin, for which the UV–visible spectrum is determined by subtle quantum mechanical effects involving solvent reorganization and protonation processes [674].

Just as the protein environment tunes the flavin's photoactivity [674], so too can the flavin cofactor influence the process of protein folding as a newly-transcribed amino acid chain envelopes it [675]. These biophysical processes are essential to innumerable aspects of molecular-scale physiology, and critical to all major forms of biological energy generation (aerobic and anaerobic respiration, photosynthesis, denitrification, etc.) [676]. Yet the nature of the environmental interactions that determine flavin biophysics remains ambiguous even after many decades of flavoenzyme research, while its environmentally-dependent functional properties continue to elude prediction [677]. For example, research elucidating the mechanism of flavin oxidation by O_2 remains very limited, and the interaction of flavin with oxygen remains poorly understood [678]. This is due in part to the fact that there are no structural rules to predict how a flavoprotein will react with oxygen. Rather, a number of subtle factors such as electrostatic pre-organization, charge distribution, protein dynamics, and active-site solvation contribute to the balance of interactions that control reactivity with oxygen in flavin-containing proteins [679]. Flavin reactivity can also be adjusted by covalent modifications, either synthetically, or naturally by structural modifications found in about 10% of native flavins [668]. Simulations of flavin photophysics require complicated quantum mechanical approximations represented in an assortment of computational methods, and there is currently only limited consensus regarding which methods are appropriate [676].

Accurate quantum mechanical modeling of flavin photophysics therefore hinges on the need to rigorously establish an appropriate set of modeling techniques with robust physical benchmarks. Flavins represent challenging candidates for quantum simulations for the same reason that they are important to study: Their versatile and tunable photophysical, spin-chemical, and redox properties emerge from the highly correlated and delocalized electronic structure of their distinctive isoalloxazine ring systems. Quantum mechanical properties of biomolecules like flavin are modulated by complex interactions with the surrounding cellular matrix including the protein, solvent, and electromagnetic field in which the flavin functional group resides. This amounts to a fundamental quantum control task, as the protein environment manipulates the delocalized electronic state of the biomolecule to achieve specific physiological goals. This molecular-scale quantum control is critical to maintaining metabolic regulation and homeostasis. Thus, modeling the biophysical control of key molecular properties such as the redox activity of organic cofactors can require in-depth quantum mechanical analyses which account for the electron correlation and delocalization effects that determine the molecular properties involved in enzymatic processes [668, 669, 676, 678].

Quantum mechanical electronic delocalization plays a central role in many enzyme-mediated biological processes, including individual electron or proton transfer events, structural modifications to molecules during anabolic metabolism, and fundamental energy transduction processes. Electronic delocalization provides the basis for key concepts in organic chemistry such as resonance and aromaticity [680], yet even seemingly-simple chemical concepts like the idea of the covalent bond remain open to scientific study and debate because of the subtle nature of the underlying electronic delocalization [681]. Electronic delocalization is particularly relevant to the chemical effect of charge-resonance stabilization, wherein the structure of a molecule is stabilized by non-local resonance interactions [682]. However, delocalization and resonance stabilization are not identical concepts [683], and electronic delocalization can stabilize or destabilize molecular systems with important chemical implications in either case [684].

Electronic delocalization can be particularly important in the reaction centers of photosynthetic antenna, where strong coupling between the chlorophylls comprising the photosynthetic "special pair" enables the initial formation of a delocalized excitonic state prior to charge separation [685]. Electronic delocalization across the chlorophyll special pair is rationalized in terms of a charge-resonance interaction between the photosynthetic special pair and the formation of coupled excitonic and charge-transfer quantum states [686]. This theoretical treatment is closely related to Fano's seminal work on quantum resonances [687–689], where charge separation is facilitated by excitonic delocalization and environmental interactions such as excited-state vibronic coherences [92, 690].

Long-range charge-resonance stabilization is also implicated in the catalytic action of the diheme enzyme MauG [691, 692]. MauG plays an important role in the tryptophan tryptophylquinone (TTQ) biosynthetic pathway of several gram-negative bacteria [693], where it employs a charge resonance stabilized pair of positively-charged iron atoms (i.e., iron cofactors in a *bis*-Fe(IV) oxidation state) to perform radical-based biocatalytic reactions [694]. The unusual stability of MauG's highly-reactive *bis*-Fe(IV) catalytic cofactor is attributed to its exploitation of a type III charge-resonance phenomenon in which the high-valent iron radical is delocalized across both hemes via an intervening tryptophan residue [694]. A complex network of surrounding amino acid side chains and structured water molecules control the enzyme's reactivity by coordinating the multiple proton and electron transfers steps that enable charge–resonance stabilization of the high-valent catalytic state [695]. The catalytic mechanism of MauG is unusual because its heme cofactors do not interact directly with the reactant substrate, instead mediating catalysis via long-range electron transfer through another intervening tryptophan [695]. Although the catalytic properties of MauG are extraordinary, the protein nonetheless exemplifies a broad class of metalloporphyrin enzymes that carry out a vast assortment of oxidative processes in living organisms [696–698].

Another broad class of enzymes, known as electron-bifurcating enzymes, allows crucial yet thermodynamically-costly reactions to occur spontaneously using a process known as electron bifurcation [699, 700]. In an electron-bifurcating reaction, a pair of electrons (from a two-electron donor) is distributed separately over two distinct electron transfer pathways which correspond to distinct electron acceptors (e.g., an iron–suflur cluster and a *b*-type heme) [701, 702]. During photosynthesis or respiration, for example, an aromatic quinol (either ubiquinol or plastoquinol, respectively) is oxidized by a cytochrome complex that separates and distributes the two electrons from the quinol oxidation site via electron bifurcation [703, 704]. Electron-bifurcating enzymes optimize the use of free energy by coupling thermodynamically-unfavorable "endergonic" reactions to favorable "exergonic" ones, enabling a variety of chemical reactions necessary to cell physiology [705, 706]. Electron bifurcation is therefore considered one of the primary energy conversion mechanisms in biology [699], along with ATP hydrolysis and ion gradient-driven processes. These three fundamental energy transduction mechanisms provide the biochemical driving forces in all living systems.

Studies have revealed an intricate mechanism governing essential electron-bifurcating processes, enabling them to occur reversibly and spontaneously [702]. As a consequence of this complexity, the essential mechanism underlying bifurcating enzymes eluded characterization for decades, precluding the design and synthesis of artificial electron-bifurcating enzymes [707]. The thermodynamics of respiratory electron bifurcation puzzled chemists for many years because of its efficiency, enabling the thermodynamically-unfavorable reduction of cytochrome *b* by coupling it simultaneously to the more favorable reduction of cytochrome *c* via an iron-sulfur (FeS) cluster [708]. The key to the whole process is its reversibility, allowing it to operate either forward or backward interchangeably (without significant energy loss) by coupling an "uphill" reduction reaction directly to a "downhill" one [707, 709].

Electron bifurcation processes have drawn attention from researchers interested in modeling quantum thermodynamic effects [88,205], notably using the enzyme cytochrome b_6f as a model system because of its role as the primary electronic coupling site between photosystems I and II during photosynthesis [704,710]. Other simulations derived from crystallographic data and electron transfer models were employed to predict the electron bifurcating function of the cytochrome bc_1 complex using stochastic dynamics [711]. A diverse assortment of chemical gating schemes were proposed to rationalize the absence of any electronic short-circuiting during the charge-separation step of electron bifurcation, but those schemes could not explain the existence of charge bifurcation because they did not address its reversibility (a crucial aspect of short-circuit suppression in the bifurcating systems) [707]. More detailed studies shed light on characteristic features of the free energy landscapes that enable high-efficiency electron bifurcation [707, 709], leading to the development of a kinetic theory of bifurcation processes [712].

Correlated electronic motion is an essential feature of electron bifurcation processes because mean field theories fail to capture essential correlations that enable biological energy transduction processes [713]. Inherently quantum mechanical effects such as spin-spin coupling and electron transfer are also implicated in the fundamental mechanisms of electron bifurcation that underpin energy transduction in all domains of life [714]. Protein structural rearrangements serve as a control mechanism for electron dynamics along electron transfer pathways through many diverse electron bifurcation enzymes [715]. Hence, the underlying dynamics of the correlated electron motion are complicated by a multitude of protein conformational states which synchronize transitions between electron transfer and bifurcation states of redox enzymes, coordinating structure-function relationships to optimize catalysis [716].

Electron bifurcation had been considered unique to Mitchell's Q cycle for forty years before Buckel and Thauer discovered that flavin-based electron bifurcation is carried out during anaerobic metabolism in microbes [708, 717]. Inspired by Keilin's ground-breaking work on cytochrome systems in the cellular respiratory chain [219] and building on the Wikström–Berden model of electron transport through complex III [718], Mitchell was the first to identify electron bifurcation as the mechanism underpinning the Q cycle during oxidative phosphorylation — the process by which 95% of all energy is obtained in aerobic organisms [719,720]. Mechanistic models of how electron-bifurcating enzymes carry out highly-efficient bifurcation processes became prominent after electron bifurcation was recognized as a key function of enzymes that perform thermodynamically-costly reduction/oxidation (redox) reactions [714, 721], although many key details of electron bifurcation mechanisms still remain unclear [714].

Similar to the quinol-based electron bifurcation schemes of the Q cycle [717], flavin-based bifurcation couples free energy from a favorable redox reaction to power a thermodynamically-unfavorable one by coupling the common enzyme cofactor nicotinamide adenine dinucleotide phosphate (NADPH) to a pair of electron acceptors. ETF is a dimeric enzyme comprising two distinct subunits, ETF-α and ETF-β, conventionally containing the electron transfer cofactor FAD and the non-redox-active cofactor adenosine monophosphate (AMP), respectively [722]. ETFs are best known for their role in transferring electrons from the oxidation of amino and fatty acids to the respiratory ETC [723]. However, electron-bifurcating forms of ETF instead carry a second FAD instead of the usual AMP cofactor in a molecular arrangement that is correlated with a conserved amino acid sequence motif in those electron-bifurcating ETFs [699]. During flavin-based bifurcation, NADPH reduces FAD to its anionic hydroquinone form ($FADH^-$), effectively sending one electron "uphill" to a low reduction-potential acceptor (such as ferredoxin or flavodoxin) using energy from another electron traveling "downhill" to a high reduction-potential acceptor (e.g., benzoquinone, menaquinone, heterodisulfide, or crotonyl-coenzyme A) [724].

Quantum mechanical interactions were revealed between the FAD cofactors of the α and β subunits of bifurcating ETF (i.e., α-FAD and β-FAD, respectively) by electrochemical observations which showed that NADH reduced β-FAD by hydride transfer about 1500 times slower in the presence of a distant α-FAD$^{\bullet-}$ radical [724] than in the presence of α-FAD. This dramatic difference in observed hydride transfer rates was confirmed to be due to the radical nature of the α-FAD$^{\bullet-}$ species (and not its ionicity) when the rate of β-FAD reduction by NADH in the presence of α-FADH$^-$ was found to be identical to that of β-FAD reduction in the presence of α-FAD. The spin dependence of the α-FAD$^{\bullet-}$-inhibited reaction indicates the presence of a latent quantum effect, raising the question of how a bystander radical could impede hydride transfer (from a distance of at least 14 Å).

Kinetic isotope effect (KIE) measurements were carried out to help address that question, revealing a normal KIE for the reduction of β-FAD in the presence of α-FAD, and an inverted KIE in the presence of α-FAD$^{\bullet-}$. This suggested that a radical within no less than 14 Å of β-FAD could slow the tunneling-mediated hydride transfer reaction sufficiently to invert its KIE [724]. In this case, KIE experiments performed on bifurcating ETF from the bacteria *A. fermentans* revealed the presence of radical-controlled KIEs. The unprecedented findings of radical-controlled KIEs in ETF from *A. fermentans* are also expected to occur in other similar bifurcating ETFs commonly found in archaea and anaerobes [724]. Observations of KIEs are relevant to quantum biology because the dependence of a reaction rate on its isotopic components is a fundamental effect of quantum mechanics [725]. As such, KIEs designate many established examples of quantum effects in biology, including many vital processes such as biological electron transfer [726], oxygen-activation [727, 728], and enzyme-catalyzed hydrocarbon-cleavage reactions [729].

Spectroscopic observations of the partially-reduced form of bifurcating ETF revealed a sharply-peaked band around 726 nm which gradually appeared and disappeared during FAD reduction [730]. That unexpected finding indicated the presence of a delocalized charge-transfer species involving both FAD cofactors (α and β), distributing electrons coherently across the flavins in a degeneracy-breaking effect that is very similar to the charge-resonance mechanisms identified in the chlorophyll dimer of the photosynthetic "special pair" [685, 686], and also the heme dimer of the enzyme MauG [691, 692]. The two-flavin arrangement found in bifurcating ETF is therefore likely to contribute to the signature efficiency of the electron bifurcation process. For comparison, experimental observations have also associated fast, efficient charge separation with delocalized electron transfer in organic electronics [731]. In those systems, electron delocalization breaks the system symmetry to create a level-repulsion effect that divides the system into characteristically-distinct energy states with counter-intuitive consequences for charge separation [732]. The profound impact of a distant radical on the rate of hydride transfer in bifurcating ETF [724] is comparable to the discovery that the RPM is suppressed by a nearby electronic spin [482], both in the case of an interaction with an additional non-reacting radical [482] and in the case where there is coupling to a radical scavenger [503]. The involvement of a radical mechanism in the formation of the 726 nm signaling state is consistent with observations of the formation of a charge-transfer state between the two flavins (α-FAD and β-FAD) [733, 734]. The dramatic decrease in the hydride-transfer rate found in bifurcating ETF containing an α-FAD$^{\bullet-}$ radical indicates that the electron-bifurcation reaction may incorporate a spin-chemical mechanism involving a delocalized electron pair.

Key functional aspects of biocatalytic mechanisms remain poorly understood even after more than a century of research [735, 736]. Organic coenzymes like flavins and metallorganic cofactors such as chlorophylls and hemes are excellent candidates for studies designed to develop physical models of the mechanisms that govern these catalytic systems. Electron-bifurcating flavoproteins, in particular, constitute an exemplary class of enzymes for use in studies of quantum biology because of their high tunability, nuanced reactivity, air tolerance, and relative simplicity [706, 736–738]. Exploratory studies have revealed a wealth of possible quantum mechanical effects in bifurcating flavoproteins with evidence for electron tunneling, delocalization, resonance stabilization, spin dynamics, and electron exchange effects [714,724,730]. The wide range of quantum processes exhibited by bifurcating ETFs present an exceptional staging ground for investigations into the nature of quantum effects in biological systems and the interplay between the various quantum mechanical phenomena which govern their action and stability in a complex system of quantum control.

The trend that we find in all quantum biological systems considered in Part I is the nature of quantum mechanical coherence and delocalization in determining biophysical structure and dynamics in relation to the environment, whether it be in the elementary delocalization effects that produce "incoherent" tunneling in the electron transport chain, vibronic delocalization in photosynthetic apparatus, non-adiabatic photochemical dynamics in rhodopsin, singlet-triplet interconversion in cryptochrome, related forms of intersystem crossing in photolyase, or evidence for electron pair delocalization in the reversible kinetics of electron bifurcating enzymes. The foundational investigations of the physical principles of open quantum systems underlying these effects provide the basis for a conceptual framework universally applicable to all biological systems.

Part II

Coherent Quantum Effects in Biology

"It can scarcely be denied that the supreme goal of all theory is to make the irreducible basic elements as simple and as few as possible without having to surrender the adequate representation of a single datum of experience."

— Albert Einstein, 1933 [739]

Chapter 8

Ultraweak Photon Emission & Cell Processes

Although the treatment of disease with sunlight is as ancient as medicine itself, even a century ago there was no real understanding of how sunlight influences health and disease [740]. As such, the formal study of the role of radiation in biology is purely a modern development [741], sparked by the discovery of the photoelectric effect and the characterization of the Planck spectrum that together gave birth to quantum mechanics [121]. In his seminal work [742], Planck showed that the temperature-dependent radiation spectrum of a material deviates dramatically from the light spectrum predicted by classical physics [743]. This laid the foundation for the quantum theory of light [744], and provided one of the first demonstrations that seemingly-classical phenomena (i.e., the spectrum of incoherent light) could emerge from fundamental quantum effects. Today it is well known that light is an essential environmental factor for regulating circadian rhythms, metabolic rates, and cell growth in living organisms [745].

Light is broadly divided into three categories centered around the 380–720 nm spectrum of visible light, where wavelengths of light longer than 720 nm constitute infrared (IR) spectra and wavelengths shorter than 380 nm define the UV regime. IR light is widely considered to have health benefits [746], offering possible protection against a number of chronic diseases which may be linked to improvements in mitochondrial function and ATP production [747]. Visible light influences many aspects of vitality with an impact on retinal function, sleep, cancer, and the persistence of mental health disorders [745]. Visible light from artificial lighting has been shown to have photon energy-dependent toxic effects which can reduce longevity while increasing oxidative stress, neurodegeneration, and chronic conditions such as obesity and diabetes [745]. Though exposure to UV light is known to induce cancer, promote aging, and decrease longevity, moderate doses of UV light can be beneficial to human health [745].

The study of the impact of UV light on living cells was pioneered by Gurwitsch in his definitive onion root experiments of 1923, which first revealed that living cells emit a faint spectrum of radiation [748, 749]. These faint radiation emissions, sometimes described as biophotons [750], are now widely recognized as ultraweak photon emissions (UPE) [751, 752]. The wavelengths of UPE span the UV, visible, and IR light spectrum, where the UVA-visible-IR range of UPE is documented especially well [753, 754]. However, what most distinguished Gurwitsch's discovery of UPE was his claim that certain wavelengths of the faint light could promote cell replication. According to Gurwitsch [755, 756], UV UPE primarily in the 190–250 nm wavelength range [757, 758] could also promote growth in nearby cells in a phenomenon that he described as the "mitogenetic effect" [759, 760]. Although the mitogenetic effect described by Gurwitsch still remains to be decisively established [759], there is an ever-growing body of evidence that a diverse range of photon energies are involved in biological processes [749, 761–765].

Quantum effects of light are distinguished by their wavelength dependence, where different colors of light can produce different effects in living organisms with substantial implications for health and medicine [766]. This is due to the discrete nature of the light quanta themselves (i.e., the photons) [119]. Consequences of the influence of different wavelengths of light on living processes were studied and applied quite famously by Finsen who came to be regarded as the founder of modern phototherapy after he pioneered light-based treatments for smallpox and *lupus vulgaris* (skin tuberculosis) in the 19th century, in turn winning the Nobel Prize in Physiology or Medicine for his efforts in 1903. That same year, Rollier established the first clinic for solar therapy, known as heliotherapy, to treat tuberculosis using high-altitude sun baths which were implemented in clinics around the world [767].

The implications of quantum theory for biology and medicine were considered so great that, by 1932, Bohr himself was invited to give the opening address to the *International Congress on Light Therapy* in Copenhagen that year [768]. His lecture marked his first concerted attempt to extend the concepts of quantum physics to the life sciences, challenging the notion that the principles of life itself could be reduced to pure physics and chemistry [769]. Bohr's sentiment was not new, but had been anticipated a quarter century earlier in the philosophy of Bergson [770] (who too received a Nobel Prize in Literature in 1928 [771]). Bohr's lecture was so impressive that it provided the impetus for Delbrück, whom Bohr had invited to attend, to switch fields from physics to biology [772].

Bohr's sentiment was echoed in more technical terms by Wiener who extended the line of reasoning in his work on *Cybernetics* [773], drawing on principles of quantum theory while incorporating insights from Szent-Györgyi [774] and Haldane [775]. Wiener noted that the stability of living matter must be a consequence of its high internal resonance or "degeneracy" [773], concluding that the stability of a quantum state depends on its resonance structure insofar as the most stable states will be those which can be transformed into themselves by the largest number of isomorphic transformations with "the ability to make small causes produce appreciable and stable effects" [773]. Wiener's identification of resonance phenomenon in living matter [773] reflects Gurwitsch's original conceptualization of the mitogenetic effect in terms of fundamental resonance processes [749]. Reflections of those proposals were recapitulated in subsequent ideas developed by Fröhlich later in the 20th century [776, 777].

Delbrück went on to study the impact of X-ray irradiation on genetic mutations, concluding that the fundamental nature of genetic material must be molecular [778]. He would go on to receive the Nobel Prize in Physiology or Medicine with Luria in 1969 for discovering the genetic structure and replication mechanism of viruses. Drawing on insights from Delbrück and Bohr [769], Schrödinger prepared the

1943 lecture series "What is Life?" [42] in which he put forward the premise that the preservation of genetic inheritance could not be accounted for by classical statistical laws because of the ever-present statistical noise found on the molecular scale. He reasoned that any microscopic genetic information would be washed out by ubiquitous and disorderly thermal motion, precluding the possibility that life could rely on the function of classical genetic material. In his own words [42]:

> "All the physical and chemical laws that are known to play an important part in the life of organisms are of this statistical kind; any other kind of lawfulness and orderliness that one might think of is being perpetually disturbed and made inoperative by the unceasing heat motion of the atoms."

Schrödinger's work was so influential that it inspired Watson and Crick to pursue research on the underlying structure of life, with Watson himself noting that Schrödinger's lectures "very elegantly propounded the belief that genes were the key components of living cells and that, to understand what life is, we must know how genes act" [559]. It is a matter of history that Watson and Crick — drawing on data from Franklin and Wilkins [779] — discovered the structure of the DNA double helix in 1953, for which they received the Nobel Prize in Physiology or Medicine in 1962. That same year, Delbrück invited Bohr to deliver the opening lecture at the dedication of the Genetics Institute in Cologne. He took the opportunity to revisit the theme of his address to the Light Therapy Congress in Copenhagen thirty years before, delivering a lecture titled, "Light and Life Revisited," which was to be his final public address [772]. In that speech, he re-framed his original thesis, reflecting that [780]:

> "...it appeared for a long time that the regulatory functions in living organisms, disclosed especially by studies of cell physiology and embryology, exhibited a fineness so unfamiliar to ordinary physical and chemical experience as to point to the existence of fundamental biological laws without counterpart in the properties of inanimate matter studied under simple reproducible experimental conditions. Stressing the difficulties of keeping the organisms alive under conditions which aim at a full atomic account I therefore

> suggested that the very existence of life might be taken as a basic fact in biology in the same sense as the quantum of action has to be regarded in atomic physics as a fundamental element irreducible to classical physical concepts."

In that 1962 address [780], Bohr updated his stance to convey the idea that the regulation of living organisms exhibits a kind of sensitivity that is not found in conventional physical or chemical systems. Like Schrödinger, he believed that this pointed to fundamental biological laws without counterparts in the properties of inanimate matter. Thus, he proposed that the task of biology could not be that of accounting for the motion of every individual atom in an organism, as no clear distinction could be made between the life-sustaining mechanisms in an organism and the functions they fulfill. Rather, he proposed a kind of complementarity between the description of the individual atoms of a living system and its activity as a whole. In this specific sense, he proposed that the action of the whole organism in its natural setting could not be effectively reduced to that of its particular molecular components.

Delbrück carried the legacy of Bohr's search for complementarity in biology in an endeavor to express the concept rigorously and in concrete terms [781]. In what would become the third in a trilogy of talks initiated by Bohr, Delbrück delivered an address at the Centennial of the Carlsberg Laboratory in Copenhagen in September of 1976 titled, "Light and Life III." That lecture was dedicated "literally to Light and Life, to wit, to the question of how and when the various basic photochemical reactions involved in life have come about and what they do" [772]. Beginning by introducing a colorful cast of photo-enzymatic "characters" (chlorophyll, protochlorophyll, retinal, phytochrome, cryptochrome, and DNA photolyase), he quickly and adeptly proceeded to survey a list of fundamental topics that remain prototypical research subjects in quantum biology today, including the biomolecular photoexcitations, biological electron and proton transfers, reactive oxygen species, and organic molecular-bridge systems [772].

By the 1990s, after five decades of research on biological light emissions using photomultiplier tubes (PMTs), UPE from living organisms had been demonstrated conclusively [782, 783]. This body of work included observations of UV UPE in the 190–250 nm wavelength range predicted by Gurwitsch [784], although attempts to reproduce the effect of UV UPE-induced mitosis were set back by methodological failures from research groups that did not follow the necessary experimental protocols laid out by Gurwitsch [749, 759]. Meanwhile, biological photon emissions in the IR range were also brought to light as having potential effects on living systems by way of a rudimentary form of cellular "vision" characterized by a comprehensive series of experiments by Albrecht-Buehler [785–789]. That body of work was exemplified by the finding that the seat of cellular IR detection is contained in the centrosome [786] and observations that pulsed near-IR signals could perturb the stability of radial microtubules around the centrosome in fibroblast cells [788]. The long-range attraction between aggregating mouse embryonic fibroblasts was also found to be mediated by near-infrared (NIR) light [789], revealing a key role for long-range electromagnetic interactions between living cells. Those findings were corroborated by evidence that NIR laser light can induce the formation of cell protrusions in fibroblasts and neurons by enhancing cellular actin recruitment [790].

Dynamical cellular interactions in the terahertz (THz) regime have been known for decades, highlighted in the work of Albrecht-Buehler on cell behavior in response to NIR [787]. Long-range electrodynamic (dipole-dipole) interactions have also been shown to mediate attractive forces between distant proteins when driven by an external light source [791]. These findings provide a proof-of-principle that non-equilibrium collective oscillations can enable long-range dynamical effects to promote distant protein-protein coupling for aggregation, and has been reproduced using THz resonance spectroscopy [792]. The discovery followed from lines developed in previous work on non-equilibrium phonon-condensation [793, 794]. This is consistent with other observations of long-range orientation in solvent ordering involving molecular dipole-dipole interactions [795].

The production of UPE has been demonstrated in practically all forms of life [754]. It is widely understood that mitochondria produce the majority of UPE in eukaryotic cells [796], although DNA was also recently proposed as a significant UPE source [797]. The ubiquity of UPE in living tissue and its link to oxidative biochemical reactions has made it a recurring subject of interest for medical diagnostic and therapeutic applications [751]. For example, the optical stimulation of white blood cells has been shown to induce respiratory bursts, not only in target neutrophils, but also in a second population of chemically-separated but optically-coupled cells, supporting claims of the existence of long-range optical coupling between living cells [798]. Endogenous (i.e., non-thermal) UPE from living organisms are well-documented [799] in the context of oxidative stress. Stress-induced UPE carries a spectral signature that is distinct from that of spontaneous UPE [799], implicating biophotonic modulation under physiological conditions in an organism and presenting a potential role for photonic signaling. However, the links between UPE, cell metabolism, ROS generation and microtubule cytoskeletal dynamics remain far from clear [800]. This leaves a critical gap in contemporary knowledge of cytoskeletal structure and function, where cooperative photonic interactions may be essential to coordinate collective behavior [74].

To consider how UPE are likely to affect cellular order and disorder, one may examine certain key characteristics of cancer. Cancer cells exhibit unique UPE signatures [802], as well as disturbed microtubule structural dynamics [803, 804]. Furthermore, microtubule-stabilizing drugs are commonly used as chemotherapeutics [805,806]. Taken together, evidence for UV and IR microtubule photodynamics suggests that microtubules may provide a mechanism for optical (up and/or down) conversion between IR and UV frequencies [758]. Further research is needed to decipher the relationship between optical-frequency radiation, microtubule structural dynamics, and diseases like cancer for which the primary pathology is the failure of cellular communication, regulation, and control.

Microtubules were found to reorganize structurally under the application of UV light [807, 808], suggesting the possibility of photochemical control of microtubule dynamics [806,809,810] which are likewise associated with ROS regulation in living cells [811,812].

ROS were shown to control cytoskeletal dynamics with glutathione [813], and studies of anticancer drugs have linked pharmacological activity with increased ROS production and altered microtubule dynamics [814, 815]. Microtubules exhibit superradiance [74] and make good light harvesters [82], similar to quantum optical phenomena observed in light-harvesting nanotubes [335, 816]. This is technologically promising, because the presence of superradiance can be accompanied by superabsorption and subradiance effects which are advantageous for light harvesting and precision sensing applications [296, 817]. Structural arrangements of chromophores have important functional implications in macromolecular complexes found in biological light-harvesting systems [818]. Ring-like arrangements of chromophores, typical of biological light-harvesting complexes, are structurally similar to arrangements of quantum emitters used in quantum optical experiments on cooperative quantum phenomena using advanced light–matter platforms [819].

Strong light–matter coupling has been proposed as a means to overcome long-standing problems precluding efficient optical emissions from carbon nanotubes by enabling a way to suppress light-quenching effects, presenting a viable route to exciton brightening in these systems [820]. Curiously enough, the structural organization of microtubule chromophores has been found to enable a similar brightening effect in the lowest-excitonic state [625], which has likewise been proposed to enable observations of enhanced fluorescence quantum yields in tubulin when aggregated into microtubules [801]. Nanotubes have also been proposed as ultrafast light-harvesting systems [296], echoing the discovery of light-harvesting properties in microtubules [82]. Expansive applications of carbon nanotube-based chemical sensors have also been proposed [821], similar to the established sensory properties of biological microtubules *in vivo* [822, 823]. Realizations of strong light–matter coupling in these systems have important prospects for quantum optical control of biochemical processes (because of the exceptionally-large vacuum Rabi splittings of aromatic organic molecules — in the hundreds of meV — due to their large transition dipole moments) [321, 824–830].

Although light does not typically interact with other electromagnetic fields, the formation of polaron-polariton quasiparticles can enable the direct manipulation of hybrid light–matter states using electric and magnetic fields [831, 832]. The prospect of light-mediated biochemical interactions is especially promising in light of proposals for optical-frequency communication and information processing in biochemical systems [833, 834]. In spite of much anticipation regarding the possibility of using THz frequencies to generate the quantum mechanical condensates predicted by Fröhlich [794,835–837], biomolecules like microtubules and bovine serum albumin have yet to exhibit non-equilibrium Fröhlich condensation [838]. Nevertheless, non-equilibrium THz-frequency interactions have now been demonstrated between biological protein-pigment complexes in a triumph for the theory of electrodynamic processes in living matter [791–793, 839, 840]. Nonlinear optical effects involving UV-frequency UPE have been proposed to play a significant role in metabolic redox reactions and mitogenesis as hypothesized by Gurwitsch [758]. Biophotons have also been proposed to play an essential role in brain function [841] with myelinated axons potentially serving as photonic waveguides [833]. Neuronal biophoton emissions are facilitated by high potassium (K^+) ion concentration and can be impeded by the removal of extracellular calcium (Ca^{2+}) ions, indicating that measurements of UPE could be useful for monitoring redox physiology, metabolism, and stress [842, 843].

Observations of strong light–matter coupling in nanotubes and microtubules highlight the implications of quantum electrodynamics (QED) effects and suggest a key role for hybrid light–matter polariton states in these systems [816, 820, 844]. Exciton-polariton lasing has been demonstrated using fluorescent proteins trapped between laminate sheets [845]. Polaritonic states hold special promise for the prospect of biologically-generated coherent light because of their capacity to enable electrically-pumped inversionless laser-like beams at a fraction of the power consumption needed for conventional laser mechanisms [846]. Effects of QED have been observed in chlorophyll placed in a plasmonic nanocavity [847], and methods have been proposed to observe cavity-mediated excitonic correlations in photosynthetic quantum aggregates using phonon-driven dephasing as a means to probe correlations associated with long-range exciton migration in light–matter systems [848].

UV light inhibits the assembly of tubulin into morphologically normal microtubules [849], and UV radiation was found to induce breaks in cytoskeletal microtubules in the cell cortex [850]. Advances relating the dynamics of cell mitosis and meiosis to the electromagnetic fields generated by cell organelles and supercomplexes have revealed how synchronized oscillations of microtubules, centrosomes, and chromatin fibers may facilitate cell division and reproduction, with key implications for the diagnosis and treatment of cancer [851]. Prospects for photochemical control of microtubule dynamics [806–810] have been extended to demonstrate optical control of the microtubule cytoskeleton [852] using light-induced microtubule stabilization techniques [853]. Demonstrations of photochemical control of cytoskeleton morphology and dynamics may have substantial implications for biotechnology, including cytoskeletal manipulation for agriculture [854], embryology [855], immunology [856], and medicine [857].

Chapter 9

Electromagnetic Oscillations in Biostructures

Microtubules are main elements of the eukaryotic cytoskeleton along with actin and intermediate filaments [858, 859]. As the structural scaffold of the eukaryotic cell, their dynamics enable cell motility, cytoplasmic transport, mitosis and meiosis [860]. Microtubule dynamics are crucial to cytoskeletal regulation for tissue homeostasis, where microtubule destabilization has been correlated with cellular dysregulation and carcinogenesis [861]. Microtubules form complex networks of bundles in neuronal tissue, where the reordering of cytoskeletal microtubules is essential for learning and memory [862–864], and cytoskeletal dysregulation can lead to the occurrence of neurodevelopmental disorders and other pathologies [865,866]. Although textbook depictions of microtubules may appear static (like microtubule samples stabilized by taxol [867]), actual microtubule dynamics are vibrant, dynamic, and transient [868, 869]. This inherent dynamism is referred to phenomenologically as "dynamic instability," wherein microtubules undergo a continuing process of growth, collapse (a.k.a. "catastrophe"), and recovery (a.k.a. "rescue") [870].

The microtubule cytoskeleton has attracted attention for its mechanical and electrical properties [871, 872], as well as its capability for ion transport [873], charge storage [874–876], and signal amplification [877]. Microtubules have been considered as electric field generators [878,879]. Ion condensates along microtubule surfaces have also been identified using local pH measurements [874]. Tubulin, the protein from which microtubules are made, has an unusually negative charge on its C-terminal tail, imparting it with a range of signaling capabilities [871]. Given these features, microtubules likely operate as intracellular electromagnetic computing elements [878, 880], in line with proposals for biochemical information processing schemes [881] and known computational abilities of single cells [882]. Plant microtubules, for example, exhibit sensation relevant to the perception of mechanical stress [883] and sensory microtubules are integrated into plant cell architectures to optimize stress responses [822]. Microtubules also mediate mechanoreception in the cytoskeletal networks of insect cells to enable a sense of touch [884]. Likewise, microtubules have been proposed to mediate mechanotransduction in striated muscles [885] and neurons [886].

Microtubules are composed of repeating arrays of tubulin subunits in ordered arrangements of heterodimers, with each dimer containing a pair of tubulin monomers labeled α and β [871]. Tubulin is a mechanically durable and electrically tunable protein suitable for sensing, sorting, and information processing applications [887–890], as well as flexible optoelectronics [875]. Post-translational modifications can modulate microtubule structure and function via polymerization dynamics [858], but these do not fully explain the dynamical properties of microtubules (such as resistance switching [891]). Microtubule polarity is crucial to the formation of complex cellular structures including neuronal axons and dendrites, although the mechanisms that govern the formation of these hierarchical cellular structures remain largely unknown [892–895]. The inherent dynamics of microtubule organization processes are essential for maintaining the morphology and function of cells [896].

Tubulin monomers aggregate into microtubules by a non-equilibrium self-assembly process characterized by its dynamic instability [870]. Guanosine 5′-triphosphate (GTP) hydrolysis drives dynamic instability during microtubule self-assembly [897], resulting in rich collective behavior and stochastic switching between polymerization by self-assembly and depolymerization by sudden collapse (catastrophe), as well as its reversal (rescue) [898]. This enables cell reorganization and remodeling for mitosis, differentiation, transport, exploration, and apoptosis [899]. Atomistic models of dynamic instability have been implemented in multi-scale simulations [900] and there has been progress controlling non-equilibrium microtubule assembly and disassembly using programmable dynamics [901]. Microtubule disassembly can be initiated by pulses of THz frequency radiation [902], and repeated assembly and disassembly of microtubules by thermal hysteresis has been shown to induce protein aging and degradation [903].

Spontaneous electrical oscillations are an innate property of microtubule bundles found in the brain [904], which likely govern such cellular features as cytoskeleton-regulated ion channels and neuronal electrical activity. Thus, microtubules can operate as biological transistors with nonlinear transmission capabilities to support electrochemical signal amplification and propagation [905]. Microtubule oscillations are also likely to contribute to cellular electric fields, which in turn facilitate cell function [878]. Electromagnetic fields generated by microtubule oscillations [906] synchronized in concert with chromosomes are likely to affect key cellular processes during mitosis and meiosis [851]. Honeybee brain oscillations were observed to be mediated by microtubules, revealing a general oscillator mechanism to enable brain function and synchronization [907]. Likewise, bundles of mammalian brain microtubules have been used to demonstrate oscillatory entrainment and synchronization, revealing the capacity for microtubules to generate, propagate, and amplify coherent electrical signals in vitro (when the bundle preparation method avoided the use of microtubule-stabilizer and anti-tumor drug taxol) [908, 909]. Simulations of microtubule oscillations have been used to reproduce experimental data based on a kinetic model of microtubule nucleation, assembly (from GTP-loaded tubulin), disassembly, and dissolution [910].

The decoherence timescale in microtubules was estimated to be 1 fs–100 fs [911] (or even as much as 400 fs [310, 879]), which may be used as a reference for other decoherence processes in biology [912, 913] and is consistent with exciton decoherence time estimates of up to about 100 fs in photosynthetic systems [317]. This implies an ultrashort timescale for coherent processes in cells, where signals traveling at the speed of light nevertheless have adequate time to span the length of a typical eukaryotic cell (with a cellular diameter ranging from 10 to 100 μm). Coherent electrodynamic interactions may therefore be able to play important roles in cellular signaling and regulation over the complete range of length and time scales available in the cell. Biological synchronization must be continually renewed due to the presence of decoherence in the thermal environment of the cell, however noise can also be beneficial for transport and sensory-motor activities in biology [914–916]. It has become evident that microtubule dynamic instability provides a highly-optimized noise source in the cytoskeleton [917] — which may also incorporate such diverse phenomena as ionic flows, excitonic waves, and solitons [918, 919] — in a complex system of adaptive sensory response and control during cell homeostasis, division, differentiation, and growth [855, 920, 921].

Cytoskeletal microtubules became a major topic of interest in quantum biology after nonlinear electrodynamic phenomena in microtubules were theorized to enable the concerted organization of biological activities in the pioneering work of Hameroff, Smith, and Watt [922]. Since that time, microtubules have been proposed as a potential source of quantum effects in biology [923–925]. Microtubules are linked to a wide range of biological information processing functions including ciliary and flagellar control, axonal transport, and awareness [868,926] — insofar as "awareness" encompasses the whole set of mechanisms underlying the acquisition, storage, and processing of information at all levels of cellular organization [927–929]. This attention has motivated a picture of the microtubule as a quantum electrodynamical cavity with potential quantum information processing capabilities [879, 930]. Models of nonlinear dynamics in microtubules have been developed to inform the understanding of the functional properties of microtubule networks involved in motor protein dynamics and cellular information transfer processes [931]. Characteristic aspects of microtubule relaxation dynamics such as the relaxation of the system to a mixed or pure quantum state have been shown to depend on critical environmental parameters such as temperature [932].

Microtubules have permanent longitudinal electric dipole moments which cause them to align parallel to strong electric fields [933], whereas the application of an oscillating electric field has been shown to increase microtubule polymerization rates [934, 935]. The self-assembly of tubulin into microtubules was found to be enhanced by the application of MHz-band oscillating electromagnetic fields, where a cavity electrodynamic effect was employed to enhance microtubule growth by inducing synchrony and arresting spontaneous disassembly [936]. Beyond their properties as electrodynamically-oscillating sensors and electro-mechanical actuators, microtubules have been proposed to operate as optical cavities according to principles of cavity QED (cQED) [925] similar to other findings of confinement-mediated photochemical effects in organic nanocavities [937]. Cavity-induced electromagnetic interactions have been used increasingly to modify reactive, conductive, and relaxation pathways in organic chemistry [938], making microtubules an ideal testbed for investigations of cQED effects. This cQED microtubule model has been developed to propose mechanisms for generating mesoscale coherent quantum states [879, 939]. In particular, long-range "laser-like" quantum phenomena are proposed to enable optical signalling in microtubules, free from losses and thermal noise [924]. These proposals reflect established atom-optics experiments that realize strong light-matter coupling effects with fiber-optic cQED setups to enable networked quantum information processing schemes [940–942]. This picture of a microtubule network as a cQED system is accentuated by an ongoing debate about the experimental value of the refractive index of tubulin [943].

Discrepancies between estimated values of the refractive index of tubulin are intriguing because substantial variations in observed values of tubulin's index of refraction could be attributed to plasmon and/or exciton effects in tubulin [930, 943], particularly in light of demonstrations of plasmonic [944], polaronic [945, 946] and excitonic [947] control of refractive indices in photonic matter.

Prominent studies have placed the refractive index n of tubulin in the range of 2.4–2.9 [930, 948, 949], suggesting that tubulin may be at least as refractive as diamond. Alternatively, established dielectric theory predicts the refractive index of protein to be within a range $n = 1.5$–1.7, which is supported by critical-angle dispersion-refractometer experiments indicating a value of $n = 1.6$ [943]. Birefringence experiments using oscillating electromagnetic waves have shown that the microtubules respond to electromagnetic fields differently from tubulin [950]. The substantial difference between the predicted refractive index of $n \approx 1.6$ and multiple independent experiments finding $2.4 \leq n \leq 2.9$ presents the possibility that the refractive index of tubulin may be greatly altered by plasmonic resonances *in vitro* [930, 948] or by native dynamic effects *in vivo* [949].

Motivated by work on the excitonic properties of microtubules [625,807,951,952], Nganfo *et al.* have examined the dynamical properties of exciton-polaron pairs in microtubules. When an electron in a tubulin dimer is excited, the electronic excitation itself leaves a charge hole where the interaction between the excited electron and the hole form an exciton. Consequently, this results in an exciton-phonon quasi-particle known as a polaron [951]. Although a large refractive index $n > 1$ is preferable to confine electromagnetic modes to an optical cavity (by total internal reflection), a smaller refractive index $n \approx 1$ can enhance electronic dipole-dipole coupling in the cavity (by its $1/n^4$ dependence). Balancing this trade-off may be critical for quantum bio-polaritonics where the quality (Q) of cavity confinement must be weighed against the strength of excitonic coupling, in contrast with common experimental cQED set-ups that achieve strong light-matter coupling with lone atoms or dilute gas. Typical cQED set-ups require very large $Q \gg 1$, unlike condensed matter systems where the sheer number of molecules coupled to the field can ensure strong coupling even for modest Q. Moreover, recent studies have shown that nonlinear cQED effects can be dramatically enhanced through the tailored use of exciton-polaron interactions [953], presenting profound implications for the use of resonant electronic or vibrational oscillations to control chemistry [954]. Polarons can form easily in polar materials like microtubules by coupling extra electrons and/or holes with polar vibrations, and can lead to profound changes in material properties and functions [955].

A diverse range of factors is known to govern polarization effects in microtubules, although a comprehensive picture of their interactions remains lacking [956]. Studies have shown that exciton behavior is modulated by polaronic interactions with phonons in an electronically excited medium [957], where modified transport properties and lattice deformations due to exciton-polarons have been studied in depth in the context of structural dynamics [958]. Experimental support for the presence of polaronic coupling between phonon vibrational states and excitonic modes was provided by observations of Fano resonances in tubulin and microtubules by Raman spectroscopy [959]. This is consistent with previous experimental findings that linked observations of Raman spectra of metabolically active cells with coherent electric oscillations and vibrational solitons [960]. Electrical oscillations of microtubules are well documented [851, 878, 904, 906, 961–965], and the dynamical complexity found in microtubule networks has inspired proposals that microtubules can exhibit exotic physical and topological phenomena [966–968]. Microtubules play essential roles in cell motility, signaling, and transport [969, 970], and a cQED model of microtubules has been put forward in order to hypothesize a role for dissipationless energy transfer and computing in living cells [939].

Collective oscillations are found ubiquitously in biological systems where they govern many developmental processes [135]. Living cells coordinate cooperation between thousands of active molecules that collectively carry out tasks which range from cell differentiation and mobilization to motility and replication [971]. Collective spontaneous movements exemplify emergent phenomena on the molecular scale where self-organized dynamics generate active responses in living cells [972]. In multicellular systems, cells, tissues, and organisms are nested functionally as well as structurally, prompting the idea that collective cellular decisions are coordinated primarily by biochemical, bioelectric, and biomechanical signals [973, 974]. Microtubule ensembles, in particular, are able to exhibit collective spontaneous movements [971,975]. Membrane potentials are believed to play critical roles in biological patterning processes [976], where bioelectrical signals complement chemical and mechanical signals governing embryonic development, morphogenesis, and healing [977]. Cellular ion channels, known for regulating tissue morphology in planarian flatworms [978, 979], have also been characterized using non-Hermitian quantum theory [980]. This presents the possibility that the inherent topological features of living organisms may emerge from underlying quantum mechanisms which give rise to characteristic forms of biological synchronization [981].

Chapter 10

Functional Chemical Dynamics in Living Cells

Cellular homeostasis is contingent on maintaining a matrix of metabolites and signaling molecules in the cytosol (i.e., cellular fluid), which includes various metal ions, reactive oxygen and nitrogen species, kinases, and hormones [982]. Specific combinations of these molecules form the signatures of stress responses and other metabolic changes within the networks of chemical potentials in living cells [822]. To this effect, microtubules monitor calcium and ROS signals that in turn control microtubule dynamics in a regulatory feedback loop in collaboration with calcium ion (Ca^{2+}) channels. In this process, crosstalk between microtubules, ROS, and Ca^{2+} ion concentrations enables the propagation of electrochemical signals in response to stress [982]. Although many studies have revealed wave-like flows of electrons and ions along cytoskeletal elements, it was only recently shown that ion flows along the cytoskeleton are optimized for signal transmissions whereby microtubules carry environmental information to the centrosome and nucleus with minimal data loss [890]. Similarly, the stochastic chemical dynamics of membranous ion channels are known for their weak signal detection and amplification abilities [983].

Stochastic fluctuations in microtubules have been associated with the phenomenon of stochastic resonance (SR) [984], a counterintuitive physical effect found in bistable systems subject to both periodic and random fluctuations — where an increase in the input noise can improve the output signal-to-noise ratio [985]. Although noise is often considered an enemy of signal detection, random fluctuations play a beneficial role in enhancing weak signal detection in nonlinear systems that possess delicate sensory thresholds [986, 987], such as precision photodetector systems [988]. Fundamental noise — *true randomness* — is an essential resource for computing tasks in many scientific fields including cryptography and optimization, machine learning, chemistry, and materials science [989–991]. Noise is also a key component in many natural decision-making processes, where it enables self-organized systems to adapt to dynamically changing environments by striking a balance between stable and flexible behaviors [992]. SR is common in many chemical, electromagnetic, mechanical, and optical biosensory systems [983, 986, 987, 993–997]. SR was investigated as a noise-driven signal amplification mechanism in tubulin, where the amplified signal-to-noise ratio was derived for an effective potential governing delocalized electron dynamics in a tubulin dimer [984]. Microtubules, comprised of tubulin, are cellular developmental regulators that function at the intersection between biochemical and mechanical growth control [883], responding to various forms of chemical, physical, and mechanical stresses [998, 999]. Microtubule fluctuations are crucial for sensing, learning, and memory [1000, 1001], and label-free microscopy has revealed that stochastic switching between disassembly and assembly phases of microtubule growth is stochastic [1002].

Microtubules provide the structural framework for a cylindrical cytoskeletal structure known as the axoneme in cellular appendages such as motile cilia and flagella, where a dynamic influx of Ca^{2+} ions initiates and synchronizes the action of the dynein motor proteins which drive the motile power stroke to propel the cell or circulate surrounding fluid [1003, 1004]. Ca^{2+} ions are crucial to many signaling pathways in cilia and flagella, even though very low concentrations of Ca^{2+} ions will disassemble microtubules (unlike Mg^{2+} ions which were found necessary for microtubule polymerization [1005]). Calcium signals also control bundle motility in actin filaments [1006]. Microtubules are linked to a wide range of regulatory functions including transport, positioning of cell organelles, and ciliary and flagellar control [868]. In contrast with motile cilia, primary cilia are solitary, non-motile extensions of the centriole found in most eukaryotic cells, where they operate as specialized calcium-signaling organelles [1007]. Also known as sensory cilia, primary cilia transduce optical, chemical, and mechanical signals that enable cell signaling pathways [1008, 1009]. Their role as the "antennae" of the cell may explain why primary ciliary protein mutations are linked to severe developmental defects [1007, 1009]. A growing body of research is beginning to uncover the complex relationship between cellular mechanical force sensing and chemical signaling via Ca^{2+} ions, hormones, ROS, and the immune system [1010]. Synchronized interstitial calcium oscillations, in particular, are crucial to enable tissue functions which have synchronization properties determined by the dissipation of coupled cellular oscillators [1011].

Non-equilibrium microtubule dynamics are fueled by hydrolysis of GTP into guanosine 5′-diphosphate (GDP) in a process where practically all of the energy freed by hydrolysis is stored in the microtubule lattice [1012]. GTP hydrolysis is involved in regulating most cellular processes, yet the mechanism of enzyme-catalyzed GTP hydrolysis has been controversial [1013]. Free energy input is required for the reaction-diffusion process of microtubule self-organization [1012, 1014–1016]. Coupling between the reactive and diffusive processes governing microtubule self-organization are associated with the emergence of complex phenomena in the form of macroscopic traveling waves and microtubule "treadmilling" effects that can enable long-range transport [1017, 1018]. Self-organizing reaction-diffusion dynamics of microtubules are also believed to be responsible for functional deformations of the cell membrane during the growth of microtubule networks [1019]. Their rich nonlinear dynamics have made microtubule networks a fertile subject for research in solitary waves [1020], including kink-like energy transfer mechanisms [1021] and bell-shaped solitons [1022], as well as dark, bright, and singular solitary waves [1023]. Solitary waves were proposed to enable essential information processing mechanisms in a nonlinear model of microtubule dynamics [1024]. Conditions for the amplification, chirping, and damping of solitonic ion waves along microtubules have also been predicted using the nonlinear Schrödinger equation [1025, 1026].

Nonlinear amplification is an essential ingredient in the formation of solitons, self-reinforcing solitary waves which move with constant velocity and shape [1027]. Microtubule networks [1020, 1022] have been proposed to support the propagation of charge-transfer solitons [878] and soliton-like "clouds" of Ca^{2+} ions [1028], similar to models of solitonic electrical impulses in actin filaments [1029]. Electrical solitons have been observed in neuronal dendrites, where they were proposed to influence ionic flow generation [1030]. More importantly, solitonic "wetware" of this kind has been proposed to work as a liquid core waveguide for use as a polariton transmission line in a polaritonic network for applications of strong light–matter coupling [1031, 1032]. Coherent interactions of this nature are proposed to have central importance for coordinating the shape and movement of the cell [836,1033], which is in turn a determining factor controlling tissue maintenance and processes such as embryological development, angiogenesis, bone formation, collagen synthesis, immunity, inflammation, and metastasis [1034–1038].

Similar to microtubules [984], membrane-bound ion channels are understood to employ SR to enable coherent collective dynamics at the biomolecular scale [986, 993, 1039, 1040]. The identification of this coherent collective phenomenon, known as coherence resonance [993], has inspired nanoengineering research efforts to use coherent oscillations generated from inherently noisy stochastic components for precision sensing, ion transport, and energy transduction applications [996, 1041]. Quantum coherence is believed to enable the high ionic conductance and selectivity in biological ion channels [1042, 1043], where quantum resonances have been proposed to explain ion conduction and selection processes [1044, 1045]. Cellular ion channels are recognized as active linear amplifiers in the cell [1046] which also have been proposed to enable the amplification of electromagnetic fields using gating effects that are modulated with SR [1047]. In general, SR is possible in bistable systems with inherent noise [1048] (such as tubulin dimers assembled into microtubules [984]). SR is also theorized to exist in single biomolecules and individual ion channels at physiological temperatures [1049]. This has motivated research on coherence resonance effects in carbon nanotubes and nanopore systems for ultrasensitive detection systems, ion-based computers, and single-molecule chemical reactors [996]. As a biological equivalent to carbon nanotube structures [950, 1050], microtubules have been investigated for their potential properties as advanced electromechanical biosensors [1051].

A recent meta-analysis revealed that low frequency magnetic fields are likely to influence calcium homeostasis in cells *in vitro* [1052], where moderate-intensity static magnetic fields may influence physiology by varying the action of Ca^{2+} flux through ion channels [1053], indicating a possible role for ion fluctuations in the cellular responses to magnetic fields [1054]. Cytosolic calcium oscillations have been linked to low frequency (50 Hz) electromagnetic field effects in living cells [1055]. Simulations indicate that the most sensitive step in a calcium oscillation chain reaction is the release of stored Ca^{2+} ions by calcium-induced calcium release (CICR) [1056], similar to mitochondrial ROS-induced ROS release (RIRR) [1057]. This finding reveals how experimentally observed magnetic resonance effects may be explained using nonlinear dynamics, where the dynamics' extraordinary sensitivity to periodic waves of cellular Ca^{2+} oscillations indicates a role for chaos in cells' detection of weak electromagnetic signals [983]. Various models have been proposed to explain the mechanisms of ionic signal transduction and chemical bifurcation by which calcium oscillations arise in cells [1058].

Similar to the role that Ca^{2+} and K^+ ion concentrations play in mediating cellular signaling [881], the messenger molecule nitric oxide (NO) mediates communication in various tissues. Interestingly, NO levels were found to modulate firefly respiration and bioluminescence in a light-dependent switching effect [1059]. Furthermore, synchronized action potentials were found to originate from heterogeneous subcellular subthreshold Ca^{2+} signals in the cardiac sinoatrial node, i.e., the chief pacemaker of the heart, during the entrainment of spontaneous excitations in the individual pacemaking cells [1060]. The interplay between ion and oxide oscillations, physiological synchronization, and biological luminescence [1061] presents an interconnected series of biochemical reactions that suggest the need to develop a new paradigm for characterizing synchronization effects in biology [1062].

The synchronized control of ROS and other free radicals is essential to homeostasis, as a sufficient number of ROS are needed for metabolism, signaling, and immunity [194]. However, an excess amount can spell death for the cell. The problem is complicated by the "open shell" electronic structure of molecular oxygen. Diatomic oxygen (O_2), as it is found in the air we breathe, is uniquely configured with a triplet electronic ground state (i.e., $^3O_2$). This means that the lowest energy state of molecular oxygen has non-zero spin, in contrast with the usual singlet (i.e., zero spin) ground state of "closed-shell" chemical systems. Cellular ROS are known regulators of microtubule dynamics [1063], and studies have shown that microtubule dynamics are diminished in the presence of ROS *in vitro* [811]. Microtubule network destabilization, depolymerization, and re-assembly were also found to be triggered using an oscillating magnetic field [1064]. Taken together, these findings indicate that microtubule dynamics entail magnetically-sensitive chemical processes that are quenched by ROS. This was anticipated by the hypothesis that perception of the geomagnetic field could be enabled by magnetosensitive chemical reactions that compete with radical scavenging to enable chemical amplification [1065].

An extensive body of research has emerged to associate physiological ROS modulation with both static and time-dependent MFEs at field strengths on order of mT or smaller. ROS are critical for self-renewal in stem cells, but the ways that ROS signals are integrated into self-renewal mechanisms remain unclear. Although ROS are a well-known by-product of mitochondrial respiration, recent insights have also revealed specific biochemical regulatory pathways that invoke NADPH oxidase, initiating and amplifying ROS signals that activate transcription factors in a positive feedback loop that drives cellular regeneration [1066]. ROS signals also contribute to the proliferation and survival of many cancers [1067], making the precision modulation of mitochondria/cell redox communication a promising avenue for therapy and oncology. Moreover, the multitude of roles played by ROS in homeostasis, carcinogenesis and cellular renewal is revealing a subtle and cryptic system of redox regulation of cellular function.

Growing evidence indicates that cancer cells suffer increased ROS stress — over and above that of normal physiological ROS-generating processes — due to oncogenic stimuli, mitochondrial dysfunctions, and metabolic dysregulation that can all contribute to high ROS levels in cancer cells [1068]. Further research on the influence of magnetic fields on ROS levels in biological tissues will be needed before magnetic field-based treatments can become available for oxidative stress-related diseases such as cancer [1069]. Although moderate amounts of ROS mediate cell signaling [1070], proteolysis, and microbial defense [1071], excessive amounts can lead to mitochondrial and DNA damage, lesions, and death [1072]. Redox stress is correlated with age-related diseases [1073, 1074] including cardiovascular disease, chronic obstructive pulmonary disease (COPD), kidney disease, neurodegeneration, sarcopenia, fibrosis, frailty, and ischaemic stroke [1074–1076]. ROS also influence pathogenesis in a wide range of other diseases that include autoimmune disorders, congenital defects, diabetes, and cancer [1077, 1078].

The appropriate balance between chemical reduction and oxidation is essential to healthy cellular physiology, where dysregulated redox balance often occurs with the production of excess oxidative species, leading to DNA damage, lipid peroxidation, and abnormal post-translational protein modifications that cause injury, disease and cell death [1063]. Despite the inherent risks associated with molecular oxygen [1079], it is critical to cell function where it regulates cell dynamics including growth, healing, and regeneration. Monitoring of cellular ROS levels has become a topic of growing biomedical interest [1080–1082] because ROS are essential to cell function and immunity [1083, 1084]. ROS include superoxide ($O_2^{\bullet -}$) and hydroxyl ($HO^{\bullet}$) radicals as well as non-radicals including hydrogen peroxide (H_2O_2), singlet oxygen ($^1O_2$) and ozone (O_3) [1085]. NADPH oxidase is the prevailing source of ROS in many cells [194, 1074, 1086] where different forms of NADPH oxidase have similar ETC structures with comparable NADPH and FAD-binding sites in the enzyme complex's cytosolic C-terminal tail [1087]. Of the many types of NADPH oxidase, phagocytic NADPH oxidase has been called the prototype of the NADPH oxidase ETC systems [1087]. Phagocytic NADPH oxidase is a superoxide-generating enzyme complex that constitutes an ETC connecting NADPH to the phagocytic vacuole where microbial killing occurs in phagocytes (e.g., white blood cells) [1087]. Phagocytes are incapable of killing bacteria and fungi without NADPH oxidase and a deficit in this enzyme can result in life-threatening infections [1088]. NADPH oxidase complexes have been found in various plants, fungi, sea urchins, worms, and flies, as well as in several different organs of higher animals [1087].

In eukaryotic cells, mitochondria produce ROS, regulate cellular redox stress, maintain Ca^{2+} homeostasis, synthesize and break down high-energy metabolites, and regulate cell death — all while producing about 90% of the cellular energy needed to survive [1089]. In white blood cells, for example, ROS concentrations are also modulated by biochemical mechanisms that influence NADPH oxidase activities [1090]. Hypoxia-induced ROS amplification has been found to trigger mitochondria-mediated apoptosis, accelerating the rate of intracellular ROS production under oxygen-starved conditions [1091]. Sodium (Na^+) ions have also been found to control hypoxic signaling by the mitochondrial ETC in an unanticipated interaction with phospholipids [1092]. Ionic signaling between cells has now been observed in many different cell types to coordinate cellular regulation and controlled cell death [769]. This form of intercellular chemical signaling can become a double-edged sword in light of the fact that cancer cells create a microenvironment of oncometabolites that perpetuate cancer growth in a vicious chemical-signaling cycle [1093].

Calcium ions are essential to maintain homeostasis and coordinate function in mitochondria [1094], which in turn produce most cellular energy, control oxidative metabolism, maintain cellular Ca^{2+} levels, regulate high-energy metabolites, and mediate apoptosis [1089, 1095]. All this relies on a complex system of Ca^{2+} sensors which mitochondria use to regulate their internal and surrounding Ca^{2+}

concentrations [1096–1098]. Calcium regulation also plays a key role in gene expression [1099]. Although calcium homeostasis is usually modeled in terms of an intricate system of chemical machinery made up of pumps, ion channels, and auxiliary proteins [1100, 1101], some recent proofs-of-principle have demonstrated photosensitizer-based mechanisms that use light to control cellular calcium regulation [1094, 1102–1104]. Light-based Ca^{2+} signaling methods involving endogenous calcium reserves and native photosensitizers (such as flavin-containing compounds [1103]) are particularly significant, as they suggest the possibility of light-based homeostatic regulation. Mitochondria have been shown to demonstrate light-based interactions [1105], consistent with documented examples of nonlinear cellular responses to low-intensity radiation [1106, 1107].

Mitochondrial redox stress is a key regulatory factor in immune responses controlling ion homeostasis, ROS signaling, cytoskeletal dynamics [1063], mitophagy, and apoptosis [1108]. When redox homeostasis is included in the cellular model, it extends the essential problem of regulating electron tunneling for respiration in the ETC to encompass a vast undertaking in chemical sensing and control. ROS regulation thus entails an enormous control scheme to regulate chemical dynamics in the many-electron systems of cells, tissues, and organs [1109]. In the NADPH oxidase ETC, electron transfer cofactors are arranged to tune the transfer of O_2-reducing electrons from inside the cell to the ROS-generating site on the outside of the cell membrane in cooperation with structural elements that control NADPH binding, tightly managing ROS generation [1110]. In turn, NADPH oxidase action is coupled to the activity of voltage-gated ion channels [1111] and reactive chemical intermediates such as peroxynitrite [1112], highlighting its influence on osmotic balance and redox homeostasis. NADPH oxidase regulation is associated with hypertension, making it a major source of cardiovascular and renal redox stress [1113]. For example, neutrophils sequestered in the pulmonary vasculature can release ROS which can in turn induce endothelial dysfunction and disruptions responsible for the primary clinical manifestations of acute respiratory distress syndrome

(ARDS) [1114]. These factors suggest a promising prospect for the pharmacological and dietary manipulation of cellular NADPH oxidase activities, which are commonly implicated with oxidative stress in neurodegenerative disorders [1115].

Mitochondria function as critical integrators of cell signaling cascades involving calcium, ATP, and ROS due to the central role of mitochondria in metabolism, redox chemistry, and cellular life/death decisions [1095, 1116–1118]. Although calcium regulates mitochondrial function by stimulating ATP synthesis and ROS generation, the mechanisms by which Ca^{2+} ions elicit mitochondrial dysfunction have been challenging to identify [1119]. Dysregulated mitochondria can undergo structural and dynamical changes via carcinogenesis [1120,1121], and various forms of tumors have revealed dysregulated mitochondrial dynamics [1122]. As a result, cancer cells suffer from a wide range of impaired functions that include metabolic regulation, energy transduction, calcium homeostasis, ROS production, and apoptosis [1123]. Mitochondria have thus been designated as the "metabolic switch" of oncogenesis [1124] in pathophysiological models that treat mitochondrial ROS as endogenous oncogens [1125, 1126]. Aberrant mitochondrial morphology may lead to increased ROS production and consequently decreased mitochondrial health, further exacerbating a self-perpetuating cycle of escalating oxidative stress [1127]. Mitochondrial dysfunction and disturbed cell coherence have thus been described as a "gate to cancer" [804], prompting interest in cellular exposure to coherent electromagnetic fields as a potential cancer therapy [1128, 1129]. Indeed, tumors in rodents exposed to a constant $60\,\mu$T magnetic field combined with an alternating 100 nT magnetic field (with six frequencies) experienced tumor growth suppression with increased survival rates and enhanced average lifespans [1130]. This has prompted a range of proposals for new mitochondria-targeted cancer therapies [1131]. Nevertheless, the exact link between mitochondrial dysfunction, cellular electromagnetic oscillations, and cancer remains unclear [1132].

The bioenergetics of mitochondrial respiration have emerged as important factors in determining whether cancer cells either succumb to or evade apoptosis [1133]. A cell's stress response to the harsh tumor environment is associated with the acquisition of common malignant traits [1134]. Two hallmarks of cancer are the use of glycolysis to support energy production, known as the Warburg effect, and dysregulated mitochondrial oxidative phosphorylation (OXPHOS). Both of these contribute to carcinogenesis [1120] in a process that has been exacerbated by the emergence of drug-resistant cancer stem cells [1135–1137]. Cancer stem cells allow cancers to relapse following initially-successful treatments by conventional radiation or chemotherapy treatments. The theory of ROS-as-oncogen is corroborated by recent work showing how cells placed in a magnetic field weaker than the Earth's undergo metabolic shifts towards dysregulated oxygen use; this is similar to the Warburg effect found in the progression of cancer metabolism [1138]. Comorbid stiffening of the extracellular matrix (ECM) during the reverse Warburg effect [1139] results in inadequate mechanical transduction and may promote the development of further malignancy [1140].

About half of human mitochondrial disorders are associated with complex I mutations [123]. Dysregulated complex I activity is correlated with defects in its assembly which lead to metabolic disorders in humans [1141]. Complex II has also gained attention as a ROS generator [1142] with mutations correlated with cancer and neurodegeneration [1143, 1144]. Thermodynamic analyses of mitochondrial complexes have shown that vulnerabilities in complexes III and IV can emerge even under resting conditions, highlighting the importance of metabolic control of these complexes [1145]. Mitochondria support innate immunity by producing ROS in response to cell damage and stress [1146], shaping the orientation and character of the cell's response to stimuli [1147]. Control of the electron transfer dynamics which produce ROS and free radicals is therefore crucial to maintaining cell homeostasis [125, 1148].

Mitochondria play a central role in determining cellular health [1149] and impaired mitochondrial function can cause debilitating disease [1150]. In eukaryotic cells, up to 90% of the ROS are generated in the mitochondria during ATP synthesis [1151] wherein mitochondrial complexes I and III are considered primary generators of $O_2^{\bullet -}$ and other ROS [1152–1155]. Disruptions in the electron transfer efficiency of respiratory carriers such as electron transfer flavoproteins (ETFs) and their partner enzymes are an important source of ROS in cells [722]. Salient features in the structures of the quinone binding sites of the ETC reveal an ideal framework to inspire programs in rational drug discovery [1156]. ROS play a central role in autoimmune diseases, inflammatory conditions, cancers and bacterial infections. This has made ROS amplification an attractive effect for use in the design of therapeutic interventions, but practical implementations of ROS amplification are problematic because interventions that systemically increase oxidative stress can cause severe side effects [1090]. Targeted, cell-specific control of redox chemistry is therefore a preferable method to enable biomedical applications, but the cellular regulation of ROS, metal ions, and metabolites is complex and interconnected [1119]. Quantum mechanical effects are implicated in many aspects of the vast network of mitochondria, microtubules, ion channels, and ETCs that enable biochemical regulation and immunity in living cells. This understanding of cellular signal integration between nuclear, mitochondrial, and cytoskeletal organelles inspires a new vision of the multifunctional and responsive nature of the cellular microenvironment [1157–1159].

Chapter 11

Magnetic Biomodulation: Biodynamic Control

The Earth's magnetic field influences various animal behaviors, ranging from migrations to herd behaviors [1160–1162]. Many creatures across various taxa have demonstrable abilities to perceive and respond to magnetic fields, especially birds (known for their long migrations) including European robins, silvereyes, garden warblers, homing pigeons [1163] and zebra finches [1164]; but also cockroaches [449], fruit flies, honey bees, mice [1165], mole rats, lobsters, salamanders [1166, 1167], salmon, sea turtles, sharks [1163], and humans [1168]. Moreover, detailed studies have revealed the importance of the Earth's magnetic field to biological function. Geomagnetic field deprivation was found to reduce neurogenesis and cognition in mammals [1169], to disrupt appetite and feeding in insects [1170], and to induce abnormal morphological development in amphibians [1171, 1172]. Insights into mechanisms underlying these physiological disruptions have come from studies showing that microtubule assembly is disordered in the absence of the geomagnetic field (GMF) *in vitro* [1173], and that geomagnetic field shielding leads to disordered actin assembly kinetics that results in reduced cell motility of human cells [1174]. These findings have inspired a proposal to use hypomagnetic field effects as a means to test the RPM in biology [1175].

Long-term deprivation of the Earth's GMF was shown to impair nerve growth and development in the mammalian brain [1169]. This impairment was demonstrated to be ROS-mediated when normal neurogenesis and cognition were recovered by pharmacologically restoring ROS levels in mice subject to a magnetic field much weaker than the Earth's [1169]. The careful controls used in that study further demonstrated the functional relationships between magnetic fields, ROS levels, neuron growth and cognition [1169]. Defective neurogenesis and cognition were reversed in the GMF-deprived mice by returning them to the GMF which restored neuronal ROS to normal physiological levels. The cognitive benefits of normal physiological ROS levels could again be eliminated by chemically lowering hippocampal ROS levels in mice shielded from the GMF, effectively demonstrating a key role for Earth-strength magnetic fields in regulating physiology via biochemical modulation effects [1176, 1177].

A hypomagnetic field (HMF) — i.e., an ambient magnetic field much weaker than the GMF — can have unfavorable effects on biological processes including embryonic development, morphogenesis, and behavior [1178, 1179]. Despite the body of evidence demonstrating the influence of subtle magnetic fields on physiology, substantial progress developing magnetotherapeutics has been stymied by a lack of widely accepted mechanisms to explain how electromagnetic fields induce physiological changes in living cells [505, 1180]. Without a unified theory of MFEs on biology, a range of hypotheses have emerged based on diverse mechanisms involving radical scavenging and recombination reactions, voltage gated ion channels, kinase signaling pathways, and/or heat shock proteins [1152]. The finding that low-intensity magnetic field-induced ROS generation requires the presence of the flavoprotein photoreceptor cryptochrome in human cells [1181] supports the hypothesis that carcinogenesis associated with electromagnetic field noise, electromagnetic field-induced ROS production, medical treatments using pulsed magnetic fields, and magnetic sensing *in vivo* share a common mechanistic basis [534]. Alterations to cellular free radical activities were also reported to affect DNA integrity, immune and inflammatory responses, cell proliferation and differentiation, wound healing, neural activities and behavior [1152, 1245].

Questions about the influence of $\sim$50 μT geomagnetic fields on physiology have been motivated by models of animal magnetoreception and suggestions that even weaker anthropogenic low frequency electromagnetic fields ($<1\,\mu$T) may have vital implications for human health [505, 1182]. Magnetic fields much smaller than 1 mT can influence rates of radical pair reactions in chemistry, although the resulting product yields are often on the order of a few percent or less [1183]. These marginal product yields have left important questions open about a role for weak field effects in biochemistry that pertain to sensing, metabolism, and cellular growth [1184, 1185]. Thus, substantial work has been done to characterize the chemical amplification of MFEs due to the presence of radical scavenging molecules [501, 552] interacting through radical termination reactions [519]. This work has further motivated the question of whether a small magnetically-induced change in the rate of one step in a multi-step reaction could disproportionately change the overall biochemical reaction rate [1183], inspiring subsequent investigations of chemically-amplified MFEs using a spin-dynamical reaction–diffusion model of ROS oscillations in complex III [1186].

The effect of magnetic fields on reaction rates and kinetics have been known to chemistry for decades [480], as electronic and nuclear spin polarization effects were shown to influence chemical reactions in the 1960s [483, 484]. The influence of a weak magnetic field (WMF) on a chemical reaction has also been called the low field effect [495]. Speaking broadly, the presence of a low field effect is the hallmark that an otherwise spin-forbidden chemical process has been "turned on" and is now allowed by a magnetic field-mediated radical-coupling mechanism [498, 1187]. Typically, this occurs by the lifting of a degeneracy between reactant and product energies that would otherwise render the effect negligible. The effect is to modulate the rate of an ISC process that is symmetry-forbidden in the absence of radical coupling or strong magnetic fields [1188].

The biological reception of weak magnetic signals is widely associated with spin-chemical radical-pair reaction mechanisms [432, 443]. Radical pair reactions have also been proposed to play a role in general anesthesia [1189], implicating radical pair dynamics in microtubule-based cellular perception [1190]. Microtubules have been proposed to play an essential role in mediating cognition, as anesthetics bind to the microtubule building-block protein tubulin and have been linked to microtubule depolymerization [1191], and hence may account for selective action of anesthetics on consciousness and memory [1192]. The influence of quantum effects in non-equilibrium microtubule dynamics was implicated in the discovery that the assembly of tubulin (into microtubules) is disordered in a HMF [1173]. The failure of microtubule growth in the absence of the GMF led to the hypothesis that radical pairs may be involved in microtubule assembly [1190]. However, that hypothesis is problematic because it involves a singlet-born radical pair with a weak-field MFE that would be reversed in the high-field limit [495], which is inconsistent with observations of well-organized microtubule growth in strong magnetic fields [1193, 1194]. Although the RPM is widely believed to rationalize the influence of WMFs on biological systems [432], the explanatory model offered by the RPM remains limited by fundamental theoretical challenges [482]. Alternative models of biological magnetoreception have also been proposed based on a magnetite-based sense [1195] and the magneto-hydrodynamic effect [1196]. Challenges associated with identifying common biophysical mechanisms of magnetoreception across different organisms have also motivated efforts to designate innate physical properties (such as gyromagnetic ratios and thermal relaxation times) to characterize specific aspects of magnetoreceptive mechanisms in biology [1197].

Weak fields of less than 1 milliTesla (mT) have been shown to influence various biological processes in hundreds of studies [443]. Despite the wealth of evidence demonstrating that magnetic fields can influence ROS levels in living cells, the mechanism(s) underlying cellular ROS modulation has been challenging to characterize due to the high complexity of the effect [1069]. ROS are implicated in HMF effects on neurogenesis [1179] and stem-cell mediated cell regeneration [548]. Magnetic fields were shown to influence redox metabolism in human neuroblastoma cells [1138], where the application of magnetic shielding induced cancer-like metabolism by upregulating anaerobic glycolysis and repressing ROS production (i.e., the Warburg effect). As far as ROS regulation is concerned, singlet oxygen has attracted considerable attention for its toxic and carcinogenic effect in the oxidation of lipids, proteins and nucleic acids [1198]. Redox stress due to increased ROS in tumor cells is essential for cancer progression and recurrence [1244]. Paradoxically, singlet oxygen has also attracted interest for its anti-tumor effect [1199], and the production of mitochondrial singlet oxygen has provided a rationale to explain the effect of metabolic therapies in treating cancer [1200].

The application of a HMF via magnetic shielding was also shown to reduce ROS production in blood cells [1201, 1202], whereas applying a modest magnetic field could produce a moderate increase in ROS levels [1203]. Exposure to GHz frequency electromagnetic waves was also shown to modulate ROS in human kidney cells as a function of the signal amplitude [1204]. The application of a low-frequency, pulsed magnetic field was shown to stimulate collagen production in fibroblasts cultured from chicken tendon, indicating that pulsed magnetic fields can increase vertebrate collagen production [1205]. A subtle balance of homeostatic ROS levels is essential for effective wound healing [1206], and changes in static magnetic fields can alter both ROS and calcium ion concentrations, acidity (pH), and growth rates of cancer cells [1207]. Magnetic field treatments have thus been proposed as a possible method of wound therapy based on results from experiments on animals and clinical trials [1208].

Static magnetic fields have been shown to have an impact on human physiology and sensation [1209]. Blood cells have been shown to orient diamagnetically in strong magnetic fields, where a significant contribution to the total cellular anisotropic diamagnetic susceptibility was attributed to the microtubule cytoskeletal network [1210]. Static MFEs on cellular ROS levels have attracted particular interest in studies because of their apparent ubiquity in cell physiology and presumed connection with the RPM [1211]. The applications of static mT-scale magnetic fields were shown to have an effect on the growth of three types of bacteria, decreasing growth rates of *Escherichia coli* and *Staphylococcus aureus*, while increasing the growth of *Bacillus subtilis* [1212]. A combination of static and oscillatory magnetic fields produced a stimulating effect on the fission and regeneration of planarians, where removing the static component could reverse the effect [1213]. Static WMFs $<$1 mT were shown to be able to modulate stem-cell mediated growth and regeneration in planarian flatworms [548]. Endogenous ROS concentrations were shown to be modulated by control of the ambient magnetic field, where ROS levels and associated growth rates were together either augmented or diminished depending on the strength of the applied magnetic field. The correlation between ROS variations and growth was confirmed by demonstrating that growth-inhibition by the magnetic field could be counter-acted by restoring ROS levels biochemically [548]. This provides further evidence that static WMFs can influence redox homeostasis and cell metabolism, with a significant impact on tissue growth after an injury.

Magnetic manipulation of ROS levels can also alter redox homeostasis, metabolism, and growth of white blood cells [1214]. Coordinated sequences of pulsed WMFs have been applied to control metabolism in neutrophils to enable chemical ROS amplification via metabolic resonance using respiratory burst synchronization [1071]. In these experiments, electromagnetic resonance with NADPH oscillations in human neutrophils was achieved by synchronizing externally-applied pulsed magnetic fields with the innate frequency of flavoprotein redox oscillations in the neutrophil cells, where magnetic pulses were phased to coincide with the NADPH fluorescence minima. Production of NO was found to follow the production of ROS closely as the NADPH oscillation increased cumulatively with the application of each pulse. Production of ROS and NO could likewise be terminated by pulsed WMFs, leading to neutrophil depolarization, reminiscent of NO-mediated control of firefly flashing [109].

In fact, an extensive body of work has been carried out by Novikov and coworkers varying magnetic field conditions to modulate respiratory (i.e., ROS) bursts in neutrophils (i.e., white blood cells). These include studies of the following effects: the effect of hypomagnetic conditions on neutrophil ROS production [1178], the effects of weak combined static and oscillating magnetic fields of neutrophil respiratory bursts [1215, 1216], the influence of oxygen on neutrophil respiratory bursts upon exposure to WMFs [1217], the role of calcium ions and hydroxyl radicals on respiratory bursts in neutrophils [1218], the impact of combined static and oscillating magnetic fields on lipid peroxidation in neutrophils [1219], the comparison of the responses of human and mouse neutrophils to weak combined static and oscillating magnetic fields [1220], the action of weak combined static and low-frequency oscillating magnetic fields on luminol-mediated chemiluminescence in mammalian blood [1221], and the quenching of magnetic field-enhanced chemiluminescence by free-radical scavengers in human blood [1222].

The complexity of these effects suggests that several receptors may be responsible for WMF effects in biological systems [1223]. For example, very low-frequency electromagnetic fields can accelerate wound healing by modulating cytokines and metalloproteinases [1224] or induce increases or decreases in cellular DNA methylation levels [1225], suggesting that low frequency magnetic oscillations correspond to energy separations between molecular states [1226]. Whole-body exposures to WMFs could induce seizures and reduce learning of rats, or improve performance at a learning task, depending on the specific magnetic field pattern employed [1227]. Studies of physiologically-patterned weak magnetic-field treatments indicated that the intensity of the field required to ensure treatment efficacy may decrease substantially with the degree of pre-programmed field patterning [1227]. Oscillating fields have been used to alter relative yields of ROS products in living cells, indicating coherent singlet-triplet mixing at the point of ROS production and concurrent metabolic changes [230]. Even nanoTesla-scale oscillating magnetic fields have been shown to induce biological changes, and the exposure to appropriately tuned fields can significantly increase metabolic ROS levels in biochemical systems [1228]. To this effect, new classes of physiologically patterned electromagnetic signals are being investigated for magnetobiology research and therapy [1229].

Many examples of the influence of magnetic fields on physiology also involve illumination [467, 515, 1163, 1230–1233], including the human magnetic sense [1168]. This reveals that cellular behavior, and in some cases animal behavior, can reflect the complex interplay between light and magnetic fields. Cellular processes involving oxidative metabolism and redox signaling can involve light-dependent magnetic sensitivity [1234], and photochemical pathways that mediate MFEs are generally linked to ROS modulation, cell growth, and tissue repair. Magneto-optic interactions of this kind have motivated hypotheses that connect electromagnetic fields to carcinogenesis, pulsed magnetic field therapies, photobiomodulation, and other areas of regenerative medicine and biology. New results in optogenetics hold promise to circumvent conventional limitations on cellular ROS control (via system-wide applications of ROS generating-or-quenching molecules) by enabling targeted applications of light to activate genetically-encoded effector proteins [1235]. Quantum sensing techniques, such as magnetic relaxometry, have also emerged to enable the detection of chemical radicals with subcellular resolution [1236, 1237].

Quantum mechanical methods are well-developed in biochemical and molecular electronics using biomolecules such as proteins, bacterial nanowires, and DNA [1238]. Biological processes are known to respond to WMFs [443, 1246], but there is still no viable hypothesis to rationalize all of the observed effects of electromagnetic fields on cellular radicals. Efforts to leverage the quantum underpinnings of spin chemistry for quantum information processing applications have only just begun [1239–1242] and ROS physiology is emerging as an important factor in health and disease on many levels [1243].

Chapter 12

Molecular Fluctuations: Solvent Effects & Dispersion

The identification of biological systems as merely "warm, wet, and noisy" grossly underestimates the complex and non-local quantum mechanical structure of water. Despite evidence for the existence of coherent quantum effects in biological systems, the aqueous conditions of the biological environment are often presumed to suppress non-trivial quantum phenomena in the cell [1247]. This is ironic because warm, wet, and disordered cellular environments reproduce the conditions for Brownian motion [1248, 1249] — the topic of Einstein's second *annus mirabilis* paper of 1905 [1250] — in which those same features (i.e., warm, wet, and noisy) were used to quantitatively predict the existence of atoms. Einstein's theory of Brownian motion became a major triumph for quantum theory because the atomic model was still under dispute at the turn of the 20th century [1251]. Prominent chemists and physicists including Berthelot, Ostwald, and Mach opposed the theory of the atom so vehemently that the teaching of the atomic hypothesis was practically banned in France [1251,1252]. It was Perrin's confirmation of Einstein's theory of Brownian motion — the random movements of microscopic particles in a liquid — that convinced many leading scientists of the reality of atoms while winning Perrin the Nobel Prize in Physics in 1926 for his work on the "discontinuous structure of matter."

Far from being trivial, the structural and dynamic properties of liquid water remain among the most challenging of material properties to predict [1253]. This is primarily due to the ubiquitous presence of non-local quantum mechanical effects [68] with such complexity that several elementary properties of water (and ice) have only recently been estimated theoretically [69, 1254]. Liquid water can exhibit complex emergent phenomena [1255], such as the formation of a solute exclusion zone (EZ) in the vicinity of a hydrophilic surface [1256, 1257]. The formation of so-called "EZ water" has been reproduced extensively by multiple independent research groups [1258, 1259] and is most likely the result of a combination of ion diffusion and diffusiophoresis effects [1260]. QED are necessary to rationalize the existence of quantum mechanical van der Waals forces in liquid water [1261, 1262], but the question of whether the more complex, coherent effects of QED can emerge in water is contentious [1263]. This contention exemplifies many other contradictory results in water research, as water has numerous anomalous physico-chemical properties which are only partially understood [1262, 1264]. Water's anomalous properties are often attributed to its high propensity to form hydrogen bonds, but this attribution produces more questions than answers due to the lack of a formal distinction between the electrostatic and van der Waals contributions to hydrogen bonding [1265, 1266].

Electrostatic and van der Waals interactions are primarily responsible for the structure of biomolecules, where stacking interactions between planar molecules in particular are due to the London dispersion forces that arise from the quantum mechanical interactions between fluctuating electron densities [1267]. Solvent effects further complicate this picture, as the hydrogen-bonding network of the aqueous environment adapts continuously to accommodate the presence of solute molecules through hydrophobic effects. Although dispersion interactions arise from spontaneously correlated electron-density fluctuations [1268], solvent models and molecular mechanical (MM) force fields often include them only phenomenologically as a set of classical pairwise potentials that cannot capture the collective nature of dispersion effects [1269]. Accounting for the long-range effects of correlated electronic fluctuations poses a substantial challenge due to their collective and ubiquitous properties which scale nonlinearly with the system size [1270, 1271]. Solute/solvent polarization effects are fundamentally quantum mechanical [1272], and as such an atom's van der Waals radius is directly related to the quantum-mechanical atomic dipole polarizability [1273]. Beginning with investigations of the anomalous properties of liquid water [1274, 1275], a growing number of landmark investigations have revealed the failures of the classical pairwise approximations to account for dispersion effects in materials [1276].

This makes the study of biomolecular systems problematic for conventional methods in quantum chemistry like DFT which typically rely on mean-field approximations that do not account for long-range electronic correlations [1277, 1278]. The need to include dispersion in realistic simulations of condensed matter systems has motivated research on methodological approaches to correctly incorporate dispersive forces in DFT simulations [273]. Modeling biological systems thus presents a challenge because of the ubiquitous presence of weak binding and charge transfer effects which are not properly described by conventional DFT (i.e., using local electron densities or generalized gradients [1279]) [94, 1280]. Although some approximations developed for biological systems have successfully determined the structure and dynamics of assorted peptides, ligands, and DNA [1280–1282], these approximations fail to describe many long-range dispersion effects in biomolecules [1282], where noncovalent molecular interactions such as dispersion determine many important biochemical properties [1283].

Historically, accurate quantum chemical studies of biological systems have not been accessible to DFT calculations because the large sizes of biological molecules prohibit the elucidation of biological processes using quantum-mechanical calculations [255,1284]. To scale DFT-based methods efficiently to large systems, atomistic tight-binding approximations were developed [1280,1285,1286] and applied to biological systems using empirical corrections to partially account for dispersive forces between molecules [1281, 1282, 1287]. Although many-body dispersion forces may only have a minor influence on the low-energy geometries of molecular van der Waals clusters (where vibrational frequencies of breathing modes can decrease by as much as 10% [1288]), non-covalent van der Waals interactions constitute key aspects of many important biological processes such as protein folding, aqueous solvation, allosteric transformations, and electric dipole oscillations in biomolecules [1283, 1289, 1290]. Whereas covalent bonds are primarily responsible for the formation of individual molecules, non-covalent interactions can act over long ranges, extending from several Ångstroms (in the case of van der Waals, electrostatic, hydrogen-bonding, and π–π interactions) to hundreds of nanometers (for hydrophobic interactions) [1291, 1292]. The participation of van der Waals interactions in non-polar attractive forces, the formation of molecular clusters, the assembly of macromolecular structures, and other complex chemical effects has motivated a proliferation of atom- and molecule-based approaches to modeling quantum mechanical many-body dispersion effects in chemistry, biology, and material science [1269, 1292, 1293].

Although pairwise corrections to classical molecular models can account for dispersion effects in biological systems of interest such as proteins [1294], those schemes are not generally sufficient to address energetic aspects of van der Waals interactions that arise directly from the underlying quantum mechanics [1268, 1295]. Collective wavelike fluctuations that arise from the non-locality of the charge density in a wide range of biologically-important nanostructures can result in non-trivial variations in the power laws that govern non-covalent interactions at the nanoscale [1296]. Due to substantial variations in the complexity of dispersion effects, models that account for dispersion can range from simple semi-empirical or classical electromagnetic corrections to higher-order quantum mechanical approximations [1297]. As a result, developing DFT-based methods that accurately account for these collective dispersion effects has become a major goal in computational chemistry and materials science [1276, 1298].

Accurate representations of dispersive forces are particularly important for the study of proteins and DNA in an aqueous environment [1268, 1299, 1300]. The complex relaxation dynamics of the hydration layer surrounding DNA play a key role by influencing its flexibility and stability [1301], where dispersion forces allow the DNA double helix to form by balancing the electrostatic repulsion between protons bound to the base pairs [1298,1302]. Dispersion forces similar to those found in stacked DNA base pairs are also found in RNA–protein complexes where the stacking of aromatic molecular planes is achieved using a balance of classical electrostatic and non-local electronic interactions [1303], as well as in the stacking interactions between ribophosphate groups and nucleotide bases found in RNA molecules [1304]. Although they are often mistakenly attributed to classical electrostatic forces in standard texts, the π–π interactions that enable stacking effects between aromatic molecules (such as the base pairs that make up the "rungs" of the DNA double helix [1305]) define a fundamentally quantum-mechanical form of non-bonded force [1299,1306–1311]. Experiments have shown that the π-stacking of nucleotide base pairs facilitates long-range charge transfer along DNA strands [1312]. There is also broad research interest in quantum transport in DNA assemblies [1313, 1314], coherent charge transfer along sequences of DNA base pairs [146,1315], long-lived delocalized charge-separated states in DNA strands [1316], and electronic properties of DNA assemblies [1317,1318].

Charge transport mechanisms in nucleic acids can differ markedly from those found in proteins [1319], and DNA repair by the enzyme photolyase has been shown to depend on the intensity and direction of an external magnetic field [1320]. Long-distance charge transport depends on the stacking arrangements of base pairs in DNA, making it sensitive to the presence of DNA lesions, mismatches, and binding proteins [1321]. Long-range charge transfer along DNA has been proposed as a mechanism for biological sensing and signaling that may be critical for cell function [642, 1322]. DNA-repair enzymes that contain FeS clusters have also been shown to exploit DNA charge transport's remarkable sensitivity to DNA integrity in order to search the genome quickly and efficiently for errors [1323]. Thus, charge transport plays an essential role in DNA synthesis, where it is critical to detect and repair DNA mismatches [1324]. These mismatches can occur as a result of quantum tunneling effects which produce mismatched base pairs via mutagenic tautomerism [1325–1327]. Studies elucidating the role of quantum mechanics in genetic mutations have motivated the theory that random point mutations may occur through directed adaptation by organisms according to the principles of open quantum systems theory [1328], in line with proposals by McFadden, Al-Khalili, and Ogryzko [1329–1331]. Quantum mechanical processes have also been proposed to play a role in epigenetic mechanisms based on the charge transfer properties and CISS of DNA [1332]. Metabolism and genetic regulation are intertwined in the cell, where the coupling between gene expression and other metabolic processes must be coordinated dynamically to manage the fluctuating conditions of life [1333]. Despite the enormous gain in the understanding of cellular metabolism that was made through characterizations of the cellular reaction network during the 19th and 20th centuries [1334], little is still known about how small molecules and metabolites control gene expression [1335].

Beyond their role in maintaining the integrity and function of RNA and DNA, collective non-covalent interactions influence a wide variety of biochemical processes such as protein folding, molecular self-assembly and recognition, and the formation of supramolecular structures [1336]. Dispersion forces are implicated in water-mediated correlations in DNA-enzyme interactions [1337, 1338] and may also play a key role in some drug interactions [1339]. Likewise, water molecules are involved in determining the specificity, kinetics, and thermodynamics of molecular DNA–drug complexes [1340]. Dispersion forces are also predicted to play a key role in the dynamics of intrinsically disordered proteins (IDPs) [1294] that transcend the classical biomolecular structural-functional paradigm (i.e., the lock-and-key model [1341]). Unlike folded proteins that have definite, compact three-dimensional structures, IDPs do not naturally exhibit distinct folded states [1294]. Instead, IDPs exhibit glassy dynamics with a broad distribution of relaxation times under protein extension [1342], making them suitable testbeds to investigate models of dispersion in protein–solvent systems (similar to investigations of dispersive forces in other glassy media [1343, 1344]).

In contrast with IDPs, most globular, fibrous, and membrane-bound proteins are well-folded, with functions that are determined by a hierarchy of primary, secondary, tertiary, and quaternary structural motifs [1345, 1346]. Protein folding is primarily governed by hydrophobic forces which drive the protein into its stable folded state, where dispersion forces are enhanced by the tightly-packed protein interior [1347]. Protein folding is a central problem in computational biology [1348], where it is still unclear how the newly-synthesized peptide sequence relaxes into its folded state when the accessible states scale exponentially with respect to the sequence length [1349, 1350]. The process by which folding occurs is unclear because the thermal relaxation to the optimal folded state through the conformational "landscape" of the protein is hindered by a plethora of topological folding restrictions [1349]. Nearest neighbor interactions in the protein strand cannot facilitate its timely reduction to a compact conformation, precluding a resolution to the problem based on relaxation by nearest-neighbor interactions. This apparent contradiction has prompted a hypothesis that life relies on quantum-mechanical interactions and entanglement for development and survival [1351, 1352]. Efforts to solve the protein-folding problem have circumvented this issue (known as Levinthal's paradox [1353]) by employing machine learning (ML) techniques to predict protein structures empirically [1354, 1355], for which Baker, Hassabis, and Jumper received the 2024 Nobel Prize in Chemistry [1356]. Anfinsen's dogma dictates that the stable folded state of a protein must correspond to its thermodynamic free energy minimum [1357]. The problem of finding the protein fold that minimizes its free energy has therefore motivated interest in the use of quantum computing to solve hard optimization problems in molecular biology [1348, 1358–1360].

Unlike proteins, for which folding is primarily governed by hydrophobic interactions [1348], single-stranded RNA folds by way of a hierarchical process controlled by the hydrogen bonds formed between nucleotides [1361]. As a result, relaxation to the folded RNA tertiary structure is a relatively slow process which produces a definite, thermally-stable structure with little flexibility [1362]. In the absence of accurate classical force fields to use for structure prediction, intuitive physical arguments have been combined with phenomenological considerations to develop coarse-grained models with simplified effective potentials for simulating folding in DNA, RNA, and proteins [1363]. Conformational changes can also have a dramatic effect on the electronic properties of DNA, necessitating the use of multi-scale modeling and quantum calculations to predict electrical transport properties of stretched DNA [1364].

Photoexcited electronic states of DNA have attracted scientific interest because of their role inducing genetic damage in the form of DNA lesions [1365], as well as the possibility that photogenerated radical pairs could be used in quantum information science applications [1366]. Proof-of-principle demonstrations have shown that pulsed electron paramagnetic resonance (EPR) can be used to coherently manipulate entangled spin states of chromophores incorporated in strands of DNA for quantum information processing applications [1366], and quantum mechanical effects have been proposed to enable synchronization in DNA–enzyme interactions [1367, 1368]. Infrared light was also proposed to guide the seemingly random motion of DNA chromatin in collective dynamics that provide context to the cell about how to translate cellular chemical and electrical gradients into the activation of specific genes. In that model, infrared light would mediate collective interactions between the chromatin segments of DNA in the cellular nucleus via a superradiance effect in a form of quantum optical information processing, enabling the cell to engage its entire genome in electrochemical context-based decision making and regulation [1369]. That idea was motivated by previous findings that the enhanced quantum optical effects observed in microtubules using fluorescence spectroscopy could be attributed to cooperative superradiant interactions exhibited by arrangements of Trp residues [74].

The structure and function of biomolecular systems are influenced by collective many-body van der Waals interactions involving quantum effects that cannot always be mimicked effectively by classical pairwise interactions [1370]. The challenge associated with simulating these collective quantum effects is the development of fundamental models of van der Waals interactions from first principles [1276] to enable the calculation of long-range correlation energies in large structurally-complex molecular systems [1371]. The development of a comprehensive theoretical framework for modeling long-range electronic correlations will be necessary to account for van der Waals interactions and London dispersion forces in aqueous biomolecular systems such as proteins, DNA, and RNA [1270].

Chapter 13

Multiscale Modeling of Biomolecular Systems

The central dogma of molecular biology — that DNA provides the code-script for the synthesis of proteins that make up the body of life [1372] — does not solve the problem of how the greater structure of a living organism can emerge from the warm and wet environment of the cell [1373]. To the contrary, the statement that genetic information flows in one direction only (i.e., from DNA to RNA to protein [1374]) begs the question of how biological macromolecules like RNA and DNA arose in the first place [1375, 1376]. Schrödinger had posed the question of how life should overcome the statistical tendency to devolve from order to disorder a full decade before Watson and Crick identified the structure of the double helix [1377] (but still more than 70 years after Miescher's original discovery of DNA [1378]). To the classical physicist, the noisy statistical nature of the biological microenvironment can be overcome only in the limit of large numbers, where the gross anatomy and physiology of the organism are determined by the macroscopic structures and functions of cellular biophysics [42]. This ignores the underlying biophysical processes that overcome ever-present thermal noise and Brownian motion of the molecular scale to give rise to biological structure in the first place. To the classical physicist, life should not exist at all [42].

As Schrödinger realized in his time [42], distinct biological features are not limited to the gross structure of the organism and in fact can be identified all the way down to the molecular (and nearly atomic) scales [42]. To overcome the irreversible degradation from order to disorder imposed by the second law of thermodynamics, Schrödinger proposed that living organisms must take in a constant influx of thermodynamic free energy [42], which he described in terms of a need to harvest orderliness (i.e., thermodynamic free energy or "negative entropy") to stave off one's inevitable decay into an inert mass in equilibrium with its surroundings. However, Schrödinger was not the first to characterize the struggle to survive in this light. It was rather Boltzmann — who's fundamental definition of entropy remains engraved on his tombstone [1379] — who wrote in 1875 that, "The general struggle for existence of animate beings is not a struggle for raw materials ... nor for energy which exists in plenty in any body in the form of heat, but a struggle for *entropy*" [769]. In this respect, Boltzmann pondered the question of why fluctuations away from universal equilibrium should allow life to flourish transiently on worlds like ours [1380].

In the 1870s, Boltzmann published his celebrated "H-theorem" [1381] to provide a statistical foundation of the second law of thermodynamics, but similar to the Maxwell–Boltzmann equipartition principle, Boltzmann's H-theorem relied on the unsubstantiated hypothesis that the velocities of colliding particles should be uncorrelated and position-independent [1382]. In the 1920s, von Neumann set forth to correct the circular reasoning in Boltzmann's arguments by developing his own quantum mechanical "ergodic theorem" [1383], but much the same as Boltzmann's theorem, von Neumann's relied on the statistically-independent action of a classical measurement device to provide the source of randomness necessary to bring about thermodynamic decay [1382]. Although much work was carried out over the ensuing century to find a theoretical proof of the second law, the problem remains as intractable today as it was in Boltzmann's time. The *a priori* assumption of inherent randomness is necessary to both Boltzmann's and von Neumann's theorems to recover the principles of equilibrium thermodynamics [1384]. Although the assumption that the motions of component particles are essentially uncorrelated may be reasonable for large material bodies, it is not true in general, and it certainly falls into question in the highly coordinated microenvironment of the living cell [1385]. Nevertheless, great care is taken to recover the approximations of statistical mechanics in numerical models of molecular biology, where physical and chemical processes occur in or out of thermal equilibrium with the surrounding biological milieu [1386]. This is a natural way to consider living organisms that adapt to the temperature of their surroundings, as well as warm-blooded animals which can control their body temperatures internally.

In addition to replicating biomolecular structures [1387], it is therefore necessary for molecular simulations to reproduce thermodynamic properties of interest to biology such as temperatures and free energies [1388]. As such, a variety of molecular modeling methods have been developed to control simulation temperatures using Langevin dynamics [1389], as well as Anderson [1390], Berendsen [1391], and Nosé-Hoover thermostats [1392]. Physical simulations of this kind obey Newton's laws of motion, so that even though molecules can bend, stretch, and twist, the molecular bonds themselves cannot be broken or formed [1109]. To account for quantum effects such as the formation and cleavage of chemical bonds, molecular simulations must include quantum mechanical models [1393]. Yet it is not simply the electronic structure of atoms and molecules that lays the underpinnings of life [243], but life's distinct ability to direct key photo- and biochemical processes that render a classical description of it inadequate.

Semiclassical MM simulations have been used successfully to model molecular conformational changes and protein folding in countless biological structures which range from small peptides to large macromolecules in solution, membranes and membrane-bound proteins, and even macromolecular complexes such as ribosomes and nucleosomes [1394,1395]. A wide range of physical properties can also be calculated classically, including thermodynamic quantities such as free energies and binding affinities [1396, 1397], as well as material properties such as densities, viscosities, and thermal conductivities [1398]. Physical interactions between atoms and molecules in classical molecular simulations account for electrostatic forces between both non-bonded atoms and bonded atoms (with the appropriate equilibrium bond lengths, angles, and elastic force constants) by using effective force field parameters obtained from quantum mechanical calculations. These semiclassical force field parameters must also be supplemented by quantum mechanical corrections that typically take the form of a Lennard-Jones potential which approximately accounts for the van der Waals attractions and very short-range repulsions between atoms and molecules [1399]. Polarizable atomic models based on the classical Drude oscillator approximation may also be used in classical simulations where it is necessary to account for the polarization dynamics of charged particles in polar media [1400].

Quantum mechanical phenomena must be incorporated explicitly into semiclassical simulations to account for chemical reactions [1401–1403], photoinduced processes [1404, 1405], and macromolecular interactions [1338, 1406, 1407]. Combined quantum–classical methods have become useful for understanding the transition state structures and free energy profiles of enzymes [1109, 1408–1411], although it remains an outstanding challenge to fully account for the bioenergetics of many-body quantum biological systems [1412, 1413]. Consequently, applications of quantum chemistry methods to biological systems are supplemented by advanced numerical methods in quantum mechanics, typically beginning with Møller–Plesset perturbation theory [1284, 1414–1417]. For example, fragment molecular orbital methods have allowed efficient quantum mechanical calculations of electronic states of macromolecules [1418, 1419] to enable predictions of many life science phenomena *a priori* with applications to drug design, molecular recognition, and structural biology (such as characterizations of metalloprotein interactions) [1420]. Fragment-molecular-orbital density-functional tight-binding calculations have also enabled structure-based analyses of molecular interactions associated with the deadly SARS-CoV-2 virus [1421]. Far from trivial, these calculations are carried out on vast super-computers using high-order quantum mechanical corrections (e.g., fourth-order Møller–Plesset theory [1421]) to account for the contribution of quantum coherence and electronic correlations to vital biological processes [1413, 1422–1424].

Accounting for explicit quantum mechanical effects poses a dilemma to molecular modelers, because whole biological molecules are much too large to simulate quantum mechanically — even on present-day supercomputers — but classical simulations alone cannot account for the most critical steps that occur during biological catalysis and anabolic synthesis. For this reason, experts in molecular simulation have developed multi-scale modeling techniques to enable the simulation of molecular dynamics (MD) that occur over multiple scales in biology by modeling the quantum mechanics explicitly where it is absolutely necessary (e.g., at the active sites of enzymes) and treating the quantum mechanics only implicitly in the surrounding molecules and solvent where the classical approximations described above are suitable. For their work developing the molecular simulation techniques needed to span multiple biological scales, Karplus, Levitt, and Warshel were awarded the Nobel Prize in Chemistry in 2013 [1425].

Early in the 1970s, Warshel and Karplus developed calculations to study excited states of conjugated organic molecules where the electrons are quantum mechanically delocalized across multiple bonds [1426, 1427]. That research was developed in subsequent studies of protein folding [1428], culminating in 1976 with a method to simulate the lysozyme enzyme reaction that demonstrated the role of electrostatic effects in governing enzyme catalysis [662]. In their general method, Warshel and Levitt used a multi-scale approach in which a semi-empirical quantum mechanical (QM) model of the lysozyme enzyme active site was coupled to a classical model of the surrounding ions, polarized atoms, and water molecules [662]. Those calculations relied on self-consistent field (SCF) models of the active sites based on molecular orbital (MO) theory using the Hartree–Fock (HF) method with corrections due to configuration interaction (CI) effects [1426]. Since the 1970s, molecular dynamics simulations employing coupled quantum mechanics/molecular mechanics (QM/MM) methods have become the *de facto* standard for carrying out computer simulations of biophysical processes in biomolecular systems [1429, 1430].

The integration of QM methods into classical models continued over the ensuing decades with the use of continuum solvent approximations to predict properties of molecules in solution in tandem with the coupling of MM to QM methods in hybrid approaches [1431]. Simulations based on mixed QM/MM models gradually became the "method of choice" for studying enzymatic reactions involving electronic excitations or charge transfer [1432]. Based on the enormous success of these techniques [1402,1433–1435], numerical methods for simulating enzymatic reactions are now routinely taught to students in classrooms and laboratory tutorials [1436]. Despite this success, researchers in this area have yet to establish a consensus on the best way to implement QM/MM techniques for any given application. Rather, the performance of a particular QM/MM model is likely to depend on the details of the system and the physical property of interest [253]. The accuracy of a QM/MM method generally reflects the choice of QM method, size of the QM region, and details of the embedding potential, where the inclusion of a sufficient amount of water at the QM region can be critical to reduce errors [1437].

QM/MM methods are not only applicable to the effects of quantization on ground state structures and thermal energies [248], but are also increasingly being applied to study photochemically-excited states, ultrafast chemical dynamics, and spectroscopic properties of enzymes and other biological molecules [1432, 1438]. Quantum chemical dynamics of this kind have been treated in depth by Nitzan [76], and include many common biological processes such as electron transfer reactions which occur during photosynthesis and vision, solvent-controlled and bridge-mediated electron transfer, and other excited-state relaxation effects [1439]. For example, QM/MM simulations were employed to investigate the photodynamics of vitamin D photosynthesis, providing mechanistic insights into the photodynamic processes involved in synthesizing vitamin D and presenting a theoretical explanation for experimentally observed excited-state decay rates [102]. QM/MM simulations have also been used to simulate the complete photocycle of phototropin, a blue-light photoreceptor found in green algae [1440]. Those findings indicated that the key bond formation step during phototropic photoreception relies on radical pair dynamics initiated by the coordinated action of electron and proton transfer steps, suggesting that photo-induced signal transmission in phototropin may be mediated by the altered dynamical properties that result from covalent bond formation [1440].

Although fast non-adiabatic processes are ubiquitous in biology where they enable many photo-induced reactions [1439], the simultaneous use of classical and quantum methods in QM/MM techniques can result in incorrect descriptions of quantum processes such as wave packet branching and tunneling [1441]. In fact, traditional methods for simulating enzyme catalysis using QM/MM can fall short even in electrostatic systems if they contain strong quantum correlations [284,1442]. Strongly-coupled quantum resonance processes have also recently been proposed to play a possible role in certain aspects of cellular growth and development [758]. Several recent proposals have been aimed at replacing the QM portion of QM/MM simulations with classical ML models for ML-guided enzyme engineering [1443–1445], but biological systems pose a formidable challenge to classical ML due to the ubiquitous presence of quantum mechanical correlations and long-range interactions in biomolecules [1406,1446]. The failure of classical computing methods to simulate strongly-correlated quantum systems has revealed the practical necessity of using quantum computing for studies in molecular biology and biomedicine [288,1447].

Simple structural models fail to provide accurate information about the biochemical transition states, protein domain movements, and conformational changes that enable enzymatic reactions [716]. In fact, even conventional QM/MM methods (based on SCF theories) often do not capture the non-local electronic correlations underlying electronic and photophysical properties of enzymes [1448–1450]. Furthermore, classical simulations of highly-correlated quantum systems come with severe computational costs that scale poorly with biomolecular system size [1413]. To overcome these issues, multi-level embedding schemes have been proposed to allow biochemical reactions to be simulated using quantum computers [284,1451,1452]. Multi-level schemes of this kind generally incorporate a virtual embedding strategy comprising an inner level where the strongly-correlated quantum subsystem is simulated using a quantum computer, surrounded by a weakly-correlated quantum region that is represented using conventional (mean field) QM methods on a classical computer, which is finally surrounded by an outermost simulation level using classical continuum approximations and/or conventional MM methods [284, 1451, 1452]. To this effect, a number of hybrid methods have been proposed, such as density matrix embedding theory [1452] and dynamical mean-field theory (DMFT) [284] following previous work using continuous-time quantum Monte Carlo (CT-QMC) impurity solvers [1453]. Advanced numerical methods may be particularly valuable for calculations of ligand screening effects used for drug discovery, which have often employed parameterized (semiclassical) energy functions that are computationally inexpensive but unsuitable to determine chemical properties [1448]. Quantum computing methods will be critical to model biological molecules with active sites that are not simulated efficiently using classical computers [284, 1454].

Chapter 14

Quantum Correlations in Biological Cofactors

Quantum theory is necessary to describe the active sites of enzymes, where the presence of electronic correlations make the governing quantum mechanical equations computationally hard to solve [288]. This is especially true in the strongly-interacting electronic systems found universally in the active sites of metalloenzymes [1455]. This has inspired the imaginations of quantum information scientists interested in simulating strong quantum correlations in realistic physical systems [1456, 1457]. The strong quantum correlations found in certain biocatalytic cofactors make these systems molecular analogues of strongly-correlated materials [1458], an important class of quantum matter [1459–1462]. As a consequence, these highly-complex problems in quantum biochemistry have influenced the development and benchmarking of massively-parallel classical computer architectures [1454, 1463, 1464], as well as large-scale quantum mechanical simulation methods for use on advanced quantum computing architectures [1451, 1456, 1465].

Metallic active sites carry out key roles in many crucial biological functions including light harvesting, electron transport, and enzyme catalysis [1466, 1467]. Iron-containing enzymes, for example, play a central role in biochemistry where they are only partially understood even after decades of study [1468–1470]. Heme proteins, known for the characteristic porphyrin ring which binds a single iron atom at the heme active site [1471], define one of the most versatile classes of proteins with functional roles in electron transfer (ET), substrate oxidation, ion storage, oxygen transport, transcriptional regulation, chemical sensing, circadian regulation, and RNA processing [1472, 1473]. Their high versatility has made hemes a major topic of interest in areas of protein structure prediction and design for medical and clean energy applications [1473]. In-depth studies of heme protein biophysics face challenges due to the presence of quantum many-body effects at the heme active site, where the spin-state energetics of the system depend strongly on small changes in the active site geometry [1470]. These many-body effects have led some to conclude that the functional properties of hemoglobin may result from the presence of magnetic correlations [1468], suggesting that magnetic effects may also be involved in the function of other heme proteins [1474, 1475].

Iron is also present in many non-heme proteins [1476, 1477], such as the alkylation B (AlkB) family of metalloenzymes that repair methylated DNA (known as DNA repair enzyme AlkB [1478]) and convert straight chain alkanes into alcohols (via alkane monooxygenase AlkB [1479]). DNA repair AlkB enzymes play an essential role protecting the genome from alkylation damage, where they employ molecular oxygen to demethylate damaged DNA and RNA chains by hydroxylating the methylated nucleic acid base. QM/MM studies of the oxygen binding and hydroxylation steps of the reaction have enabled a detailed characterization of the electronic and structural changes that occur during the catalysis, revealing important insights into the nature of the nucleotide- and oxygen-binding channels of the enzyme [1469]. Simplified quantum mechanical models based on HF and DFT have proven effective at modeling chemical reactions as electrostatic phenomena involving short-range correlations [653, 1432] — with examples that include simple proton transfer events and cleavage reactions such as phosphate hydrolysis [1480]. However, functional aspects of biomolecules often depend on non-local electronic correlations [1277, 1481]; particularly when there is a narrowing of the energy gap — either between the highest occupied and lowest unoccupied molecular orbitals, or between the singlet and triplet spin states — that determines enzyme reaction kinetics [1482, 1483].

Iron-sulfur clusters, like the Fe_4S_4 cluster contained in the DNA base-excision repair enzyme endonuclease III [1484], perform a diversity of biological functions related to redox chemistry, electron transfer, and oxygen sensing which rely on electronic configurations with strongly-interacting (entangled) quantum states which are not described by HF or DFT [1455,1485]. Symmetry-breaking introduces non-trivial physical effects [1486], and broken-symmetry DFT has been used to produce detailed and accurate computational analyses of structural models of FeS clusters in rubredoxin and nitrogenase enzymes by incorporating weighted sums of spin multiplets in calculations [1487,1488], but this method only provides an average of the system energy spectra and does not allow for direct simulations of individual electronic states [1455]. The problem is exacerbated by the large exchange-correlation functional dependence in calculations of the open-shell electronic structure of FeS clusters which is not well understood [1489]. Extending DFT beyond descriptions of static magnetic moments associated with spin symmetry-broken states requires the inclusion of spin-orbit interactions to account for dynamical correlations involved in magnetic coupling [1490] and altered spin states due to quantum mechanical and thermal fluctuations [1470,1491].

Single electronic configuration methods can significantly miscalculate estimates of reaction barriers even when perturbative corrections are applied [1492], and the accuracies of DFT-based techniques are notoriously hard to predict [1493]. These issues arise primarily as a result of the mean field approximations used in HF and DFT which assume that the quantum system of interest can initially be approximated as a single (exchange-correlated) non-interacting electronic configuration [1494]. HF and DFT methods are most appropriate where the single configuration model is exact or where the interactions between multiple electronic configurations are small enough to be treated using perturbation theory [1482]. Quantum mechanical projection-based techniques have been developed to allow the exact quantum mechanical simulation of an enzyme active site to be embedded in a DFT-based model [1495] which in turn is embedded in a MM representation of the greater protein environment [1492].

Correlated electronic effects in the exactly-embedded quantum subsystem can then be simulated using higher-level correction techniques such as coupled cluster methods [1496, 1497] or spin-component scaled second-order Møller–Plesset perturbation theory [1464, 1498]. As the CI between multiple electronic reference states increases, correlation-corrected methods based on HF and DFT may fail due to the resulting "multi-reference" character of the quantum mechanical system [1499].

Nevertheless, single-reference methods based on HF and DFT have enjoyed unparalleled success in computational chemistry research [1500], where they have become workhorses of molecular simulation [1501, 1502]. Many techniques have been developed at varying levels of approximation, and the resulting menagerie of computational chemistry methods are arranged in increasing order of complexity [275, 1503, 1504]. These begin with single-reference approximations and the perturbative corrections thereof, and ascend a "ladder" of methods to account for strong multi-reference correlations which include multi-reference CI, complete active space self-consistent field (CASSCF) theories, multiconfigurational perturbation theories, multi-reference coupled-cluster theories, and density matrix renormalization group (DMRG) methods [1505, 1506]. Although some of these techniques have been extended to account for quantum electrodynamic effects inherent to polariton chemistry [1507–1509], approximate methods often still fall short of exact *ab initio* theories [1510–1512] — necessitating the use of experimental reference data against which to benchmark the quality and applicability of the many different quantum mechanical approximations [1513].

Metalloenzyme-catalyzed reactions provide a natural testbed for benchmarking the accuracy and efficiency of quantum chemical approximations [1503], particularly for metalloenzymes with complex electronic structures due to the presence of transition metals [1514] (i.e., atoms configured with partially-filled $3d$ or $4f$ electron shells [1466]). Conventional methods for simulating these systems using DFT with LDA or GGA techniques fail to describe transition metal correlations, resulting in predictions which can differ significantly from experiments [1466,1503]. For transition metal proteins containing iron (in hemoglobin), copper (in hemocyanin), or manganese (in oxygen-evolving complex), semilocal methods involving semiempirical "hybrid" orbitals or the Hubbard U correction provide only limited improvements over standard quantum chemical models because they do not account for electronic correlations due to quantum fluctuations [1466, 1515] or for subtle quantum effects that occur near energy-level splittings [1516]. As a result, computational studies of biological transition metal complexes often require advanced quantum chemical methods to account for highly non-trivial quantum effects [1482,1492].

A sophisticated approach to the problem of simulating strongly-correlated electronic behavior is provided by DMFT, which can describe electronic interactions by introducing local quantum fluctuations that extend beyond the single-electron picture assumed in conventional quantum chemistry methods (HF and DFT) [1517, 1518]. Combining DFT and DMFT, the DFT+DMFT approach is suitable for simulating magnetic nanostructures and metal clusters similar to those found ubiquitously in biological systems [1519–1521]. Combined DFT+DMFT has been used successfully to model entangled spin state transitions at the active site of hemoglobin, consistent with experimental observations of the transition between high and low spin states of the hemoglobin molecule [1522]. In principle the use of DMFT is limited to high-dimensional systems (i.e., as the system dimension scales to infinity), but in practice it is an effective approximation for crystal lattices and ligand systems with large coordination numbers [1519,1520]. Use of DMFT in high-dimensional systems was developed in pioneering work by Metzner and Vollhardt [1523, 1524] who showed that the method could be used to simplify studies of strongly-correlated quantum mechanical systems by taking the average of the spatial quantum correlations in a scenario where the quantum mechanical time correlations remain non-trivial [1525,1526]. This unexpected result revealed that quantum mechanical systems could exhibit strong local electronic correlations even in the absence of non-local spatial correlations, resulting in complex phase diagrams and exotic physical effects that make their study both challenging and exciting [1527].

Quantum embedding theory provides a natural framework in which to consider the interaction of a strongly-correlated quantum system with its molecular environment [1528]. Initially developed to address the infinite limit of large coordination-number lattice systems [1529], DMFT in its original form was well-suited for the study of local dynamic correlations in quantum mechanical systems (as its name implies [1525]), and was developed particularly to model strongly-correlated materials where DFT fails to predict even basic material properties [1530]. Although DMFT offers tremendous predictive power by mapping the dynamics of a strongly-correlated, infinite-dimensional quantum system onto a frequency-dependent one-particle Green's function [1526], DMFT in itself cannot fundamentally account for the non-local static correlations which lie at the heart of quantum mechanics. Practically since its inception [1526], DMFT has been refined and extended to address finite-coordination systems where static and dynamic correlation effects coexist [1525, 1527]. Cluster-based extensions of DMFT are important to bridge the gap between the single-site mean-field approach and the exact non-local many-body expansion, describing a wide range of symmetry-breaking effects which include magnetism, superconductivity, and charge and orbital order [1526].

Cluster DMFT has been used to show the importance of many-body effects of ligand binding at the active sites of hemoglobin and myoglobin [1531, 1532]. Those studies identified the metal-to-ligand bond length as a key control parameter for spin transitions in metalloporphyrin systems [1470], consistent with previous reports of entanglement in the ground state of the oxygen–myoglobin complex [1470, 1474, 1533]. Although Pauling and Coryell proposed that magnetic effects could play an important role in enzymatic systems like hemoglobin in the 1930s [1534], the role of magnetism in the functioning of heme proteins has only recently been defined as a research topic in quantum biology [1468]. Many-body DMFT has proven itself as a powerful method in studies of biomagnetism, where it simultaneously accounts for both electron valence and spin fluctuations when combined with large-scale DFT calculations [1474]. Indeed, large-scale DFT simulations have had a major impact on biomolecular studies where it is common for large simulations of 500 atoms or more to be required to predict biological properties from first principles [1277].

Large scale quantum mechanical simulations are likely to continue to play an important role in investigations of complex organometallic structures found in countless biological catalysts, such as the tetramanganese–calcium (Mn_4Ca) cluster at the core of the oxygen-evolving complex of photosystem II (PSII) [1535]. Metal-ligand systems with multiple metal centers [1536] are likely candidates for investigations of magnetic effects in systems with properties that are controlled or enhanced by quantum magnetic effects such as superexchange [1537, 1538]. In spite of their wide success in modeling metabolically-important biochemical processes, the accuracies of conventional QM/MM methods are still not well understood [253]. Complex biochemical systems present a challenge to investigate using electronic structure methods because of these systems' large sizes, numerous subtle interactions, coupled dynamical timescales, and non-adiabatic electronic effects [1495]. Strong quantum mechanical correlations are present in many biological photosystems and catalysts which make them inherently different from systems which have been successfully modeled using conventional quantum chemical theories [1539]. This makes it necessary to employ comprehensive models of static and dynamic electron correlations involving the use of sophisticated multi-reference (MR) quantum mechanical calculation methods to obtain efficient and quantitatively accurate descriptions of many biochemical processes [1540–1544]. Noteworthy examples of methods which extend the conventional QM/MM formalism include quantum embedding methods based on projector techniques, as well as hybrid, multilayered and fragment-based methods, not to mention density functional- and density matrix-based embedding schemes [1452, 1495, 1545, 1546].

Sophisticated computational methods are accompanied by the need for increasingly-advanced numerical techniques to implement them, including high-level parallel computational packages designed to carry out DMFT and CT-QMC algorithms on classical supercomputing architectures [284, 1547, 1548]. The vast computational resources required to perform these simulations have motivated interest in the use of quantum computing to study biochemical systems [284]. Although the computational armory of numerical quantum chemistry is well-stocked with methods that range from primitive Hartree–Fock techniques to highly accurate coupled-cluster and CI methods, exact treatments of quantum systems with more than a few dozen electronic orbitals remain intractable on classical computers [1452]. Fortunately, embedding methods developed for classical computer simulations of quantum systems are readily translated to hybrid quantum/classical computations executed on architectures that combine classical and quantum computers [284, 1451, 1452]. These methods suggest a practical and systematic path to multiscale descriptions of quantum systems for use in quantum simulations of large, realistic biomolecules [1452, 1549, 1550].

Quantum dynamics play a critical role in the excited state properties of magnetic molecules and materials [1551], such as the FeS clusters found in the enzyme nitrogenase which exhibit strong correlations between its eight transition metal atoms and have been called a holy grail of quantum chemistry [1452, 1458]. Quantum information processing will be critical to predict the dynamics of complex quantum systems such as these, which are beyond the reach of simulations that employ even the most advanced classical algorithms [1552]. For example, the molecular fragmentation pathways of the photo-dissociating sulfonium ion H_3S^+ have been simulated on a superconducting quantum processor, demonstrating the usefulness of hybrid quantum/classical algorithms at simulating molecular decay spectra [1553]. The ubiquity of ultrafast non-adiabatic processes in biological systems has made investigations of quantum non-adiabatic molecular dynamics an important topic of study for small molecular systems using high-level electronic structure calculations which account for the quantization of nuclear motion [1441, 1554]. Whereas a decisive demonstration of an advantage at simulating quantum dynamics would appear to require achieving the elusive goal of developing a fault-tolerant quantum computer, present-day and noisy intermediate-scale quantum (NISQ) devices may provide as-of-yet unclear benefits in efforts to simulate molecular dynamics [1555].

Nuclear quantum dynamics beyond the (quasi-static) Born–Oppenheimer approximation are also becoming recognized in a growing number of biochemical processes [72]. Quantum simulations of non-adiabatic dynamics involved in photophysical internal conversion and intersystem crossings, the Jahn-Teller effect, and vibrationally-assisted energy and/or electron transfer reactions provide natural starting points for investigations of this kind [1441]. Simulations of the conical intersections which couple the diabatic energy surfaces during non-adiabatic reactions have been carried out on trapped ion quantum computers, providing insights into the entangled nuclear and electronic (spin) dynamics which occur around quantum singularities as the geometric phases accrued there become non-trivial [1556]. Non-adiabatic dynamics can also become significant in photoinduced PCET which are crucial to long-range signaling in cells [212]. Quantum dynamics are particularly relevant in photo-activated phenomena which are key to essential aspects of biological systems including photosynthesis and light-harvesting, the photostability of DNA, and various forms of photocatalysis [1555, 1557–1559].

Quantum computers offer a diverse range of potential advantages in biochemical and biophysical computations, where even constructing the numerical representation of a suitable initial state (such as a molecular ground state or a thermal Gibbs state [1552, 1560, 1561]) may constitute a computationally hard simulation task [1481, 1562]. It is widely believed that quantum computers will ultimately enable simulations of strongly-correlated chemical systems [1483], although the exact degree to which advances in quantum computing will benefit the simulation of quantum chemistry remains to be shown. The widespread success of heuristic approximations in quantum chemistry has prompted some authors to doubt whether quantum computers can provide an asymptotic (i.e., super-polynomial) speed-up over classical methods for calculating ground-state properties of molecules [1458]. Furthermore, it is unlikely that the forthcoming generation of NISQ computing devices will be able to emulate chemical reaction rates, due to the exponential relationship between the rate of the reaction and its free energy of activation [1563]. Nevertheless, quantum algorithms are increasingly being used to simulate open quantum systems [1564], and analogue quantum computers are predicted to offer a superpolynomial advantage over classical algorithms when simulating open quantum system dynamics [1565]. Quantum computing has important implications for computational biology [1360], with broad applications which range from quantum mechanically embedded simulations of enzyme catalysts and reaction dynamics to protein folding, molecular recognition, nucleic acid research, and functional genomics [287,1348,1359,1566]. The development of quantum computing algorithms for biochemistry is expected to have major applications in pharmaceutical discovery, material design, and catalyst optimization [1452,1567–1571].

Part III

Nanomedicine & Biotechnology

"Physics constitutes a logical system of thought which is in a state of evolution, whose basis cannot be distilled, as it were, from experience by an inductive method, but can only be arrived at by free invention."

— Albert Einstein, 1936 [1572]

Chapter 15

Photobiomodulation & Electromagnetic Therapies

While the healing power of sunlight has been known since antiquity [1573–1575], light-based therapy was established for the modern world when the 1903 Nobel Prize in Physiology or Medicine was awarded to Finsen "in recognition of his contribution to the treatment of diseases, especially lupus vulgaris, with concentrated light radiation, whereby he has opened a new avenue for medical science" [1576]. In addition to solar (UV) treatment for dermal tuberculosis, Finsen also developed phototherapy using red light to treat smallpox [1577, 1578]. However, subsequent work on phototherapy lapsed until the mid-1960s when the regenerative powers of red light were rediscovered [1579].

The first research demonstrating that ruby laser light could accelerate wound healing and hair growth was carried out in 1967 by Mester at Semmelweis Medical University in Hungary [1575, 1580] in an attempt to reproduce research by McGuff *et al.* ablating malignancies in rats with the (recently invented) ruby laser [1580, 1581]. Cancer mitigation was unsuccessful because of the low power of the laser he used, which produced enhanced tissue generation instead. The resulting treatment came to be known as low level laser/light therapy (LLLT) for this reason, garnering interest from National Aeronautics and Space Administration (NASA) scientists for the prospect of accelerating wound healing in space [1582]. The therapeutic application of light to modulate tissue function (by activation or inhibition) is potentially important for many physiological processes such as angiogenesis, collagen and cartilage production, muscle and nerve regeneration, and neovascularization, as well as decreased inflammation, edema, and pain [1583]. More than 4000 laboratory studies have been performed on LLLT, as well as more than 400 Phase III randomized, double-blind, placebo-controlled trials [1584].

It is now evident that photophysical forces applied at low energies and specific frequencies can have profound effects on physiological processes [1585]. The scope of these photobiomodulation (PBM) effects has been broadened to include other light sources such as light-emitting diode (LED)s, mainly in the red to near infrared spectrum [98]. Moreover, use of the term LLLT to refer to PBM techniques has become discouraged, because the broad connotations of "low level light" could also be applied to implementations of photodynamic therapy (PDT) and optogenetic techniques wherein low-doses of laser or LED light can be used to activate photochemical switches, rendering the term LLLT overly vague and misleading for precise usage in medical physics [1574]. The need for a more specific term to characterize the use of red through infrared light for the (nonthermal) biological stimulation of tissue repair and vascularization, as well as pain reduction and mitochondrial ATP production, has led to the wide adoption of the term PBM [1581, 1583], which is formally defined as any "nonthermal process involving endogenous chromophores eliciting photophysical (i.e., linear and nonlinear) and photochemical events at various biological scales" [1586].

Hence, PBM therapy encompasses both conventional low intensity laser and LED-based treatments [1583, 1586], as well as the growing potential for high-intensity LED-based irradiation in hard tissue regeneration therapy [1587]. PBM effects depend on the dose (intensity, duration, and wavelength) of light. During PBM therapy, light is usually applied from an emitter in the 10 mW to 500 mW power range [1584] using monochromatic light at a wavelength in the 600 nm to 1000 nm region of the optical spectrum (i.e., red to near infrared) [1588], affording the added clinical advantage of treatment within the 650–1000 nm tissue-transparency window [1589]. These power levels and wavelengths have the ability to penetrate skin as well as other soft and hard tissue structures, and were demonstrated to produce beneficial effects on pain, inflammation and tissue healing in various clinical studies [1584].

For the treatment of musculoskeletal injury and pain, an optical power density of $\sim 5\,\text{W/cm}^2$ in red to NIR spectrum (660–905 nm) may be applied to the injured or painful site for up to a minute a few times a week for a period of weeks to reduce inflammation and pain while accelerating tissue regeneration [1584]. PBM using far red and NIR light was found to reverse the effects of neurotoxins in neuronal cell cultures, supporting the hypothesis that PBM up-regulates cytochrome *c* oxidase to increase energy metabolism in a signaling cascade that prevents poisoned neurons from undergoing cell death [1589]. PBM using red and NIR light has also been shown to have therapeutic effects on neurological diseases and nerve injuries including stroke, traumatic brain injury, Parkinson's disease, Alzheimer's disease, and major depressive disorder [1590–1594]. Therapeutic benefits have been reported for PBM using blue light [1595, 1596], particularly for use in dermatology and wound healing [1597]. PBM therapy is also being studied for its capacity to improve skin health, joint mobility, and athletic performance [1598].

PBM mechanisms are primarily believed to operate via photoactivated retrograde signaling pathways whereby mitochondria relay their functional status to the cell nucleus [1596, 1599]. This was proposed by Karu as a result of the appearance of a 300–860 nm action spectrum for DNA and RNA synthesis of cell cultures, even though cell nuclei do not have dedicated chromophores to absorb in this region [1588]. Instead, those action spectra data revealed PBM-active chromophores residing in the respiratory chain, implying the existence of a "retrograde" signaling mechanism from the mitochondria to the nucleus. Retrograde signaling is also a key factor in cellular homeostasis and regulation of neuronal function [1600]. Cytochrome *c* oxidase, hemoglobin, and myoglobin are the three main photoacceptor molecules known to absorb NIR light in mammalian tissues. Of these three, only cytochrome *c* oxidase is directly associated with energy production [1589]. Thus, PBM is believed to improve cell viability by stimulating mitochondrial function [1582] via light–matter interactions with mitochondrial chromophores [1580].

Although there is ample evidence that PBM is mediated by mitochondrial retrograde signals in irradiated cells, cellular responses to PBM initiated in the nucleus have also been documented, and a number of PBM mechanisms have been proposed that do not rely on cytochrome *c* oxidase [1588, 1601]. In particular, new evidence has shown that PBM at 660 nm can stimulate cell proliferation without the involvement of cytochrome *c* oxidase [1602]. This indicates that a number of other candidate molecules such as ROS, genetic transcription factors, and hypoxia-inducible factors are also involved in PBM-induced changes that enhance cell metabolism and growth. PBM using NIR light has also been shown to stimulate assembly and disassembly in tubulin and microtubules, respectively [1581]. PBM mechanisms have been linked to research on cellular UPE and the hypothesis that living cells may communicate with each other via metabolic biophoton emissions [1603].

Despite its broad range of applicability, widespread adoption of PBM treatments has been limited by sparing knowledge of its physiological basis [98, 1582, 1597, 1604]. Partly for this reason, optimal PBM treatment parameters (such as wavelength, power density and fluence, pulse shape, and treatment timing) are not known definitively [1605, 1606], although suboptimal parameter choices can reduce treatment effectiveness or result in negative outcomes [98]. Full characterizations of the effect of PBM on living cells have remained elusive because of the uncontrolled variations that distinguish different PBM protocols. Progress is being made to establish more favorable methods for promoting the proliferation and differentiation in stem cells *in vitro* [1607, 1608]. Mitochondrial function is linked to cytosolic Ca^{2+} ions, ROS, and AMP levels. Beyond influencing overall mitochondrial function, changes in these levels can also activate a diversity of retrograde signaling pathways to the cellular nucleus [1609]. PBM can initiate mitochondrial retrograde signaling by way of a Ca^{2+} ion and cyclic AMP-related cascades which modulate cellular functions (such as proliferation and migration) and can result in tissue regeneration [1610]. When functioning normally, retrograde signals are linked to mitochondrial biogenesis, cell phenotype, and metabolism [1611]. However, the exact role of retrograde signaling in mitochondrial dysfunction is complex, and the mechanism by which mitochondria influence gene expression is still unclear [1611, 1612].

PBM has been shown to stimulate growth and metabolic activity in mammalian cells, yeast, and bacteria [1588, 1599, 1613]. PBM is also proposed to influence physiological (i.e., neuronal and circadian) rhythmicity [1585], consistent with findings that PBM effects exhibit a biphasic dose–response curve and that biological stimulation for a given cell or tissue type occurs only through a therapeutic dose window [1608, 1614]. There is growing evidence that PBM may be indicated in cancer therapy, both as a primary treatment method [1615] and for chemotherapy-related neuropathies [1593, 1594]. Furthermore, preliminary studies show that PBM may be combined with other therapeutic applications including hyaluronic acid and gold nanoparticles to significantly accelerate tissue healing [1616].

As with PBM, electromagnetic effects have been ascribed to a plethora of biological processes ranging from wound healing and cell migration to gene expression and DNA synthesis [1617–1619]. Although magnetic fields have been known to interact with living organisms for centuries [1620], interest in the influence of electromagnetic fields on biology and ecology grew over the 20th century [1621] as electronic devices came to influence practically every aspect of culture — from health and medicine to science, engineering, and communications. Public interest in the health effects of ambient electromagnetic fields has grown [1622], with studies finding that these fields can exert stimulating or inhibiting effects on immune function [1623].

Pulsed electromagnetic field (PEMF) therapy influences biological systems through the application of short bursts of electromagnetic fields which are both tissue and frequency-specific in their effects [1624, 1625]. It has emerged as a promising method for treating chronic inflammatory and immunological conditions such as rheumatoid arthritis by targeting mesenchymal stromal cells through modulation of differentiation, proliferation, cell cycle regulation, and metabolism, as well as cytokine and growth factor secretions [1624, 1626]. PEMF is proposed to affect cellular function and differentiation by altering intracellular calcium ion concentrations and restoring potassium channel activity, down-regulating inflammatory cytokines, heat-shock proteins, and proangiogenic molecules to make it possible for cells to commence regeneration of healthy cartilage. The use of PEMF as an inflammation and immune system modulator is considered relatively safe in comparison with other broad immunosuppression techniques currently used to treat arthritic conditions [1624]. Likewise, extremely low frequency (ELF) electromagnetic fields have been found to accelerate wound healing [1224] and have been used to treat pain and inflammation without causing cytotoxic or genotoxic effects in human mesenchymal stromal cells [1627].

Oscillating electromagnetic fields have been shown to induce cell autophagy *in vitro*, promoting homeostasis in a cytoprotective effect [1628]. Furthermore, combinations of μT-scale static and alternating magnetic fields were found to reduce amyloid plaques in the cortex and improve memory in transgenic mice used to model Alzheimer's disease, suggesting that magnetic fields may combat neuronal degeneration early on in brain diseases with Alzheimer's-like symptoms of amyloid deposition [1629]. Pulsed electromagnetic fields have been demonstrated to have beneficial effects on an intracerebral hemorrhage (ICH) model system *in vitro*, protecting against brain injury by alleviating inflammation and promoting hematoma clearance [1630]. As such, transcranial magnetic stimulation is being clinically investigated as a potential non-invasive therapy for neurological and psychiatric disorders [1631].

While the exact physiological basis for these effects remains to be fully understood, it is clear that practically all living cells are diamagnetic [1632]. In fact, electromagnetic fields have been used clinically to aid in the healing of delayed union and non-union bone fractures for several decades [1071, 1633], and static magnetic fields have been shown to induce deformations in bone-forming osteoblast cells *in vitro* [1634, 1635]. PEMFs have also been shown to induce ROS production levels in a cryptochrome-dependent effect in mammalian cells [1181]. Those findings prompted interest in the potential use of magnetic field therapies to influence circadian clock activity to control such factors as glycolysis, lactate production, extracellular acidification, the ADP/ATP ratios, and cytosolic ROS levels — with substantial implications for progressive conditions such as inflammation, infarction, cancer and stroke [1636].

Cancer cell proliferation can be inhibited by amplitude-modulated electromagnetic fields using tumor-specific frequencies [1637–1639]. PEMF stimulation has demonstrated promise in treating a variety of cancers (including bladder, breast, colon, liver, lung, ovarian, pancreatic, prostate, skin, and thyroid), although only a few limited examples of PEMF cancer treatments have been demonstrated on humans [1640, 1641]. This is consistent with the picture of a cancer-like phenotype in which cell metabolism is upregulated in the absence of a magnetic field, accelerating anaerobic glycolysis [1138] and in turn reducing ROS production [1201, 1202]. Although relatively little is known about the potential benefits of combining PEMF and radiation therapies in cancer treatments, the radiosensitizing effect of electromagnetic field treatments holds promise to increase the efficacy of established radiotherapies by enhancing the effects of ionizing radiation on cancerous tumor cells [1642].

Studies have shown that specific PEMF protocols can induce membrane permeabilization and apoptosis in medulloblastoma cancer stem cells *in vitro*, where ROS generation was observed in the surviving cells together with cell-cycle arrest correlated with a senescence-associated phenotype [1643]. Those findings reveal a promising new therapeutic option to target medulloblastoma cancer stem cells, which may provide an effective means to overcome radiotherapy resistance in cancer stem cells. This finding is particularly important because medulloblastoma is a highly malignant tumor of the cerebellum that constitutes about 20% of all primary childhood central nervous system cancers — which have been characterized as a leading cause of death in children and adolescents [1643].

Transduction of electromagnetic signals to implement control of cell processes — ranging from nerve impulses to gene expression and apoptosis — have been demonstrated both *in vitro* and *in vivo* [1631]. Chronic ROS levels can promote cell proliferation and tumor metastasis, whereas a sudden burst of ROS can initiate apoptosis immediately [1644,1645]. In fact, cellular ROS levels are increased by most chemotherapeutics today, which often alter redox homeostasis in cancer cells as part of their drug action [1646]. In contrast, iron oxide (Fe_3O_4) magnetic nanoparticles delivered to the tumor can generate ROS under the influence of magnetic fields, modulating the processes governing mitochondrial catalysis, DNA damage, and (abnormal) cell growth — consequently altering cell morphology and ROS production while promoting tumor growth [1647].

Blocking cellular ROS production can also reduce cancerous cell populations in cancers that are characterized by ROS overproduction, effectively inhibiting oncogenic signaling [1648]. Extremely low frequency electric fields are widely reported to contribute to the risk of some types of cancer [851, 1649], and non-ionizing wireless radiation from wireless devices is increasingly being considered as a possible carcinogen [1650]. Under certain conditions, PEMFs have been found to exacerbate the cytotoxic action of DNA-damaging agents bleomycin and methyl methanesulfonate [1651]. Nevertheless, cancer-suppressing magnetic fields have been used to inhibit tumor growth over a broad range of tumor cell lines representing lung, kidney, gastric, and pancreatic cancers [1652]. Thus, the potential of magnetic field-based treatments for cancer and other ischemic disease is diminished by the lack of clear biological mechanisms underlying the inconsistent effects of WMFs and ROS on cancer metabolism [1646], presenting an urgent need for fundamental advances in this area.

Healthy, neoplastic, and cancerous tissue differ substantially in terms of electromagnetic and mechanical properties [1653]. While differences in tissue stiffness may be well-known (i.e., thanks to conventional medical guidance to check manually for tumorous lumps), electromagnetic methods of imaging cancer and other pathologies have become a mainstay of medical science subject to ongoing development and improvements [1654–1656]. More recently, advanced imaging methods have been developed using nonlinear resonance interactions with weak electromagnetic fields to improve the discrimination of cancer, bone disease, inflammation, and vascular disorders [1657]. The tissue resonance interaction method (TRIM) has been shown to detect malignant tumors in the bladder [1658], colon [1659], and liver [1660], holding promise for cancer diagnostics without the need for invasive procedures.

Nuclear magnetic resonance (NMR) imaging has notably been used to visualize metabolic and physiological signs of health, as well as predict and detect early responses to cancer treatment, allowing for the possibility of NMR-guided individualized cancer therapies [1661]. Electromagnetic fields have been found to affect many biological functions such as gene expression, cell fate, and cell differentiation [1625]. Studies on the effect of NMR modulation of cell proliferation and gene expression have motivated interest in the therapeutic use of NMR technologies to treat a variety of medical conditions such as nerve injuries [1662]. Therapeutic NMR has been found to interact with cell hypoxic signaling pathways through circadian clock mechanisms via cryptochrome and clock protein expression levels according to a nonlinear dose–response curve [1663]. This has prompted a hypothesis that quantum mechanical effects based on the RPM underlie the effect of therapeutic NMR on cell metabolism [1636].

While there is some broad understanding of the physiological mechanisms that are likely to be involved in mediated cellular interactions with electromagnetic fields via PBM and PEMF treatments, the physiological action mechanisms of static magnetic fields are more enigmatic [1619]. As previously outlined, radical-pair recombination reactions provide established mechanisms by which magnetic fields can interact with chemical dynamics in biological systems [1620], altering the activities, concentrations, and lifetimes of paramagnetic free radical species [1664]. Low-intensity static magnetic fields are known to influence a wide range of cellular photochemical reactions [443]. Thus, as an uncontrolled factor in biological experiments, variations in ambient magnetic fields may alter the production of free radicals and heat shock proteins, producing significant changes in rates of cell replication, differentiation, and survival [1665].

Biological electron transfer is implicated in cellular responses to WMFs because the sudden transfer of an electron can result in the formation of a magnetically-sensitive radical electron pair. Low field effects (LFEs) of this kind have been implicated in a wide range of physiological processes ranging from the development of cancer metabolism (i.e., the Warburg effect) to beneficial cellular regeneration effects. Likewise, electromagnetic fields have been found to alter redox metabolism, biological ROS signaling, and cell growth *in vitro* [1666]. Findings of this nature have inspired proposals to assess and limit the environmental risks associated with long-term human exposure to ambient electromagnetic fields [1628, 1667].

Cancer cells display distinctive bioelectrical, biochemical, and biomechanical signatures during tumorigenesis [1668, 1669]. This is known to coincide with the bioelectric decoupling of those cells from their surrounding somatic environment, where the cell's bioelectric state can either induce metastatic behavior or suppress tumorigenesis [1670]. In spite of these research efforts, the biophysical mechanisms that stem cells use to sense and interpret electromagnetic cues in their environment remain unclear [1671]. As a result, cancer patients are often faced with dismal prognoses and marginal recovery outcomes amidst a proliferation of treatment-resistance and recurrent cancer phenotypes [1125, 1672, 1673]. Given its high mortality rate, severe side effects, and low prospects for complete remission, there is great need to understand the root causes of cancer and to develop new therapeutic modalities to combat it [1640]. These prospects are brightened by a growing body of evidence that is revealing the importance of electromagnetic fields in regulating cellular processes and maintaining tissue homeostasis [1674].

Chapter 16

Photodynamic Therapy & Nanotheranostics

Energy transfer from photo-excited chemical species to cell components can result in cellular luminescence [764, 1675]. Demonstrations of biological ultraweak photon emissions reveal that the phenomenon of "photochemistry without light" is in fact a universal characteristic of living organisms. In the 1960s, a general method of carrying out photochemical transformations "without light" was proposed by White, Wiecko, and Roswell [1676, 1677]. In this method, the need for direct interaction with light in a photochemical transformation is circumvented by the use of a primary "photosensitizer" molecule which is prepared in a photophysically-excited spin triplet (or singlet) state [1677]. The resulting photoproducts can then promote high-energy biochemical transformations without light. These have immediate importance for nucleic acid photodamage and repair, flavin photochemistry, vitamin D production, lipid peroxidation, urocanase activation, chromophore excitation, and aerobic metabolism [761].

Flavin and porphyrin coenzymes are natural photosensitizers which can act as strong oxidants under photoexcitation [1596, 1678–1680]. The extraction of hematoporphyrin from dried blood in 1841 marked the first isolation of a photosensitizer molecule from its biological milieu [1680]. Its structure took the better part of a century to characterize, culminating with the 1930 Nobel Prize in Chemistry to Fischer for his research on the structure and synthesis of porphyrins. Building on the work of 1915 Chemistry Nobel Laureate Willstätter, Fischer showed that chlorophyll and heme are structurally related, being formed from the same pyrrole constituents, where heme carries a single iron atom at its center in the same way that chlorophyll holds an atom of magnesium [1681]. The structural similarities shared by chlorophyll and heme make them essential for energy harvesting and transduction in plants and animals [1682,1683]. For this reason, heme and chlorophyll have been called "the two most important pigments for life on earth" [1684].

The effects of combining specific dyes with certain frequencies of light have been known since the beginning of the 20th century, when Raab observed that acridine red dyes would kill paramecia when exposed to light but not in darkness [1685]. The term "photodynamic reaction" was later coined by von Tappeiner in his work studying the role of oxygen in the photo-sensitization of fluorescent substances [1686]. With his colleagues, von Tappeiner carried out the first tests to explore the therapeutic benefits of the photodynamic effect in a trial on patients with facial basal cell carcinoma using the photosensitizer eosin, treating the affected areas with a 1% eosin solution in sunlight or arc-lamp light. He found that two thirds of the patients treated experienced complete tumour resolution and a relapse-free period of twelve months [1687, 1688]. Systematic studies on tumor localization and the phototoxicity of porphyrins were carried out throughout the mid-20th century, when Lipson *et al.* developed hematoporphyrin derivative (HPD) by chemically modifying hematoporphyrin [1689,1690]. Injections of hematoporphyrin preparations and derivatives led to selective tumor fluorescence into oncology patients, establishing the basis for modern day photodynamic therapy as HPD became the precursor to the first photosensitizer approved by the Food and Drug Administration (FDA) [1691].

PDT has emerged as a modern, non-invasive form of therapy which can be used to treat cancers as well as many non-oncological conditions. It has been applied in areas of dermatology, oncology, gynecology and urology, and in treatments of chronic inflammation and drug-resistant bacterial infections [1692,1693]. Treatments entail the local or systemic application of the photosensitizer, a photosensitive compound which accumulates in the affected area(s) where it is activated by light to initiate the targeted destruction of pathological tissue or cells [1675, 1692]. Light-initiated ROS production constitutes the primary action mechanism of PDT treatment, so that the effects of PDT are abolished in the absence of oxygen [1694]. PDT protocols are typically painless, efficient, simple to use, free from side effects, and well-tolerated by patients due to PDT's tissue-selective action [1692, 1695].

During PDT, the absorption of a photon transforms the prescribed photosensitizer molecule from its ground (singlet) state into a relatively long-lived electronically-excited (triplet) state via ISC from a short-lived photoexcited singlet state [1696]. The excited triplet can then react in one of two ways, either by reacting with a substrate (e.g., a membrane or another molecule) by transferring a hydrogen atom to produce a radical which ultimately combines with oxygen to form an oxygenated product, or by direct energy transfer to oxygen in a quantum mechanical "triplet-triplet annihilation" reaction to form a highly-reactive singlet oxygen molecule ($^1O_2$) [1675, 1694]. Both reaction pathways can occur simultaneously, and the ratio between them depends on the sensitizer type, the oxygen and substrate concentrations, and the sensitizer–substrate binding affinity [1694]. Subsequent reactions of $^1O_2$ with organic molecules invariably produce peroxides in chain reactions which readily yield a wide range of other oxygen-based radicals, reactive intermediates, and final products [1697]. This abundance of action has made maximizing the $^1O_2$ quantum yield a priority in photosensitizer research, where molecules can be engineered to harness radical pair dynamics and symmetry-breaking effects that enhance the $^1O_2$ and fluorescence quantum yields of photosensitizers when designing high-performance applications for PDT and diagnostics [1696].

PDT, also known as photodynamic inactivation, combines light with a biological substrate and photosensitizer to generate ROS that eradicate undesired cells involved in cancers and infections [1573, 1698, 1699]. Following Photofrin's success and developing on its limitations [1694, 1700, 1701], many photosensitizer molecules have been proposed for similar medical applications [1702–1705]. PDT is just one of many tools being developed that use nanomaterial platforms to combine therapy and diagnostics in the emerging field of nanomedicine [1692, 1706, 1707]. In principle, the use of targeted nanoparticles can improve the effectiveness of anti-cancer and other therapies while eliminating harmful side effects by leveraging the non-invasiveness and high sensitivity of these systems [1706, 1708–1711]. A growing number of nanodynamic therapies employ sensitizer molecules to generate cancer-killing ROS upon the application of an external energy flow such as an electric field during electrodynamic therapy (EDT) [1712, 1713], ultrasound during sonodynamic therapy (SDT) [1714–1718], or X-rays during radiodynamic therapy (RDT) [353, 1719–1722].

Proposed in the modern context by Feynman [1723, 1724], nanomedicine encompasses the visionary idea that nanoscale structures, devices, machines and robots may be designed and manufactured to perform comprehensive surveillance, control, repair, and defense of biological systems [1725]. The many diverse applications of nanoparticle systems include diagnostic instruments, imaging methods, targeted medicines and pharmaceuticals, biomedical implants, and tissue engineering [1723]. Nanomedicine is a growing and interdisciplinary field that integrates nano engineering and technology with the life sciences where it may eventually be integrated into all aspects of biomedical science [1726–1728]. Since Feynman put forth the revolutionary proposal that the future of technology lay at the nanoscale [1729], advances in nanotechnology have revolutionized practically all aspects of human life [1730].

Nanomedicine itself has branched off in hundreds of directions that encompass all manner of materials and devices structured to carry out medical diagnostics and clinical therapeutics at the nanoscale [1724, 1731]. Similar to PDT, chemodynamic therapy (CDT) defines a therapeutic strategy to eliminate cancer using Fenton or Fenton-like reactions to produce hydroxyl radicals directly at the tumor site [1732]. Originally described by Fenton in 1894 [1733], the Fenton reaction is an iron-mediated chemical process in which hydrogen peroxide (H_2O_2) is converted to a hydroxide ion (OH^-) and a hydroxyl radical ($OH^\bullet$) by way of iron oxidation ($Fe^{2+} \rightarrow Fe^{3+}$) [1734]. Proposed by Zhang *et al.* [1735], CDT exploits the mildy-acidic, peroxide-rich tumor microenvironment to ionize amorphous iron nanoparticles, generating cytotoxic $OH^\bullet$ in cancer cells without nanoparticle photoactivation. Compared to traditional cancer treatments, CDT offers higher tumor specificity and selectivity, lower systemic toxicity, and fewer side effects [1736, 1737]. Also — unlike PDT, SDT, and EDT — applications of CDT operate independently of external stimuli (such as light, sound, or electricity) and do not depend on a source of oxygen [1738].

Although CDT has been scientifically validated and is widely recognized, clinical applications of CDT remain challenging because its is still poorly understood [1734, 1736]. Clinical performance of CDT is limited by certain key mechanistic issues, primarily the insufficient endogenous H_2O_2 content of tumors and the low catalytic efficiency of existing chemodynamic agents *in vivo*, in spite of expectations of high catalytic efficiencies for iron-based catalysts and copper-based nanomaterials used in CDT [1737]. These limitations have made CDT insufficient to completely eliminate malignant tumors, motivating the development of synergistic treatment strategies that combine CDT with methods to increase H_2O_2 levels, reduce pH levels, and eliminate native reductants that counteract $OH^\bullet$ production in many tumors [1737, 1739–1741]. More fundamentally, the current limitations of CDT raise questions about the intrinsic cell interactions that influence the Fenton reactions, biochemistry, and biological signaling pathways that govern the CDT mechanism(s) [1736, 1742]. Precision control of quantum mechanical properties of the Fenton reaction active site include its electron density, configuration, and spin state and are significant for designing optimal nanoparticles for CDT [1736]. Recognition of these issues has inspired recent efforts to comprehensively optimize the CDT Fenton reaction centers to systematically overcome the current drawbacks of CDT [1727, 1743, 1744].

Photothermal therapy (PTT) designates another emerging method of combating cancer by using photosensitizer molecules that convert light energy into heat to carry out thermal ablation of tumor cells [1745, 1746]. PTT is a short, simple procedure with a rapid recovery time, which has been tested intensively in clinical trials [1747]. Its drawbacks are primarily associated with the systemic uptake of photosensitizers into the body during treatment and the precision limits of the lasers used for photoactivation [1745]. To improve its selectivity, PTT can be administered with microneedles to deliver photosensitizers directly into an affected area [1748]. A wide range of nanomaterials have been proposed for use in photothermal nanotherapeutics [1749, 1750]. Heat produced during PTT can be used to initiate the release of therapeutic compounds to regulate gene transcription and enzyme activity on demand, and to enhance other chemical processes in pathological tissues [1751, 1752]. PTT can be combined with other treatment modalities [1751, 1753], such as PDT which can have a synergistic effect thanks to the increased action of ROS at elevated tissue temperatures [1746]. Nanomaterials that combine efficient ROS production with antioxidant consumption can facilitate apoptosis-inducing ROS bursts while eliminating cancer cells' defenses against ROS overproduction [1644]. Biomimetic nanoparticles can also be functionalized to combine immunosuppressive signal blocking with photothermal tumor ablation to enhance anti-cancer treatments [1752, 1754].

Although the study of photosensitive processes dates back at least to the 19th century and heliotherapy itself has been practiced for millenia [1573], characterizations of these phenomena in quantum biological terms are a recent development [1236]. This may be due to the renewed interest in medical nanoparticle systems that has accompanied recent advances in quantum dot technologies for medical applications [1755–1758]. For example, iron selenide (FeSe) quantum dots have been functionalized with a coating of glutathione, polyethylene glycol (PEG), and human epidermal growth factor receptor 2 (HER2) antibodies to enable multiphoton imaging of tumor cells [1759]. It is common for HER2 to be overexpressed in several types of cancer including breast, ovarian, bladder, stomach, ovary, colon, bladder, lung, uterine cervix, head and neck, and esophageal cancers, as well as uterine serous endometrial carcinomas [1760]. As a result, HER2 has become a predictive biomarker for certain forms of cancer, particularly in breast and gastric cancers where therapies targeting HER2 have enabled revolutionary improvements in clinical outcomes [1760]. Nanoparticles or antitumor drugs conjugated with HER2 antibodies will bind preferentially in cancer cells where HER2 is overexpressed, allowing for targeted imaging, diagnostics, and therapeutics [1761].

Quantum dots have been called "designer atoms" [1762] because of their unique, tunable optoelectronic properties, which can differ substantially from their properties in bulk. They are typically semiconductor nanocrystals with 2–10 nm diameters that contain between 10 and 50 atoms [1757]. They are known for their vibrant optical properties which include high brightness and resistance to photobleaching [1756], as well as excellent surface-to-volume ratios which enhance their optical, mechanical, magnetic, and catalytic qualities, affording them with high absorbencies for pharmaceutical applications [1723]. They are increasingly being developed for photochemical pharmaceutical applications that require photoresponse patterns to interfere minimally with imaging while maximizing responses to the wavelengths used for photo-activated drug delivery [1763, 1764]. Photopharmaceuticals have demonstrated progress in mammalian disease model systems, but the general use of photopharmacology requires precise drug targeting following system-wide administration. This necessitates a combination of high bioactivity, photoswitchability, and potency, while maintaining metabolic robustness, solubility, and distinguishability for imaging [1765].

Similar to other nanoscale drug delivery systems [1766], "nanotheranostic" techniques combine targeted nanotherapies with diagnostic imaging [1767–1769]. They have been developed steadily over recent decades using an extensive range of nanoparticles and structures [1770]. In principle, theranostic treatments enable specific diagnoses, time-resolved drug monitoring, and enhanced drug action when treating a wide range of pathological phenotypes. Nanosized magnetic particles are increasingly being investigated across a wide spectrum of biomedical fields, as functionalized magnetic nanoparticles can be situated in live cells using magnetic fields, where they can be activated to enable applications like DNA transfer, gene therapy, thermal induction, or tissue engineering [1771, 1772].

Photo-driven nanotheranostics, or phototheranostics [1773, 1774], rely on diverse excited-state energy conversion properties of theranostic elements that integrate diagnostic imaging and therapeutic treatments and employ particles with favorable biocompatibility, tuneable photophysics, and photochemical characteristics upon aggregation [1775]. These systems pave the way for the development of tailored theranostic treatments with ideal activation characteristics that may be readily prepared as all-in-one multimodal interventions that combine specialized photothermal and photodynamic properties with diverse modalities. These currently include fluorescence-guided PDT [1776], photoacoustic-guided PTT [1777, 1778], and more general multi-modal design strategies that combine high-performance photosensitizers with high photothermal conversion or other "multimodal" functional capabilities that tune the relevant excited-state energy dissipation pathways as needed on a patient-centered basis [1769, 1779].

Precision treatments are being developed to target individual (cancerous or infectious) cells as well as specific proteins within cells. Chromophore-assisted light-inactivation (CALI) is a method to inactivate (i.e., damage or denature) specific proteins with high spatiotemporal accuracy using photochemically-generated ROS [1780]. CALI can be used in cancer therapies to target cancer-associated proteins (such as integrins which are associated with tumor development, angiogenesis, invasion, metastasis, and treatment resistance [1781]), or in various protein-targeted research applications (such as studies of actin polymerization [1782]) [1780]. Photoswitchable pharmaceuticals present a growing portfolio of biomedical applications, based on the development of photoswitchable designer molecules including oligonucleotides, saccharides, peptides, and proteins [1783]. Proposals for photo-active drugs run the pharmaceutical gamut from analgesics and inhibitors to antidepressant, antidiabetic, antiestrogenic, anti-inflammatory, antihypertensive, antipsychotic, antimycobacterial, antioxidant, and antitumor drugs [1784, 1785].

This wide range of applicability presents an open challenge to photochemical pharmacology to identify applicable reagents for therapeutic investigations *in vivo*, where photoswitchable reagents present vast new possibilities for high-precision research on cell development, division, motility, and transport [1786]. One approach has been to substitute traditional anti-mitotics with photoswitchable ones, enabling light-mediated spatiotemporal control of microtubule dynamics [1787]. Anti-mitotic cancer treatments interfere with cell growth by suppressing microtubule dynamics with a systemic drug action that produces the well-known and often-devastating side effects of chemotherapy [1788]. Photoactive tubulin-polymerisation inhibitors developed for optical control of microtubule dynamics hold promise to enable methods for the quantitative photocontrol of the microtubule cytoskeleton [852, 1786, 1789]. Thus, photochemical approaches aim to revolutionize cancer chemotherapy by opening a new horizon of targeted treatments involving photoswitchable anticancer medications.

Direct control of cancer cell redox chemistry defines a grand challenge of oncology, where it would constitute a means of inducing apoptosis or modulating cancer metabolism, if not reversing carcinogenesis itself. In lieu of direct control of cancer metabolism [1669, 1790, 1791], cellular redox control is increasingly being achieved using ROS-generating photosensitized nanoparticles — even in the absence of a definitive picture of the photopharmacological mechanisms in these systems [1699, 1792, 1793]. Magnetically-driven ROS production provides another novel method for treating cancer, tissue damage, and cellular degeneration [1794]. Functionalized magnetic nanotransducers have been shown to enable magnetic control of certain cellular processes such as neuronal firing, gene expression, and apoptosis [1631]. Weak mT-range magnetic fields can also interact synergistically with drugs to potentiate effects involving cholinergic, dopaminergic, opiate, nitric oxide, and serotonergic signaling pathways [1795]. Cobalt ferrite ($CoFe_2O_4$) nanoparticles have been used to direct the alignment of microtubules, creating ordered microtubule ensembles which may have applications for the control of cellular structures and transport mechanisms [1796].

Transformative new schemes for implementing light-based control of biological systems are being developed in the emerging field of optogenetics [1797–1799]. In optogenetics, natural and bioengineered photoreceptors are introduced genetically into the cells of interest for the purpose of making previously light-insensitive biological mechanisms both photosensitive and optically addressable [212, 1797]. These include many mechanisms involving cell cycle transitions, embryonic development, metabolic homeostasis, and stress responses [1799]. For example, optogenetic tools provide an ideal method for probing the complex spatial and temporal dynamics of gene expression signals [1800], enabling a high degree of spatiotemporal control over many biological processes [1801]. Optogenetic engineering has been used to engage protein oligomerization, protein-target interactions, and conformational switching to enable the reversible control of light-induced calcium-channel gating, dynamic protein-microtubule interactions, and membrane contact-site assembly [1802]. These methods show promise for the precise control of physiological development by patterning cell ion currents and signaling cascades [1803]. Originally conceived as a tool for neuroscience by Crick in 1979 and developed formally by Deisseroth *et al.* in 2006, the growing range of biomedical applications for optogenetics now include microbiology, cardiology, endocrinology, neurobiology, and oncology [1798, 1804, 1805].

Light-activated molecules are involved in many cellular processes and have inspired a growing assortment of nanomedical and nanotheranostic mechanisms and methods for advanced healthcare solutions with the potential to improve treatment and patient outcomes on a global scale [1806]. Minimally invasive and often straightforward to administer, these nanomaterial platforms enable precision diagnostics and drug delivery, improving the effectiveness and reducing side-effects of treatments for many common and deadly diseases [1692]. Photosensitive chromophores, quantum materials, and other biomolecular cofactors function at the heart of these new biomedicines, with quantum mechanical properties that depend critically on the physics and chemistry of the cellular surroundings.

Chapter 17

Regenerative Processes: Cells, Tissues, & Organs

Regeneration restores damaged tissue and recovers lost organ structure and function, making it the basis for healing and a key aspect of the preservation of life [1807]. Substantial loss of life can be attributed to organ system and regenerative failures, making tissue regeneration a crucial area of medical science [1808]. Hence, regenerative science has emerged as an important topic at the forefront of medical research in areas such as dermatology [1809], hepatology [1810], pharmacology [1811], neurology [1812–1814], oncology [1815, 1816], immunology [1817, 1818], cardiology [1819], and nanotechnology [1820–1826]. As the biomedical arsenal of regenerative biotechnologies continues to develop and diversify, there is growing interest in enhancing the precision and efficacy of regenerative methods with other emerging technologies such as artificial intelligence, biomanufacturing, and advanced data analytics [1827, 1828].

Circadian rhythms orchestrate fundamental regenerative processes [1829–1831], while physiology and behavior follow daily programs that are well-adapted to the alternating opportunities and challenges of day and night [1832, 1833]. Diurnal cycles of light and darkness govern rhythmic changes in the physiology of living organisms [1834, 1835] along with other environmental and seasonal cues [1836, 1837]. Light plays an essential role in modulating countless physiological processes which are influenced by the intensity, duration, rhythmicity, and colors of light signals [1838, 1839]. Many processes need to be regulated over each 24-hour cycle, making circadian rhythms essential to practically every aspect of homeostatic regulation [1840–1846]. Circadian rhythms govern tissue repair and regeneration [1830, 1831, 1847, 1848], regulating the cytoskeletal dynamics that govern wound-healing from the cellular scale to that of the whole organism [1849]. Likewise, disruptions to circadian rhythms are associated with aging and disease [1844, 1849, 1850]. These considerations have motivated the idea that the repetition of low-energy background stimuli can have profound effects on physiological function [1585], with broad implications for the mechanisms that govern biological timekeeping and the associated field of chronobiology [1834, 1851]. This is consistent with PBM therapy outcomes which depend on the wakefulness of the treatment subject and the time of day when treatment is applied [1603]. Observations that PBM treatments using red and NIR light can have a regenerative influence on cell function have inspired proposals for the strategic use of light therapy to regenerate human tissues and organs [1852].

Studies of wound healing provide insight into collective cell behaviors [1853], as local tissue responses to injury can produce lasting consequences for the organism [1854]. Cellular metabolites such as ROS have been demonstrated to play an important role in healing in several highly regenerative species of animals, including planaria, zebrafish, tadpoles, and geckos [1855]. A growing body of literature shows that physiological ROS levels can be modulated by the application of static and dynamic electromagnetic fields [443], presenting an emerging picture of the essential role played by electromagnetic effects in the morphological development of living systems [1856]. Inadequate regeneration of organ function after injury is associated with many outstanding medical problems worldwide, raising the question of why some organisms have such greater regenerative capabilities than others [1855].

Although growth and regeneration are universal characteristics of life, the capacity for healing can vary greatly and is restricted in most multicellular organisms [1857–1859]. In mammals, the regenerative ability of most tissue is limited to reparative processes that result in the permanent replacement of damaged tissue by collagenous connective tissue known as a scar [1857]. Only in certain highly-regenerative organisms is it possible to replace whole body parts, or in some cases, entire bodies [1859]. Free-swimming flatworms known as planaria are in this privileged category, making them a popular organism to research among biologists [1860]. The planarian regenerative capacity is largely attributed to the high density of stem cells found in their adult bodies, where a single type of pluripotent adult stem cell known as a "neoblast" can differentiate into the complete range of planarian cell types [1861, 1862]. Planaria are living proof of the extent of anatomical regeneration that living systems are able to achieve [978].

The healing abilities of planaria have positioned them as a model organism for research on regenerative medicine, where the focus is on developing a systems-level understanding of the morphological decision-making processes that enable planaria to regenerate their whole bodies [978]. Planaria demonstrate complex responses to different colors of light [1863], and PBM has been shown to stimulate stem-cell proliferation in planaria during regeneration [1862, 1864]. WMFs less than 1 mT have been shown to influence tissue growth and regeneration in planaria where WMF exposure can increase or decrease the growth rate of new tissue [548]. Those studies revealed that ambient magnetic fields in the 100–500 μT range could induce statistically significant changes in planarian regeneration, due to altered stem cell proliferation and differentiation mediated by the accumulation of ROS and the expression of heat shock protein 70 [548, 1865]. These results highlight the potential to use non-ionizing electromagnetic radiation and static magnetic fields to control tissue growth and regeneration [504, 1865].

ROS signals have also been shown to regulate new tissue growth in zebrafish where ROS production triggers the proliferation of cells necessary for regeneration to progress [1866]. Upon the amputation of a zebrafish tail fin, physiological ROS generation is regulated at the wound site to initiate healing and coordinate fin regeneration [1867, 1868]. Zebrafish, salamanders, and newts are some of the only vertebrates that can persistently regenerate amputated limbs throughout the course of their lifespans [1866]. Similar to zebrafish, axolotl salamanders produce ROS at the site of a tail amputation to initiate regeneration and repatterning of the lost limb [1869]. Studies on tail regeneration in tadpoles show that the influx of O_2 produced by an injury increases local ROS levels, in turn stabilizing hypoxia inducible factor (HIF) 1α while modulating heat shock protein 90 levels and electric currents critical to regeneration [1855]. ROS gradients at the wound site then attract immune cells to the area [1870]. Regeneration-organizing cells at the wound site serve as a primary source of growth factors and regenerative signals, promoting proliferation of the underlying progenitor cells [1871] and orchestrating the interplay of biochemical and physical regenerative signals during the regeneration of limbs in vertebrates and invertebrates alike [1855, 1870]. In this context, ROS have emerged as a key regulator of stem cell populations in regenerative and developmental processes [1872, 1873].

Stem cell therapies have shown recent promise to aid the restoration of tissue or organs compromised by injury or disease. These can be highly effective, especially when combined with adjunct treatments such as gene-editing or anti-inflammatory therapies among others, but can be costly and labor-intensive to implement [1808]. Stem cell therapies have continued to advance, although the translation of regenerative science into standardized clinical applications has been limited by a mixture of outcomes across specialties which may be attributed to differences in clinical practice and implementation [1827]. The use of stem cells to repair, replace, and regenerate damaged tissues during stem cell therapy is itself associated with risks because it is not straightforward to isolate stem cells or to commit them to develop into a desired phenotype. Stem cells can differentiate into all cell types, including cancer cells, making tissue overgrowths (as tumors or teratomas) a main concern for stem cell treatments [1807, 1808].

Although only a relatively small number of stem cell therapies are clinically available today, stem cell engineering has the potential to revolutionize clinical practice by using tailored biomaterials with tuneable mechanical and biochemical properties to effectively recreate tissues *in vitro* [1874, 1875]. Lab grown tissues have tremendous implications for fundamental physiological studies for translational therapies intended to repair, preserve, and enhance tissue function using stem cell implants [1852, 1875, 1876]. These therapies enable the controlled migration, stimulation, and differentiation of stem cells to materially reconstruct compromised tissues [1586,1877,1878]. However, practical implementations of stem cell engineering in the form of designer tissues and engineered cartilage implants remain elusive and there is an outstanding need for hydrogel implants that incorporate growth factors into functional three-dimensional (3D) cell cultures [1879–1882]. Therapeutic and tissue engineering strategies have thus been proposed to coincide optimally with stem cell circadian rhythms for regenerative medicine [1849, 1883–1886].

Reconstitution of damaged cartilage poses one of the greatest challenges to regenerative medicine because the regenerative capacity of cartilage is restricted by its avascular composition [1887–1892]. Repairs to the articular cartilage covering joint surfaces pose a difficulty for joint surgery due to its limited capacity for self repair [1888, 1893–1895]. For this reason, current therapeutic treatments for osteochondral defects remain primarily palliative (rather than curative) [1887]. Nevertheless, a variety of innovative approaches have led to improved outcomes in the treatment of damaged cartilage and bone [1896]. Tissue engineering strategies that incorporate biocompatible scaffolds embedded with chondrocytes and/or stem cells have shown promise for the effective regeneration of articular cartilage tissue [1882, 1893]. Synthetically-prepared, layered scaffolds that mimic the structure of articular cartilage and subchondral bone have demonstrated preliminary success in repairs of osteochondral defects [1895].

Articular cartilage may suffer damage or degeneration as a consequence of aging, injury, or disease [1882, 1893]. Although damaged bone and cartilage will heal naturally in principle, functional tissue regeneration can be hampered by a wide range of pathological factors [1896, 1897]. As a result, there is a substantial need for clinical methods that promote the regeneration and repair of articular cartilage [1893]. Cartilage rejuvenation has become a major topic in orthopedic research to mitigate osteoarthritis and joint degeneration [1898, 1899]. The drawbacks of current orthopedic treatments have motivated interest from biomedical science in the development of new polymers and methods for preparing customizable 3D scaffolds and implants for osteochondral tissue regeneration [1899].

Tissue engineering methods have emerged as potential treatments to fully restore damaged articular cartilage [1891]. Combining biomaterials, live cells, and growth factors to regenerate damaged tissues, tissue engineering aims to overcome limitations of conventional treatment strategies based on surgeries and transplants [1882, 1897]. Although hydrogel implants using natural and synthetic gel-type scaffolds have demonstrated positive outcomes in preliminary clinical studies [1882, 1900, 1901], osteochondral tissue itself remains the best scaffold to restore osteoarthritic defects [1902]. Although biomaterials are now used extensively in orthopedic work, their capacity to repair tissue is still limited by their failure to fully reproduce the mechanical, biochemical, and metabolic characteristics of the tissues they replace — highlighting the need for bioactive scaffolds that differentiate into cartilage and bone [1887].

Osteochondral tissue homeostasis is maintained by articular cartilage chondrocytes and subchondral bone osteoblasts according to physico-chemical cues which influence their generative functions [1903]. Those cells synthesize the structural components of the tissue matrix and establish key features of homeostasis via the corresponding metabolic, inflammatory, and immunologic pathways. Mesenchymal stromal cells are stem cells which play a central role in maintaining the osteochondral matrix, secreting growth factors and signaling molecules which mediate tissue regeneration and repair [1897, 1904]. Despite their importance, the complex mechanisms and interactions that govern regeneration and stem cell differentiation in osteochondral tissue remain unclear [1897, 1905]. Nonetheless, it is broadly understood that biomechanical and electromagnetic factors provide key stimuli that regulate cartilage growth and ossification in line with skeletal development [1903]. Hence, the successful engineering of osteochondral tissues will require the strategic coordination of the interplay between cells, materials, and molecular factors [1906]. PBM has been found to promote cartilage regeneration for use in biomedical and tissue engineering applications [1907, 1908].

While the body has a natural ability to repair small or moderate sized injuries and tissue defects, substantial damage or damages to avascular tissue can become difficult to heal naturally [1896]. During injury, blood is typically the first tissue to respond by sealing the wound with an aggregation of platlets that are cross-linked with fibrin polymers to stop bleeding and promote wound repair [1583]. Fibrin is produced from fibrogen, a glycofibrous protein that is characterized by its quadruple α-helical coiled structure and ubiquitous presence in human blood [1909]. It is considered essential for blood coagulation, wound healing, inflammation, and the formation of new blood vessels, among other functions. As a soluble protein, it forms an insoluble clot upon its conversion to fibrin by the allosteric serine protease thrombin [1910]. The structure of the fibrin matrix provides a temporary home for cells to proliferate and organize a coordinated response to injury at an inflamed site, making it a promising candidate to facilitate tissue regeneration in surgical procedures as a hydrogel biomaterial composite [1583, 1909].

Fibrin is increasingly used in biomedical applications where it can be applied to create scaffolds for tissue regeneration or targeted drug delivery systems, such as 3D hydrogels which recreate properties of the ECM that support cell growth, differentiation, and migration [1583]. Fibrin hydrogel systems are aqueous molecular networks with tunable properties which can be used to develop complex tissue mimetics that reproduce certain prototypical features and interactions of the cell matrix [1911]. Recreation of *in vivo*-like conditions *in vitro* are naturally suited for the development of native tissue models, disease mechanisms, pharmacology, and cell therapy — but hydrogel systems suffer from restricted diffusion and lack of vascularization which can result in cell hypoxia, stress, and death [1912]. Established uses of PBM for promoting cellular resistance to hypoxia and stress have made it a candidate to improve the viability of cells cultured in fibrin scaffolds and hydrogels [1912]. Thus, PBM has been proposed as an adjunct therapy for use with fibrin compounds in tissue engineering and regenerative medical applications [1583]. Photobiomodulation therapy has been shown to enhance bone tissue regeneration [1913], particularly when combined with the surgical use of a fibrin biopolymer to repair bone defects [1914].

Fibrin scaffolds have been proposed for use in surgical procedures for neurology, orthopedics, periodontics, plastic surgery, and implantology, as well as in oral and maxillofacial surgeries [1583]. PBM is now being applied in the developing fields of regenerative medicine and tissue engineering, but the great diversity in PBM sources and parameters, cell types, scaffold compositions and geometries make it difficult to predict the outcome of any particular combination of experimental factors [1912, 1915, 1916]. Red-to-NIR range PBM has been demonstrated to be effective at enhancing cell proliferation, differentiation, and survival in 3D scaffold and hydrogel systems *in vitro* [1912]. The use of specific frequencies of light has also been explored to enhance specific cellular processes — such as the influence of 632.8 nm PBM on chondrogenesis [1917]. This work has inspired proposals to design implantable PBM devices for the photo-stimulation of interior organs (i.e., in the body cavity) *in vivo* [1852]. Despite its longstanding use and demonstrated effectiveness, there is still no clear understanding of photophysical mechanisms and chemical pathways that govern PBM. This means that research on PBM effects in 3D hydrogels, scaffolds, and implants may also provide new insights into the fundamental quantum mechanical processes governing PBM.

PBM also makes a promising choice as an adjunct treatment to stem cell therapy, because PBM has been developed as a potential treatment for controlling stem cell differentiation, proliferation, and migration *via* ROS modulation [1918, 1919]. Full characterizations of the effects of PBM on living cells have remained elusive due to the uncontrolled variations that distinguish different PBM protocols, although progress is being made to establish more favorable methods for promoting the proliferation and differentiation in stem cells *in vitro* [1607, 1608]. PBM has been found to enhance stem cell proliferation and promote differentiation *in vitro*, where specialized media can be used to induce cell differentiation into mature phenotypes that include adipocytes, chondrocytes, and osteoblasts [1920]. It has been shown to stimulate the expression of osteogenic genes in periodontal ligament stem cells, promoting enzyme activity related to bone growth and mineralization [1921]. Promising results of this kind support the idea that PBM can be used to systematically improve methods for proliferating and differentiating bone marrow stem cells *in vitro* prior to host transplantation during stem cell therapy [1922]. This presents the hopeful prospect of combining PBM with stem cell therapies in advanced tissue engineering and regenerative biomedical applications.

Cells exhibit diverse responses to their electromagnetic environment which influence respiration, photosynthesis, nutrient uptake, and other biochemical attributes related to concentrations of ROS, antioxidants, proteins, and metabolites [1923, 1924]. However, the complex interplay between electromagnetic, biochemical, and mechanical mechanisms that collectively regulate tissue response to injury, stem cell dynamics, regeneration, and cancer suppression remain unclear [1925]. To form a complete picture of the mechanisms underlying the collective behaviors of multicellular systems [1926], it is necessary to fully integrate the biomechanical and biochemical models of anatomy and physiology while extending them to incorporate electrodynamic phenomena. The development of a complete quantum biophysical picture of the underlying processes that govern cell growth and tissue regeneration will present unprecedented possibilities to advance research in regenerative medicine, bioengineering, and nanotechnology.

Chapter 18

Morphogenetic Integration & Immunodynamics

Cells must integrate the information they receive from their environment to coordinate responses to environmental cues [1927]. Cell behaviors are central to all aspects of life, including development, homeostasis, reproduction, and regeneration [1928]. Beyond the conventional range of sensory-motor behavioral responses, cells and tissues exhibit a vast array of physico-chemical phenomena during complex phase transitions [1929]. These can incorporate multiple states of matter (solids, liquids, gases, and plasmas [1930]) while exhibiting diverse quantum optical, electrochemical, and magnetic properties [515, 1931, 1932]. To create an atlas of cell types, characteristic cellular states of health and disease have been cataloged and benchmarked using various genomic, biochemical, and bioelectric signatures [1933, 1934]. This cataloging and benchmarking of cellular phenotypes does not yet provide a complete understanding of how cells interact collectively to form the tissues and organs of higher plants and animals.

Raised in the modern era by Schrödinger [42], the question of how living beings are able to generate biological structures while maintaining homeostasis in the intrinsically noisy cellular microenvironment remains one of the foremost problems of anatomy and physiology [1935, 1936]. To rationalize the high degree of functional organization found in biological systems, it has been proposed that the robust nature of living things ultimately relies on the performance of a complex ensemble of gene regulatory networks and maintenance ("clean-up") systems [1937, 1938]. Indeed, the genome specifies a cellular collective which is innately plastic, executing morphological rearrangements until a target morphology is achieved [1856]. Yet this simplistic picture of morphogenesis begs the question of how biological systems determine when to stop remodeling themselves once a preferred morphology has been reached.

Animal cells can change shape at will, responding dynamically and adaptively to many different environmental cues, by virtue of the tightly-coupled biochemical and mechanical properties of the cytoskeleton which enable cell polarization, morphogenesis, and motility [1939]. Unlike long-lived cellular structures, an individual microtubule's lifetime may be as short as twenty seconds *in vivo* [1940], so microtubule networks are renewed constantly to maintain cytoskeletal structure and organization [1941]. Cytoskeletal self-organization arises from the reaction mechanisms that control microtubule growth and shrinkage *in vitro* [1942–1944], and hence reaction–diffusion processes have been proposed to explain similar organizational behaviors *in vivo* [1945, 1946]. However, reaction–diffusion mechanisms are not sufficient to explain certain key aspects of cytoskeletal reorganization during cell migration [1947]. Cells are sensitive to their electrochemical environment and respond to a wide variety of signaling molecules including growth factors, hormones, and neurotransmitters [650, 1948]. Cellular mechanotransduction, the conversion of biomechanical forces into biochemical signals, is believed to influence important biological processes involved in embryonic development, tissue repair, and regeneration [1949–1952]. For example, the viscosity of the ECM has also emerged as a key regulator of cell and tissue dynamics [1928], with demonstrations that cells respond to the viscoelasticity of their environment in their growth, migration, proliferation, differentiation, and development [1953].

During embryogenesis, tissue repair, and cancer metastasis, cells undergo migration either as individuals or as coordinated clusters of cells [1954]. The collective dynamics of cell clusters are mediated by cell–cell junctions which transmit and respond to physiological signals between cells, as well as the single-cell interactions which mediate cells' senses of their environment [1853]. Embryonic cells within clusters may also adjust their stiffness in response to the stiffening of their native substrate when initiating collective cell migrations, where the softening of cluster stiffness can trigger collective cell migration in soft substrates [1954]. Above and beyond cell migration, a wide range of dynamic morphological processes including cell division, phagocytosis, and stem cell differentiation are known to be modulated by the external cellular environment [1955, 1956]. A number of physiological factors linked to the ECM, including cell deformations, external mechanical forces, ECM mechanics (e.g., stiffness, density, elasticity), and topographical features may all govern stem cell differentiation [1956]. Studies of collective cell migrations *in vitro* have revealed cellular dynamics similar to those of a phase transition known as a glass transition [1957].

Mechanical signals affect many biological processes in both developing and adult organisms, including cell migration, differentiation, morphogenesis, and immune responses [1958, 1959]. For example, experimental research *in vitro* has shown that MSCs cultured on stiff substrates mimicked bone stiffness, whereas MSCs cultured on soft substrates mimicked the softness of brain tissue [1960]. Research on mechanical signaling in cells is relatively new, and detailed insights into the factors that influence mechanical transitions are rare [1958]. The mechanisms that regulate microtubule nucleation in frog eggs are largely unknown [1961]. Likewise, the nature of organizational waves in the microtubule networks of zebrafish embryos remain unclear, although researchers have concluded that microtubule network organization is not fully described in static terms and must also include dynamic effects [1962]. The successful development of quantitatively-predictive multiscale models of microtubule dynamics are expected to require deep, comprehensive insights into the structural and mechanical properties of tubulin [868].

Microtubule networks control many aspects of cellular architecture, polarity, and proliferation in animal cells, but the mechanisms which govern the formation of those networks are only minimally understood [1037]. Tubulin is predicted to respond to short, intense electromagnetic pulses in simulations [964], and Cifra *et al.* have proposed that microtubules can themselves generate electric fields when excited electrochemically [851]. Electrical oscillations are intrinsic to bundles of microtubules found in brain cells, where they contribute to the fundamental oscillatory modes of neurons [905]. Although anti-cancer drugs such as taxanes are well known to have microtubule-stabilizing properties which perturb microtubule dynamics and interfere with microtubule growth, the exact nature of pharmaceutical action of those drugs remains poorly understood [909]. Proper microtubule function is crucial to cell health and thus microtubule system abnormalities can result in neurological disorder and disease [865, 1963].

Cellular mechanical responses are mediated by the cell membrane, cell adhesion complexes, and the cytoskeleton itself [1964]. Cells respond to mechanical forces during tissue development and countless disease processes as they regulate the mechanics of the ECM that maintain its physiological properties [1965]. The ECM is a highly responsive structure present in all tissues which is continuously remodeled in the process of maintaining homeostasis [1966, 1967]. It provides the essential structural foundation for multicellular organisms as well as many of the physiological cues that initiate and control cellular behavior [1968]. It regulates morphology during tissue development by governing key aspects of cell adhesion, differentiation, migration, and morphology [1959, 1969]. Cell migration, for example, is contingent on the cytoskeleton's mechanical properties which achieve structural stability and flexibility through dynamicity [1941]. Distinctive cell characteristics give rise to different forms of cell migration. During mesenchymal cell migration, cells polarize and form leading-edge protrusions, which adhere to the ECM via integrin-mediated cell-to-matrix adhesions [1970, 1971]. These dynamical processes are actuated by vast cytoskeletal rearrangements and signaling cascades [1970]. The cytoskeleton is mainly composed of microtubules, actin, septin, and intermediate filaments which each have distinctive mechanical properties [1035, 1972, 1973]. Crosslinking plays an essential role mediating the dynamics of the cytoskeleton, where microtubule crosslinking facilitates organization and transportation, and actin crosslinking determines mobility [1974, 1975]. Despite the cytoskeleton's critical physiological importance, there is still only a preliminary understanding of the mechanisms that govern its functionality [1976, 1977].

Numerical studies have shown that quantum relaxation effects in microtubule networks are likely to be important and comparable in time with other dynamical processes occurring there [1978, 1979]. Real-time microtubule imaging using scanning tunneling microscopy (STM) revealed how the application of a quantum mechanical tunneling current could be used to stimulate microtubule assembly from tubulin when the alternating tunneling current was tuned to resonance frequencies of tubulin [936]. Advanced methods for simultaneously visualizing and manipulating the cytoskeleton that combine optical tweezers or atomic force microscopy with fluorescence methods are likely to provide further insights into the fundamental mechanical properties of cellular microtubule networks [1939]. Deep insights into the physics underlying cytoskeletal mechanics are needed to inform models of the cytoskeletal dynamics associated with cell motility, migration, differentiation, development, and disease [1952, 1970].

Cell migration is necessary for physiological development, immune responses, and healing, but it is also a mechanism of cancer [1955]. During migration, cells navigate complex tissue environments, relying on chemical, electrical, and mechanical signals for direction, and traversing chemical gradients of chemokines and growth factors to reach their intended destinations [1980]. While thousands of scientific works have explored biophysical mechanisms of cells migrating on flat, 2D substrates, less is known about cell navigation in 3D environments [1955, 1981]. Chemotaxis is the movement of cells in response to their chemical environment [1982, 1983]. Although chemotaxis has been studied for over a century, research on mechanotaxis — cell movement in response to mechanical cues — is only emerging now to characterize cell responses to material substrate adhesions, stiffness, and density [1982, 1984–1987].

Cells undergoing tumorigenesis exhibit cytoskeletons with increased motility but reduced polarity [1988], and aberrant cell migration is also a characteristic of chronic inflammation, vascular disease, and metastasis [1971, 1989]. Although increased cell motility and reduced polarity are also observed in migrating cells during normal biological processes, subtle characterizations of cellular bioelectric signatures can predict the emergence of aberrant cellular behavior (e.g., in incipient tumors in amphibians [1670]). Efficient metastasis requires the coordinated operation of the cytoskeleton to form the many cellular protrusions that enable cell migrations that lead to death in 90% of cancer patients [1972]. In cancer, the ECM also becomes highly dysregulated, and the loss of proper homeostatic organization is a hallmark of solid tumors [1990]. The success of microtubule-targeting drugs has motivated a search for other pharmaceutical means of targeting mitosis to treat cancer, but this effort has been mostly unsuccessful [1991].

Collectively known as "immunotherapies" [1992, 1993], a growing range of treatments has become available which combat cancer by boosting the body's native cancer-fighting mechanisms using cancer vaccines, pharmaceutical checkpoint inhibitors, T cell expansions *in vitro*, lab-created anti-bodies and cytokines, etc. [1994]. However, these treatments are limited in scope and can include serious risks of side effects such as deadly cytokine storms [1993]. These complications have motivated research on the use of natural molecules, electromagnetic fields, and mechanical vibrations to enhance the natural human capacity for healing [1807]. Whereas direct effects of cancer treatments such as PDT, CDT, and conventional radiation and chemotherapies can eradicate the bulk of a tumor, a native immune response is still needed to eliminate any surviving cancer cells to prevent metastasis and cancer recurrence after remission [1694]. It is therefore critical to understand the underlying biophysical factors which initiate and escalate a living organism's immune response to cancers and other pathogens.

Circadian rhythms govern the immune system and synchronize endogenous oscillations in migration-promoting factors which in turn are mediated by the expression of cell adhesion complexes and chemokines [1842]. Thus, the immune system responds to biochemical and mechanical signals expressed as changes in the biophysical properties of tissue that reflect the many aspects of disease, inflammation, and aging [1995]. Immune responses orchestrate the mobilization of millions of leukocytes at the sites of active inflammation, where swarms of immune cells enact complex behavioral strategies that are now being studied using high-dimensional datasets to describe vast multi-parameter "behavioral landscapes" of immunological activity [1996]. These immunodynamics are coordinated from the lymphoid organs of higher animals, where networks of fibroblastic reticular cells inhabit the collagen mesh found at each lymph node [1997]. Each reticular cell network provides the structure to support native immune cells while distributing essential molecules that enable immune cell recruitment, trafficking, and survival [1998].

Although reticular cellular networks are primarily associated with the lymph nodes and spleen, they are also recognized for the more general biomechanical roles that they play in defining the physiological architectures of multiple organs, creating structural continuity that transcends conventional organ boundaries [1999]. Dynamic in nature, the reticular cell network relaxes during immunological activation to allow for the influx of immune cells that occurs during clonal expansion [1997]. As a "small world" network [2000], it has important topological properties that optimize communications between immune cells and provide resilience against structural damage and disruptions. This inherent topological robustness allows reticular networks to sustain losses of up to 50% of the fibroblastic cells that inhabit it without compromising immune cell recruitment, migration, and activation — demonstrating the critical role played by network integrity in mounting and sustaining immune responses [2001, 2002]. The robustness of the reticular network is an inherent feature of small-world networks [2000], similar to robust network designs proposed for entanglement-based quantum communication systems [2003, 2004]. A measure of "small-worldness" was also proposed as a topological order parameter for use in research on superconductors and topological quantum computers [2005].

Given the importance of its functions, surprisingly little is known about the structure of the reticular cell network (e.g., compared to the brain [1998]). Nevertheless, the practical role that reticular networks play, providing structural definition in higher organisms has inspired efforts to develop customizable organ grafts by "re-seeding" decellularized fascia (rich in native collagen, elastin, proteoglycans, and glycoproteins) with the appropriate cells to create structurally-organized tissue scaffolds for use as synthetic organ grafts [2006, 2007]. Fibroblastic reticular cells are stromal cells that continually remodel the reticular network in coordination with immune responses [2008], influencing all aspects of adaptive immunity and responding dynamically to infection [2009]. This is intrinsically regulated by the inner workings of the fibroblastic cells themselves, as the cytoskeletal network tension of fibroblastic reticular cells directly controls the size and flexibility of lymph nodes in the lymphatic system [2009]. Thus, it will be necessary to understand the intrinsic biophysical mechanisms that govern cytoskeletal regulation in order to harness the profound regenerative abilities of reticular cell networks for bioengineering applications.

Advances in the bioengineering of multicellular systems have driven efforts to test the limits of life's building blocks by exploring new methods of biological assembly that may reveal the fundamental biophysical principles that govern living matter [2010]. Pioneering efforts to develop organ reconstruction and manufacturing technologies for therapeutic and transplantation purposes have made steady progress despite numerous challenges [2011], and 3D organ bioprinting has stood apart as an emerging organ manufacturing technology [2012, 2013]. Although organ manufacturing techniques have had enormous success demonstrating the production of 3D-printed tissues [2013] and "organ-on-chip" technologies [2014], these synthetic methods are still far-removed from the natural growth and development processes that generate tissues and organs in living organisms [2015]. Much of the difficulty lies in mimicking the natural properties of the ECM [2016], let alone the enormous challenge of artificially reproducing the complex structure and function of whole organs [2017]. Dynamic control of the ECM, similar to that exerted by the fibroblastic cells of the reticular network, remains far beyond the reach of bioengineering and nanotechnology today. Fundamental advances in quantum biology are needed to bridge the gap between classical organ-manufacturing technologies and the natural developmental and immunological processes that create and sustain living tissues. Understanding the cells' innate control mechanisms all the way down to the quantum scale offers the potential to greatly improve these physiological processes and regenerative technologies.

Chapter 19

Quantum Biotechnology: Universal Applications

Living systems achieve a level of control over the electronic states of their internal structures that has long been sought after by researchers developing molecular electronic devices [2018–2020]. In this respect, the native biomolecular reaction networks found in living organisms harbor timeless solutions to the problem of harnessing strongly-interacting many-body non-equilibrium open quantum system dynamics [2021] — a highly-complex quantum mechanical problem that has driven intense research on molecular electronics for several decades [2020, 2022] Bioelectronic systems naturally harness the intrinsic disorder of the biomolecular microenvironment [2023]. This inherent robustness has motivated growing interest in prospects for the development of advanced quantum biotechnologies for use in quantum communication and computation [2024–2026]. Biomolecules have many advantageous functional properties [2027, 2028] which are complemented by their inherent modularity and capacity for self-assembly [2029].

Bioelectric gradients are increasingly recognized as control factors that govern growth and form in living organisms [977] and fundamental advances in bioengineering have presented the possibility of disassembling, assembling, and recombining biological structures at multiple scales to contrive living organisms in new configurations that defy many traditional conceptualizations of life [2030]. Although self-assembly is found prevalently in inanimate and biological system alike, in biological systems it primarily takes the form of self-limiting processes characterized by the assembly of smaller subunits into larger, yet finite-sized, anatomical structures [2031]. Biological assembly and organizational processes are as common as they are vital, yet by-and-large poorly understood [1611, 2032].

Although eukaryotic cells contain many specialized organelles to perform specific tasks in concert, the mechanisms that coordinate their positions in the cell with respect to their functions remain largely unknown [1611]. In fact, even the relatively unstructured cytoplasm of bacteria exhibits glassy dynamics indicative of memory effects which may be altered dramatically during cellular metabolism — suggesting a highly complex matrix that is not well approximated as a simple fluid medium [2032]. Despite the progress in the development of non-equilibrium thermodynamics to predict the emergence of self-organized behaviors in classical complex systems, the task of extrapolating those classical principles to the quantum scale remains daunting [40]. One approach, in line with theories of mitochondrial retrograde signaling [1611] and membrane dynamics [2033,2034], hypothesizes that strong electromagnetic fields generated by the mitochondrial membrane are a key factor regulating global cellular activity [2035, 2036].

Proteins play countless essential roles in nature, where they are critical for the survival and reproduction of living organisms by enabling them to adapt to continually changing environmental forces. Bioengineered protein materials have emerged recently as multi-functional media that are durable with respect to heat, light, humidity, magnetic fields, and other factors [2037]. Protein-based materials are naturally responsive and adaptable, making them choice materials for use in the development of synthetic tissues [1882], shape-changing robots [2038], and reconfigurable organisms [2039]. Efforts to re-shape living organisms for the purposes of biomedicine or bioengineering will require precision control of the structure and function of an organism at all scales [2040, 2041]. Many problems in medicine, especially, could be addressed by answering the primary question of how cells and organelles cooperate to build and maintain essential anatomical structures [2042]. More fundamentally, the prospect of building, repairing, and remodeling biological structures is attendant to the idea of generating bespoke organisms.

Controlled assemblies of physiological crystal structures produce diverse, responsive optical effects for coloration and vision across the animal kingdom [2043]. For example, investigations of the vivid patterns found on butterfly wings have revealed key aspects of the self-organizing structural dynamics that generate vibrant 3D biophotonic nanostructures [2044]. Although the formation of organic crystals is ubiquitous in plants and animals [2045], the fundamental organizational mechanisms underlying these structures remain mysterious [2043, 2044]. For example, peacock tail feathers were recently discovered to contain structures capable of laser light emissions [2046]. Research on the mechanisms that govern biological nanophotonics *in vivo* has great potential to enable applications of biomimetic optical devices using photonic crystal engineering [2044, 2045].

Single-atom photocatalysts have been proposed to overcome challenges associated with photoactive nanoparticles, nanoclusters, and bulk materials, due to the exceptionally-high photoactivity and selectivity of the electronic structures associated with atomic coordination centers [2047]. Individual atoms have the added benefits of reduced environmental and fiscal costs associated with the use of catalytic metals, highly tunable electronic properties, and comparatively simple reactive sites to ease the analysis of reaction mechanisms and structure–performance relationships [2048]. Of course, single atom catalysts are already found ubiquitously in biological systems, where porphyrin rings and their derivatives carry individual atoms of iron (in heme), cobalt (in vitamin B_{12}), magnesium (in chlorophyll), and nickel (in coenzyme F_{430}) [2049]. Biological metal–porphyrin complexes are critical to countless life-sustaining processes involving electron transfer, enzyme catalysis, oxygen transport, and photosynthesis. These complexes have provided inspiration for research on the development of artificial metalloenzymes [2050, 2051]. The many versatile physico-chemical, electronic, and spectroscopic properties of metal–porphyrin complexes have made them a subject of numerous biological studies, inspiring discoveries in multiple research areas with applications for antimicrobial drugs, biomedicine, catalysts, optoelectronics, sensing, and solar cells [2052–2055]. Yet porphyrin is only one among many examples of catalytic sites found in biological systems, be they metallic [2056] or organic [2057].

Many of these biocatalytic sites are involved in or related to the biological electron transfer processes that mediate the electronic excitations that enable "virtually all cellular activities" [801, 2058]. Ultrafast spectroscopy experiments are now uncovering a wealth of information about bioelectronic excitation dynamics [96, 2059, 2060]. Growing understanding of biomolecular interactions with light herald exciting prospects for biophotonic engineering. Genetic engineering techniques are already being applied to native molecular targets such as cytochrome enzymes to enhance photosynthetic yield [704]. Likewise, optogenetic techniques can be used to implement optical control of metabolism in bioengineered yeasts by triggering the assembly and disassembly of metabolically active enzyme clusters to control the formation of desired metabolic products [2061]. Light-based interactions have also been used to initiate chemical signaling between synthetic and natural cells, demonstrating the use of light to trigger the delivery of small molecules to live cells for biomedical applications [2062].

Research on quantum nanophotonics has undergone a recent renaissance of interest in theory, implementations, and applications of quantum biotechnologies [2063]. Biotechnological applications are bolstered by the emergence of "soft" nanophotonic devices [2064] as well as biofunctionalized nanoparticles and quantum dots [1236]. Strong light–matter interactions between nanophotonic enclosures and quantum emitters have unveiled unprecedented prospects for modifying chemical landscapes and reaction cascades with applications for biomolecular sensing, protein engineering, and control of hybrid light-matter states of living cells [2065]. Recent studies have demonstrated dynamical photocontrol of microtubule integrity, migration, and mitosis using fluorescent protein-imaging-compatible photoswitching reagents *in vivo* [1786]. Optogenetic tools have also been used to control physiological development by patterning cell ion currents and signaling cascades with high spatiotemporal precision [1803]. These optogenetic techniques can be used to leverage precise control over selected cell types — even in heterogeneous tissues [1797]. Unlike conventional optomechanical systems which required sophisticated cooling mechanisms [2066], living systems employ functional molecules with operational energies comparable to ambient thermal noise [2067]. As concurrent advances in quantum optics facilitate controlled manipulations of excitonic states at room temperature in strongly-coupled quantum systems [2068], research has revealed how the efficiency of quantum transport can be enhanced by ambient noise in natural and engineered systems designed for light-harvesting, superconduction, and ionic confinement [2069].

Biosynthetic nanostructures and assemblies have also been proposed in strategies to manipulate cell organelles *in vivo* for biomedical applications [2070]. These strategies hold great promise for proposals to carry out chemical or genetic engineering of cells *in situ*, enabling the administration of practical, cost-effective bioengineering tools to strengthen inherently-favorable cell traits or even to confer cells with new functionalities [2071–2073]. This "top down" approach has been used to endow bacteria with new abilities for food and biofuel production, cancer treatment, chemical synthesis, environmental remediation, light harvesting, and fundamental biological research [2074, 2075]. For the time being, living cells remain too complex to redesign completely, and not all new functions may be compatible with an intended host cell. These limitations have motivated an alternative "bottom up" approach [2074].

In the "bottom up" approach to synthetic biology [2076], cells and processes are dismantled and reassembled to build up new forms of life or life-like creations. This work is enabled by advances in DNA sequencing and synthesis developed in the context of systems biology [2077]. "Bottom up" synthetic biology has been used to reconstitute many natural biomolecules, cell components, whole cells, and even primitive designer organisms [2074]. It has been used to create a wide assortment of molecular-scale motors [2078, 2079] and robots [2080–2082], as well as new fundamental building blocks for life in the form of noncanonical amino acids [2083]. Synthetic cells have also been used to demonstrate luminescent signaling within and between cells, initiating photo-induced conidiation (i.e., spore forming) in neighboring fungal cells [2084]. These results define preliminary steps toward the design and construction of "bottom-up" synthetic biology by using photochemical reaction networks, biological compartmentalization, and cellular communication pathways to orchestrate biological dynamics [2085].

By merging the capabilities of "top down" and "bottom up" synthetic biology [2074], bioengineers aim to create new forms of devices, materials, sensors, and therapeutics out of living or life-like components [2072]. These emerging biotechnologies may be expected to have profound implications for agriculture, biomedicine, electronics, environmental science, and robotics [2086]. These biotechnologies raise questions about the fundamental forces or constraints that drive living systems to converge toward certain recognizable forms and functions [2087], and the emergent properties (such as multicellularity) that arise from universal self-organizing principles [2010]. Existing efforts to synthesize bionic systems (e.g., out of silicon) are prone to numerous structural and functional limitations, falling far short of the fluid and facile capabilities of native biological organisms [2088]. Nevertheless, today's rapid interdisciplinary advances in atomistic fabrication have been characterized as the beginning of a new industrial revolution [2089], and artificial nanofluidic systems are now beginning to approach a level of functionality similar to that of the biological channels that enable sensation, transport, and neurotransmission in living systems [2067].

One topic that has captured the imagination of biocomputing researchers is the prospect of molecular computing with DNA [2090–2093]. Inspired by Feynman's imperative to engineer submicroscopic computers [1729], Adleman proposed the idea of using molecular computation to solve hard combinatorial problems in 1994 [2094, 2095]. Today, computational DNA reaction networks have been programmed to perform arithmetic, digital logic, and even neural network computations for artificial intelligence applications [2096]. The applications of programmable DNA are not limited to computation, and proposals to develop DNA as programmable matter in the construction of autonomous molecular systems have emerged in line with the expanding complexity of DNA "circuit" dynamics [2096]. Programmable DNA scaffolds have been proposed to actuate spatial control over densely-packed clusters of cyanine dyes with tunable absorption spectra and strongly-coupled exciton dynamics similar to chromophore networks found in natural light-harvesting systems [618]. These efforts are representative of an emerging class of molecular-level studies intended to help mimic biological implementations of light sensing and harvesting systems [2097, 2098], and stimuli-responsive photonic materials [2099–2101]. DNA-based neural networks designed for pattern recognition have been used in disease profiling and diagnostics [2096, 2102], motivated by the need for precise, early cancer diagnoses and rapid, inexpensive, non-invasive disease screening [2103, 2104]. As the global production of digital data grows exponentially, costs associated with conventional (i.e., magnetic and semiconductor) digital storage media are becoming nonviable. For this reason, DNA-based data storage technology has been identified as a promising candidate to enable cost-effective, dense, and durable data storage solutions [2105–2107].

Molecular electronics composed of proteins, nucleic acids, and carotenoids [2027] are prototypical non-equilibrium open quantum systems suitable for use in NISQ devices [2108], sensors, nanomaterials, neuromorphic computers [2109], and advanced forms of artificial intelligence [2110]. Quantum dissipation engineering provides a method of universal quantum computation [368, 2111], and engineered dissipation has been proposed as a scalable alternative to unitary (i.e., coherent) processes for preparing entangled many-body states in noisy quantum systems [2112]. Living organisms routinely carry out dissipative many-body quantum state preparations of this kind when they synthesize strongly-correlated coenzymes for use in subsequent biochemical transformations (e.g., iron-sulfur clusters [1458]). Biocomputing systems have even been hypothesized to exceed the computational powers of quantum computers by leveraging the replicative capabilities of biological systems to solve all computational paths of hard computational problems in parallel [2113] – i.e., by constructing a non-deterministic universal Turing machine to solve the notoriously-hard class of nondeterministic polynomial-time (NP) complete computational problems. Thus, manipulation and control of biological nanostructures raises prospects for the development of radically new biotechnologies that harness quantum effects at the vanguard of physics research [758, 2114]. Bioengineered quantum technologies with characteristics of self-assembling, self-repairing, and replicating organic matter can be expected to possess powerful capabilities that harbor critical ethical implications [2076, 2116]. The advent of novel biotechnologies will also afford new and unprecedented opportunities for biomedicine and health science that engage the innate, collective intelligence of living tissues and organs [2115].

Chapter 20

Quantum Biology: Essential Further Research

Quantum biology begins wherever the energetics of life begin. All life relies on biological electron transfer, and chromophores and cofactors used to transfer those electrons are found in every living organism. The electronically active sites that enable life's electron transfer mechanisms have physiologically important quantum mechanical characteristics by virtue of their functional roles. Biological chromophores tend to be photoactive, as their name implies, with tunable photo- and electrochemical properties that are often functionally dependent on environmental conditions. These properties require further study. For example, many details of the interactions that control flavin biophysics remain unclear even after decades of flavoenzyme research.

Many structural and functional properties of respiratory and photosynthetic electron transport chains have now been characterized. There is still much work to be done to account for many-body and self-synchronization effects in those systems, and further research will be needed to fully account for the high quantum yield and thermodynamic efficiency of biological electron transporters, proton pumps, and synthases. Electron transfer reactions are often accompanied by proton transfer steps that must also be accounted for quantum mechanically in model simulations. Even developing the correct physical formalism can be challenging when modeling complex many-body systems of this kind. Further research is needed to characterize the exact roles of quantum coherence and dissipation in the mechanisms that govern photosynthesis and cellular respiration in living systems.

Biophotons have attracted interest for their potential use in evaluating redox physiology, metabolism, and cell stress for medical assessments and health monitoring. More commonly known as ultraweak photon emissions, these faint cellular light emissions are primarily a product of high-energy oxidative cellular processes. As a result, their production is linked to the reactive dynamics of various species of oxygen, and therefore also closely tied to certain aspects of mitochondrial and cytoskeletal function. Experimental observations have consistently associated weak magnetic field effects with the production of cellular reactive oxygen species, which are implicated in mechanisms of morphogenesis and regenerative healing, although these physiological phenomena also remain poorly understood.

Light–matter interactions are critical to many aspects of physiology including vision, photosynthesis, and vitamin D synthesis. Many optically-active molecules play important roles in redox chemistry, electron transfer, oxygen transport, and other catalytic processes. The flash of a firefly is just one example of the many photochemical reactions that occur in living systems. Although the importance of biological photochemistry has been recognized since photosynthetic pigments were discovered in the 18th century, the role of optical-frequency excitations in essential cellular regulatory processes such as mitochondrial retrograde signaling have only recently become topics of medical research interest. This may be due to growing interest in the development of nanomedical interventions such as photobiomodulation and photodynamic therapy, and the emerging use of optically-active nanoparticles in medical treatments. Growing use of optogenetic techniques in biological research has also brought attention to the role of optical effects in living systems, and more important discoveries may be expected in these areas.

The exact mechanism(s) underlying the effects of weak magnetic fields on biological processes remain unclear and require extensive further research. Although radical pair dynamics are implicated in many possible mechanisms that have been proposed to mediate interactions between weak magnetic fields and living cells, the nature of their role is not yet definitive. Significant efforts have also been made to characterize the chemical amplification of weak fields in the presence of radical scavenging molecules. More research is needed to assess the general role of magnetic fields in biochemistry and the physiological processes that enable sensing, metabolism, and cellular growth. More research is especially needed to determine the influence of Earth-strength ($\sim$50 μT) magnetic fields on physiology, particularly in the context of cellular magnetoreception and its implications for health. Biomagnetic theory remains to be formulated to rationalize all observed effects of electromagnetic fields on cellular radicals, and the nature of the influence of magnetic fields on cellular oxygen metabolism remains to be conclusively determined.

The microscopic mechanisms that govern the formation of the complex hierarchical structures found in eukaryotic cells are practically unknown. However, microtubules are implicated in most if not all aspects of cellular organization due to their important structural roles. Once characterized primarily in terms of its branching structural motifs, the microtubule cytoskeleton is now gaining attention for its central role in mediating the sensory-motor faculties of cells. As key integrators of biochemical, electromagnetic, and optomechanical signals, microtubules are thought to generate oscillating electric fields that coordinate core cellular functions. Nevertheless, a clear picture of the interactions that govern polarization effects in cytoskeletal networks is absent, and there is need for fundamental research to explain the growing body of evidence for photonic and quantum optical effects in microtubules.

The relationship between microtubule dynamics and mitochondrial function — including their interactions with radical species and their electron transfer dynamics — requires further attention. Microtubule assembly dynamics can be controlled by weak magnetic fields and tunneling electron currents, suggesting the importance of subtle quantum mechanical driving effects in the mechanisms that govern these systems. Tubulin is a birefringent protein with unique optical and optoelectronic material properties. It deserves to be characterized comprehensively under resting (ground state) and electronically excited conditions, which will be essential to determine its fiber-optic, optomechanical, and chemomechanical properties as a sub-cellular structural and information-carrying conduit. Theoretical and experimental characterizations of coherent quantum optical effects in microtubules require further attention to discern the role of the geometrically complex networks of biological chromophores implicated there.

Terahertz frequency radiation has been shown to influence physiological processes in living cells, including primary cellular activities, amino acid release, cell proliferation, and apoptosis. Although the mechanisms underlying Thz cellular interactions are unclear, these processes are likely to be related to other long-range electrodynamic interactions that have recently been discovered between biomolecules. Electrodynamic diagnostic techniques such as THz imaging and ultraweak photon emission measurements are likely to provide access to important metabolic information that can be used to help identify the presence and type of cancer for medical applications. However, the effects of electromagnetic fields on different types of cancerous, neoplastic, and normal tissues are not yet fully understood, and more research is needed to determine the safety, specificity, and efficacy of electromagnetic field-based therapies and diagnostics.

Although mechanical forces are traditionally in the domain of classical physics, the transduction of mechanical stimuli into electrochemical signals (and vice versa) are inherently quantum mechanical and universally present in biological systems. The systematic inventory of the various adiabatic and non-adiabatic quantum effects involved in biological processes will be useful for the ongoing characterization of biological structure–function relationships, beginning with photochemical and photophysical processes, internal conversion mechanisms, intersystem crossing effects, the Jahn–Teller effect, and various forms of vibrationally-assisted energy, electron, and proton transfer.

The future of nanomedicine depends on understanding the biomolecular basis for the regulation and control of all major cellular functions. Homeostasis requires the systematic regulation of a host of cellular processes and the associated metabolites which include reactive oxygen and nitrogen species, metal ions, lipids, peptides, and nucleic acids. Cellular regulation involves many complex interactions between governing organelles such as mitochondria, the cytoskeleton, and the cell nucleus. The complex relationship between microtubule dynamics and mitochondrial function requires further elucidation, particularly concerning the interplay between biochemical, bioelectric, and biomagnetic effects. Mitochondrial retrograde signaling, as it is implicated in the mechanism of photobiomodulation therapy, is also likely to play a critical role in cellular health and disease processes of broad biomedical interest.

Just as biological cofactors involved in enzyme catalysis respond to the electrochemical and electromagnetic properties of their environment, so do organelles and whole cells respond to their respective cellular and extracellular environments. The successful development and implementation of emerging nanomedicines and nanotheranostic tools, such as chemodynamic and sonodynamic therapies, requires in-depth research into fundamental mechanisms that determine their efficacy and efficiency. This research will also be required to realize potential applications that incorporate the full electromagnetic spectrum into treatments for wound healing and regenerative medicine.

Although the density of any stationary electronic system is quantum mechanically "trivial" in the sense that in principle it can be predicted using semiclassical theory alone, in practice the problem of finding the semiclassical approximation that replicates the exact many-electron density function of a quantum mechanical system has been recognized to be as difficult as solving any of the hardest verifiable computational problems in quantum complexity theory. This fundamental theoretical difficulty represents the ongoing quest to find the fabled universal exchange-correlation potential function of quantum chemistry. The problem is central to studies of structural biology, where functionally-critical molecular cofactors and enzyme active sites often contain highly-complex quantum mechanical electronic structures that can be intractable to solve on present-day computing hardware. Research directed toward improving approximations and numerical methods to tackle these problems is ongoing, requiring interdisciplinary efforts from molecular biologists, quantum chemists, numerical physicists, and computer scientists.

Electronically-active biomolecular systems can incorporate both local dynamic and non-local static quantum mechanical correlations which are highly-nontrivial to predict even using advanced methods and approximations. Model simulations of biomolecules present additional challenges because of the ubiquitous presence of charge transfer and weak binding effects in these systems, which require the extension of conventional methods in quantum chemistry to account for the dynamical phenomena involved in these processes. Charge transfer is fundamentally accompanied by quantum spin transfer, and spin-mediated effects are crucial to many biochemical and physical phenomena including radical pair and scavenging dynamics, spin crossing reactions, lipid peroxidations, and chirality-induced spin-selective electron transport. Yet the exact physical mechanisms underlying these processes are often unclear and subject to debate, reflecting the need for experimental and theoretical developments to resolve current ambiguities.

For example, the significance of radical pair dynamics in cryptochrome magnetosensitivity remains inconclusive. The role of radical electron dynamics is more clear in the photoactivated mechanism of the DNA-repair enzyme photolyase, where the mechanism itself is complicated and worthy of further investigation. Photosensitized processes in native biological macromolecules such as photolyase and hemoglobin have made them prototypes for biomedical research in the emerging field of nanomedicine, with the goal of developing increasingly safe, targeted, effective treatments for a broad spectrum of acute and chronic diseases. The development of novel medical treatments based on DNA-repair enzymes, for example, will benefit from comprehensive models of the complex sequence of structural rearrangements and functional dynamics that enable rapid DNA photo-catalysis in those systems. Quantum biology research heralds the possibility of dynamic control of vital spin-chemical processes for a plethora of potential biomedical applications geared toward restoring cellular regulation, communication, and control.

Non-local, wavelike quantum mechanical fluctuations play an important role in defining the form and structure of many biologically-important nanosystems. Foremost among these is the structure of DNA, where the characteristic stacking of base pairs into the iconic double helix is enabled by quantum mechanical van der Waals interactions and subtle solvent stabilization effects that are not predicted consistently using simplified classical models. The multifunctional properties of DNA have made DNA-based information processing and data storage a topic of intense and growing interest. DNA computing is representative of the broader scientific field of bioelectronic computation, where great strides are being made to develop truly molecular-scale components to use to build the next generation of miniaturized computing machines. However, current efforts to integrate molecular transistors, diodes, and logic gates suffer from various fabrication issues which may prove difficult to overcome using semiconductor devices.

Comprehensive models of the functional properties of cellular ultrastructure require analyses at all levels of approximation. However, full-scale biomolecular simulations of enzymatic systems present insurmountable difficulties to classical computational methods due to the comparatively large sizes of biological macromolecules and the ubiquitous presence of static and dynamic quantum correlations in systems of this kind. Substantial progress has been made by introducing many-body dispersion effects into simulation methods using both *ab initio* quantum and parameterized classical mechanical approaches. Computational techniques based on multi-scale models and quantum embedding schemes have had some success in simulations of enzyme-catalyzed biological processes, but these methods still entail the use of complicated approximations and the arbitrary partitioning of the system under study into distinct classical and quantum subsystems. Even small catalytic sites, such as the transition metal atom centers found in the molecular cofactors of many enzymes, can have extraordinarily complex electronic structures that defy the use of classical simulation methods even on the largest supercomputers. Quantum computing has been proposed as a means of overcoming the difficulty of simulating structural and chemical properties of complex biomolecules.

In many cases, the true significance of underlying quantum mechanical effects remains unknown, such as during protein folding, wherein it is still unclear how the newly-translated primary amino acid sequence can relax into its optimal folded conformation in finite time. The problem of folding a protein into its minimum free energy conformation amidst an exponentially-large number of folding possibilities is reminiscent of comparable fundamental problems in computer science and combinatorial optimization. This has motivated interest in the use of quantum computing and combinatorics to solve hard problems in molecular biology — and also the possibility of using molecular biology to solve the most difficult "NP-complete" problems in computer science.

Living systems overcome the problem of negotiating charge transport through an insulating medium by using tunneling-mediated electron transfer through a series of electronically active sites in the electron transport chains of the assorted photosynthetic, respiratory, and immunological engines of biology. These catalytic engines carry out the numerous electronic transformations needed to enable all aspects of anabolic and catabolic metabolism. Broadly speaking, these catalytic transformations are enabled by the many biomolecular cofactors found at the biochemical active sites of enzymes. Biomolecular cofactors are essential to carry out the countless electronic transformations that are needed to sustain growth and maintain homeostasis, and as a result they remain prototypical chemical agents in health science, pharmacology, and medical chemistry. Despite their central importance in physiology, the structural properties underpinning the functional characteristics of biocatalytic cofactors are obscured by the same highly-complex electronic configurations that enable their seemingly-endless functionalities. Addressing fundamental questions about the relationship between electronic structure and function in biology will set the stage for further research on the processes that generate, organize, and replicate living cells, tissues, and organisms.

Chapter 21

Conclusion: A Quantum Framework

Quantum biology would not exist without quantum mechanics, but quantum mechanics itself is framed by the premise that quantum measurements are performed using classical measuring apparatus. This premise is enshrined in the measurement postulate of quantum theory, made famous by the paradox that accompanies it — the so-called "measurement problem" of quantum mechanics. Yet, physical paradoxes are in the eye of the beholder, a consequence only of ill conceived assumptions. In the case of quantum theory, the paradox is that the measurement postulate requires that quantum measurements are implemented using classical measuring devices, in apparent contradiction with the fact that laboratory apparatus are made of atoms like everything else. This conundrum was not lost on the founders of quantum mechanics, who went to great lengths to formulate a quantum–classical correspondence principle.

Championed by Bohr, the correspondence principle provided the cornerstone for the orthodox "Copenhagen" interpretation of quantum physics by posing the premise that the predictions of quantum mechanics should converge to recover classical physical phenomena in some statistical limit. However, what *exactly* this statistical limit should be is no more agreed upon by physicists now than it was in Bohr's time. The correspondence principle today is not so much a principle of nature as it is a standing philosophical conjecture. Efforts pioneered by Dirac to derive quantum mechanics as a modification to classical physics — while physically appealing — remain both philosophically and mathematically unsatisfying, given that the goal of any such "canonical quantization" procedures is to derive a more fundamental theory from a less fundamental one.

Living systems are unlike either the inanimate quantum systems or the experimental apparatus found in the physics laboratory, and for this reason they may offer a unique advantage as a reference with which to consider the true nature of the quantum–classical correspondence. Rather than beginning from the assumption that living beings are essentially like other inanimate classical objects, quantum biology may begin with the founding principle that quantum theory is the rule, rather than an exception. This principle is important to adopt when considering biological systems, which are not simply chemical entities made from organic molecules, but living beings that can exert control over their own internal biochemistry, often at will.

In this work, we have provided a broad overview of the many non-trivial quantum effects and related processes found ubiquitously in biology, including electron tunneling during cellular respiration, light-harvesting during photosynthesis, and cofactor/ligand coordination during protein translation and enzyme catalysis. Although the scientific basis to recognize these as quantum effects has been established for many years, their biological implications are not widely appreciated due to the complicated nature of the quantum mechanics involved and the resulting technical difficulties associated with applying quantum mechanics to functional aspects of biology. This suggests a need to revise the current biophysical perspective to update classical assumptions about the living world.

Rather than beginning with a classical viewpoint which is then modified by quantum corrections, we propose that it is appropriate to understand the role of quantum mechanics in biological processes from first principles. This may also help to clarify the emergence of a correspondence between quantum and classical phenomena, without immediately invoking the existence of artificial classical measuring devices, yet with the final goal of explaining how seemingly-classical and quantum processes can co-exist in living systems. This approach defines a prototype framework for studies in quantum biology, where the openness of quantum systems may be accounted for using a non-Hermitian Hamiltonian formalism according to the unitary projection operator techniques developed by Fano and Feshbach (without recourse to the conventional Lüders–von Neumann projective measurement postulate).

A scientific approach that focuses on the quantum physics and maintains a quantum perspective ensures the coherence of the methodology by applying a unified set of founding assumptions throughout. This provides a natural framework in which to represent and understand quantum electrodynamic interactions in biological systems. This framework is well-posed to incorporate the influence of environmental factors on open quantum systems, which are critical to describe the complex quantum dynamics of living matter. By accounting for the interplay between the non-local electronic structures of biological cofactors, the non-adiabatic quantum dynamics of excited electronic processes, and the environmental effects of dissipation, external driving, and spontaneous decay, the approach we describe here aims to capture the essential physical features of life itself.

One goal of this shift in the fundamental research perspective is to ensure that classical reasoning is not spuriously imposed in models of fundamentally quantum phenomena, so that effectively-classical model approximations are only applied where methodologically warranted. This approach has the added benefit of acknowledging the possibility of discovering new fundamental phenomena in these systems, while maintaining a consistent theoretical framework in which to form theories, pose hypotheses, and design experiments. It emphasizes the pursuit of knowledge and understanding with the secondary intention of developing long-term applications which could be quite different from present-day technologies.

As science moves toward harnessing direct quantum control over living processes, it will enable the creation of whole new types of living systems not anticipated by previous technologies. This poses potentially-unforeseeable societal and ethical challenges. The potential risks associated with the unregulated development of these emerging technologies indicate a strong need to establish the overarching bioethical consultation, regulation, and governance needed to respond to new challenges. The implications of new forms of biotechnology may be dramatically different from their predecessors in terms of their magnitude and scope, with important consequences for biomedicine and bioengineering.

Living systems, recognized in physics as active matter, are extraordinarily dynamic. The complex phenomena associated with these systems poise quantum biology as a highly interdisciplinary field of research encompassing biology, physics, chemistry, medical science, and nanotechnology. Developing shared scientific terms, language, and methods will pose a core challenge to overcome. For this reason, it is all the more imperative to develop a common understanding of the foundational physics underlying the field. It is hoped that formalizing these shared intentions will provide the theoretical foundation needed to properly ground the work while propelling research forward. The field of quantum biology has extraordinary potential for discovery and we invite you to approach the possibilities emerging from this new scientific horizon, beginning with the physical processes and principles shared herein.

Sample Syllabus

Course Structure: This course is suitable for a multi-departmental collaboration with 22 lecture-style classes of 90 minutes each. Additional classes during the semester can be designated for research paper assistance, student presentations, question and answer periods, special topics, and review for the mid-term and final exam.

Prerequisites: For college students enrolled in physics, biology, biochemistry, biotechnology, medical science, or related studies. Specific mathematics pre-requisites are not required for this course.

Class 0 – Preliminary Introduction to Quantum Biology

I. Introduction to Quantum Biology (15 minutes)

  A. Definition and Scope of Quantum Biology

    - Overview of quantum biology as an interdisciplinary field
    - Importance of studying quantum effects in biological systems

  B. Historical Context

    - Brief history of the development of quantum biology
    - Key milestones and breakthroughs in the field

II. Course Objectives and Structure (15 minutes)

  A. Learning Outcomes

    - Discussion of key concepts and skills students will gain
    - Overview of the relevance of quantum biology to various scientific disciplines

B. Course Structure

- Outline of topics to be covered throughout the semester
- Explanation of the integration of theoretical and practical components

III. Syllabus Review (20 minutes)

A. Course Schedule

- Detailed overview of the weekly topics and class activities
- Important dates for assignments, exams, and projects

B. Grading Policy

- Explanation of grading criteria and assessment methods
- Discussion of participation, assignments, and exam formats

IV. Instructor Introduction (10 minutes)

A. Instructor Background

- Brief presentation of the instructor's academic and research background
- Overview of the instructor's interests in quantum biology

B. Teaching Philosophy

- Discussion of the instructor's approach to teaching and learning
- Importance of student engagement and collaboration in the course

V. Introduction to Key Concepts (20 minutes)

A. Basics of Quantum Mechanics

- Overview of fundamental principles of quantum mechanics relevant to biology
- Discussion of wave-particle duality: electrons, protons, neutrons, and spin

B. Applications of Quantum Principles in Biology

- Introduction to the concept of thinking from "first principles" in the sciences
- Examples of quantum mechanical influences on biological processes (respiration, photosynthesis)

VI. Conclusion and Q & A (10 minutes)

  A. Summary of Key Points

  - Recap of the course structure, objectives, and key concepts introduced

  B. Open Floor for Student Questions

  - Opportunity for students to ask questions about the course or the topics covered

Initial Assessment: Completion of a short introductory assignment (*e.g.*, a reflection on student interests in quantum biology) to be submitted by the next class.

Class 1 – Quantum Theory and the New Observables

I. Historical Introduction to Quantum Theory (15 minutes)

  A. Overview of Quantum Theory's Historical Context

  - Discussion of fundamental challenges in physics and the emergence of quantum theory
  - Key figures and events leading to the development of quantum mechanics

  B. Birth of Quantum Mechanics

  - Old Quantum Theory (Bohr-Sommerfeld quantization)
  - Matrix Mechanics (Born, Heisenberg, Jordan)
  - Wave Mechanics (Schrödinger) and Operator Algebra (Dirac)

II. Modern Quantum Mechanics (20 minutes)

  A. Scope of Quantum Mechanics

  - Discussion of coherence, superpositions, entanglement, and measurement
  - Sketch of Postulates of Quantum Mechanics including Hermitian Symmetry

  B. Contemporary Quantum Theory

  - Problem of unifying quantum theory with Einstein's relativity

  - Challenge of formalizing Quantum Field Theory (QFT) using second quantization

III. The Transition from Classical to Quantum Mechanics (20 minutes)

  A. The Gap between Quantum Mechanics and Classical Observations
    - The Measurement Problem and reliance of quantum theory on "classical" measuring devices
    - Overview of Open Quantum Systems Theory

  B. The Quantum-to-Classical Transition: Biological Complexity Spans Many Scales
    - Inadequacies of classical theories in explaining fundamental biological phenomena
    - Interplay between thermodynamic equilibrium and non-equilibrium processes in molecular biology

IV. Living Organisms as Open Quantum Systems (20 minutes)

  A. Theoretical Innovations and Quantum Biology
    - Quantum Biology and the Search for New Physics (Bohr, Schrödinger, Delbrück)
    - Stability, Excitatory, and Regenerative Properties of Biomolecules

  B. Emergence of Novel Quantum Effects from Open Quantum Systems
    - Non-Hermitian Models (*e.g.*, tunneling, Fermi's "Golden Rule," Fano-Feshbach partitioning)
    - Exotic Phenomena (superradiance, symmetry-breaking, Bose-Einstein condensates)

V. Quantum Biology and Quantum Foundations (10 minutes)

  A. Quantum Theory of Nonlinear Systems
    - Summary of linear and nonlinear processes in biology and quantum mechanics
    - The role of feedback, self-organization, and amplification in physics and biology

B. Relationship between Biology and Quantum Theory
   - Open-ended nature of living systems in contrast with closed quantum mechanics
   - Biology as a novel avenue for studying foundational issues in quantum theory

VI. Conclusion and Q & A (5 minutes)

A. Summary of Key Concepts: Biology as a Frontier for Studying Fundamental Quantum Effects

B. Open floor for student questions

Class 2 – Quantum Electrodynamics: Lighting Up Life

I. Introduction to Quantum Electrodynamics and Biology (15 minutes)

A. Quantum Electrodynamics: The Strange Theory of Light and Matter
   - Historical development of quantum electrodynamics (QED)
   - Significance of QED in understanding biological processes

B. The Role of Light in Biological Systems
   - Explanation of how photons excite electrons to drive biological function
   - Biological importance Planck's constant $h = E/f$

II. Quantum Effects in Biological Systems (20 minutes)

A. Biological Charge Transport Mechanisms
   - Introduction to coherence and decoherence in biological systems
   - Energy and charge transport in photosynthesis and respiration

B. Known Quantum Effects in Living Organisms
   - Structural Effects: Role of quantum mechanics in the structure of biomolecules
   - Dynamic Effects: Ultrafast dynamical processes in enzymes and biological cofactors

III. Quantum Tunneling: Biological Electron Transfer (20 minutes)

A. Quantum Mechanical Tunneling Explained

- Quantum tunneling as a consequence of wave-particle duality
- Discussion of resonant *vs.* decoherent tunneling in biological systems

B. Mechanisms of Electron Transfer in Enzymes

- Examination of cytochrome and its role in electron transfer reactions
- Discussion of challenges in modeling electron transfer dynamics

IV. Proton-Coupled Electron Transfer (20 minutes)

A. Mechanisms of Proton-Coupled Electron Transfer (PCET)

- Relationships between electron and proton transfers in biochemical processes
- Role of quantum coherence and electron tunneling in PCET dynamics

B. Implications for Metabolic Processes

- Influence of PCET in energy transduction and metabolic functions
- Examples of PCET in respiratory enzymes and their mechanisms

V. Quantum Spin and Biological Processes (10 minutes)

A. The Role of Chirality and Quantum Spin in Biology

- Spin-selective processes and intersystem crossing (ISC)
- Chirality-induced spin selectivity (CISS) in chiral biomolecules

B. Applications of CISS in Biochemistry

- Discussion of biological CISS effects involving radical species
- Cellular processes involving spin-polarized electron transport

VI. Conclusion and Q & A (5 minutes)

   A. Summary of Key Concepts: The Strange Theory of Light and Life
   B. Open floor for student questions

Main Assignment #1: Short 2-3 page research paper on quantum biology topic of interest related to the students field of study to be completed before mid-term study window begins.

Class 3 – Definitions of Non-Triviality for Quantum Biology

I. Defining Quantum Triviality for Quantum Biology (15 minutes)

   A. Trivial and Non-Trivial Phenomena in Physics
      - Triviality as an essential theoretical benchmark in the mathematical sciences
      - Need for consistent terminology: Phenomenological complexity in Quantum Biology

   B. Importance of Non-Trivial Quantum Effects
      - Landau Triviality: Trivial theories lack intrinsic particle interactions.
      - Dynamical Triviality: Trivial theories have trivial (stationary) dynamics.

II. Preliminary Introduction to Quantum Chemistry (20 minutes)

   A. Wave-Particle Duality of Electrons and Nuclei
      - Nuclear Particles and Electronic Waves (The Born-Oppenheimer Approximation)
      - Electron Clouds: The electron density as the basis for chemistry (The Hohenberg-Kohn Theorem)

   B. The Role of Electron Spin and Electronic Correlations
      - The Anti-Symmetry of Electrons: Exchange of Spin-Half Particles (Spin Statistics Theorem)
      - From Mean-Field Approximations to Electron Correlations (Pauli Exclusion Principle)

III. Introduction to Density Functional Theory (20 minutes)

A. Understanding Density Functional Theory (DFT)

- DFT as a fundamental method of solving for molecular energies in chemistry
- Effective modeling of many-electron chemical systems (Kohn-Sham Theory)

B. Non-Trivial Challenges in Applications of DFT

- The Exchange-Correlation (XC) Functional: A ladder of approximations
- Computational Complexity of Electronic Exchange-Correlation Effects

IV. Quantum Effects in Biological Chemistry (20 minutes)

A. Static and Dynamic Triviality in Quantum Mechanics

- Non-Trivial Static Effects: Time-Independent Quantum Correlations (adiabatic limit)
- Non-Trivial Dynamic Effects: Time-Dependent Quantum Correlations (non-adiabatic dynamics)

B. Orders of Approximation in Quantum Chemistry and Molecular Physics

- Adiabatic Orders of Approximation: Configuration Interaction and Møller-Plesset Theory
- Non-Adiabatic Approximations: Fermi's Golden Rule and Time-Dependent Perturbation Theory

V. Fundamental Challenges in Modeling Biological Molecules (10 minutes)

A. Theoretical Modeling of Quantum Correlations in Biological Molecules

- Complexities of simulating biomolecules: Limitations of semiclassical models
- Trivial Case: Stationary mean-field models of electronic structure in biomolecules

B. The Interplay of Quantum Effects in Biology: From Static Models to Correlated Dynamics
   - Inadequacy of semiclassical adiabatic approximation to reproduce correlated quantum dynamics
   - Discussion of the significance of trivial quantum effects in understanding non-trivial phenomena

VI. Conclusion and Q & A (5 minutes)
   A. Summary of Key Concepts: Quantifying Quantum Correlations in Biology
   B. Open floor for student questions

Class 4 – Photosynthesis & Open Quantum System Dynamics

I. Introduction to Photosynthesis and Quantum Dynamics (15 minutes)
   A. Physiological Importance of Photosynthesis
      - Overview of photosynthesis as the basis for life on Earth
      - The role of sunlight in biological energy conversion
   B. Quantum Coherence and Dissipation
      - Interplay between coherence and dissipation in biological systems
      - Influence of quantum coherence on photosynthetic efficiency

II. Photochemistry in Biological Systems (15 minutes)
   A. Introduction to Biological Photochemistry
      - Chromophores and Pigments: Photoactive Biomolecules
      - Chlorophyll and Heme: "the two most important pigments for life on earth"
   B. Understanding Photosynthetic Efficiency
      - Formal distinction between light-absorption and charge-separation efficiency
      - Mechanisms of non-photochemical quenching and energy dissipation

III. Mechanisms of Photosynthetic Energy Transfer (20 minutes)
  A. Role of the Fenna–Matthews–Olson (FMO) Complex
    - Introduction to the FMO complex and its function in light-harvesting
    - Overview of exciton energy transfer and its significance in photosynthesis
  B. Coherent Oscillations and Vibronic Couplings
    - Importance of excitation delocalization in natural light-harvesting systems
    - Hybrid vibronic states and their implications for energy transport dynamics

IV. Quantum Effects in Photosynthetic Processes (20 minutes)
  A. Quantum Mechanics in Photosynthesis
    - Light absorption and electron transfer in photosynthetic organisms
    - Discussion of the Marcus inverted effect in photosynthetic efficiency
  B. Implications of Quantum Coherence
    - Role of quantum coherence in electron transfer and energy transport
    - Examples of quantum dynamics in different photosynthetic organisms

V. Environmental Interactions and Photosynthesis (15 minutes)
  A. Role of Environmental Noise
    - Overview of how environmental noise can enhance light-harvesting efficiency
    - Discussion of superradiance and superabsorption in photosynthetic systems
  B. Engineering Quantum Resources
    - Exploration of how dissipative processes can be harnessed for quantum information tasks
    - Overview of difficulties in simulating photosynthetic mechanisms using classical models

VI. Conclusion and Q & A (5 minutes)

   A. Summary of Key Concepts: Harnessing Quantum Dynamics in Photosynthesis
   B. Open floor for student questions

Class 5 – Light Receptors, Spin Chemistry, & Cryptochrome

I. Introduction to Light Sensing in Biology (15 minutes)

   A. Overview of Biological Light Receptors
      – Importance of light sensing in various biological processes (vision, phototaxis, circadian rhythms, *etc.*)
      – Biological Photoreceptors: rhodopsin, cryptochrome, phytochrome, *etc.*

   B. Fundamental Mechanisms of Photosensing
      – Light-activated processes: Photochemical isomerization and electron transfer
      – Relationship between structure and function in photoactive biomolecules

II. Photoreceptors and Visual Processes (20 minutes)

   A. Types of Photoreceptor Cells
      Discussion of rod and cone cells in vertebrate vision
      – Overview of opsins and their roles in phototransduction

   B. Mechanisms of Light-Induced Responses
      – Explanation of the photoisomerization process in rhodopsin and melanopsin
      – Role of G protein-coupled receptors (GPCRs) in light-sensing and signal transduction

III. Cryptochrome and Magnetoreception (20 minutes)

   A. Role of Cryptochrome in Circadian Regulation
      – Overview of how cryptochrome functions in circadian light entrainment
      – Discussion of the relationship between cryptochrome and magnetoreception

B. The Radical Pair Model of Magnetoreception in Cryptochrome
   - Explanation of the radical pair mechanism (RPM) involving photo-generated radical pairs
   - Experimental evidence for the RPM in flavin adenine dinucleotide (FAD) and tryptophan

IV. Quantum Effects in Light Sensing (20 minutes)

A. Spin Chemistry and Quantum Dynamics
   - Overview of the significance of electron spin in biological processes
   - Discussion of how quantum coherence influences radical pair dynamics

B. Implications for Biological Function
   - Examination of how magnetic fields influence biological responses
   - The broader relevance of radical-mediated processes to physiology

V. Current Research and Future Directions (10 minutes)

A. Challenges in Understanding Magnetoreception
   - Overview of ongoing debates and challenges regarding the conventional RPM
   - Discussion of alternative mechanisms for animal magnetoreception

B. Applications of Radical Pair Processes in Quantum Biology
   - Implications of quantum oxidative processes for biomedicine and health sciences
   - Biotechnological Applications: Electronic control for light and magnetic sensing

VI. Conclusion and Q & A (5 minutes)

A. Summary of Key Concepts: Photochemical Sensing and Signaling in Biology

B. Open floor for student questions

Class 6 – Dynamic Control of DNA Repair in Photolyase

I. Introduction to Photolyase and DNA Repair (15 minutes)

A. Overview of DNA Photolyase Enzymes

- Definition of photolyases as UV and blue-light photoreceptor proteins
- Function of DNA repair in cellular processes

B. Mechanisms of DNA Repair

- Mechanism of DNA repair by photolyase using light energy
- Role of radical pair dynamics in the DNA repair process

II. Structural and Functional Dynamics of Photolyase (20 minutes)

A. Chemical Photoactivation and Electron Transfer

- Explanation of the electron transfer processes initiated by blue light
- Description of the radical pair formation and its significance in DNA repair

B. Key Steps in DNA Repair Mechanism

- Roles of FAD, pterin, deazaflavin, iron-sulfur clusters, and tryptophan functional groups
- Detailed overview of the catalytic steps in photolyase activity

III. Influence of Magnetic Fields on Biochemical Processes (20 minutes)

A. Magnetically-Modulated Radical Pair Dynamics in Photolyase

- Overview of radical pair dynamics and singlet-triplet spin interconversion
- Influence of magnetic fields on radical pair dynamics during DNA repair

B. The Influence of Magnetic Fields on Biochemical Processes

- Magnetochemical control of photolyase activity *in vitro*
- Broader implications for magnetic effects in enzymes (*viz.* Pauling-Coryell hypothesis)

IV. Quantum Effects in DNA Repair (20 minutes)

A. Quantum Spin Dynamics and Biological Processes

- Explanation of how quantum spin dynamics influence enzymatic reactions
- Discussion of the relationship between electron spin and metabolic processes

B. Implications for the Treatment of Disease

- Exploration of how photolyase mechanism(s) can inform therapeutic applications
- Overview of potential treatments for DNA damage-related diseases

V. Photochemical Research Methods and Applications (10 minutes)

A. Experimental Investigative Techniques

- Overview of spectroscopic methods used to study photolyase structure and dynamics
- Insights from time-resolved absorption spectroscopy and molecular dynamics simulations

B. Novel Applications of Photolyase Mechanisms

- Discussion of potential biotechnology and biomedical applications of photolyase
- Overview of ongoing challenges in studying and applying photolyase mechanisms

VI. Conclusion and Q & A (5 minutes)

A. Summary of Key Concepts: From Enzymatic DNA Repair to Biomedical Engineering

B. Open floor for student questions

Class 7 – Enzyme Catalysis: Quantum Fundamentals

I. Introduction to Flavins and Enzyme Catalysis (15 minutes)

A. Overview of Flavins

- Definition and importance of flavin molecules in biological systems

- Role of flavins in various cellular processes (*e.g.*, photochemistry, electron transfer)

B. Introduction to Quantum Effects in Enzyme Catalysis

- Electrostatic pre-organization: A fundamental basis for enzyme catalysis
- Photochemical dynamics: Ultrafast processes, with or without light

II. Quantum Control in Biological Processes (20 minutes)

A. Quantum Effects in Cell Function: Resonance and Delocalization

- Resonance Stabilization: From Schiff bases to aromatic molecules and coordination chemistry
- Structural Dynamics: Influence of quantum dynamics on enzyme function and signaling pathways

B. Quantum Dynamics and Radical Pair Formation

- Overview of radical pair dynamics in flavin-containing proteins
- Discussion of how magnetic effects influence enzymatic processes

III. Photophysical Properties of Flavins (20 minutes)

A. Structural Characteristics of Flavins

- Description of the isoalloxazine ring structure and its electronic properties
- Role of flavin in photo-absorption and radical pair generation

B. Experimental Techniques in Flavin Research

- Overview of spectroscopic techniques and computational methods used to study flavins
- Insights gained from time-resolved spectroscopy and molecular dynamics simulations

IV. Mechanism of Electron Bifurcation (20 minutes)

A. The Process of Electron Bifurcation

- Role of flavins in facilitating electron transfer pathways

  - Detailed explanation of how enzymes utilize electron bifurcation
  - Mechanism of electron bifurcation and its significance in energy transduction

B. Implications for Enzyme Catalysts in Biomedicine

  - Exploration of potential applications of flavin-based enzymes in medical treatments
  - Current research and unanswered questions regarding flavin, porphyrin, and tryptophan cofactors

V. Conclusion and Q & A (15 minutes)

A. Summary of Key Concepts: Quantum Enzymology from Electrons to Energy

B. Recap of Course Material from Part I (Classes 1-7)

C. Open floor for student questions

Class 8 – Ultraweak Photon Emission & Collective Cellular Behavior

I. Introduction to the Role of Light in Biology (10 minutes)

A. Historical Context: Light and Life

  - Heliotherapy: Ancient treatments using sunlight
  - Advent of Modern Photobiology and Phototherapy

B. Importance of Light in Regulating Biological Processes

  - Circadian Rhythms: Regulation of metabolism and development
  - Cellular Responses to Light: Photoreception, phototropism, phototaxis, and photomorphogenesis

II. Historical Discoveries in Light and Life (20 minutes)

A. Influential Figures in Phototherapy

  - 19th Century: Discovery of Photosensitizer Dyes (Vogel, 1873)
  - Phototherapy and Solar Clinics: Finsen and Rollier (1903)
  - Bohr and Delbrück's lectures on Light and Life (1932, 1962, 1976)

B. Gurwitsch's Experiments and Ultraweak Photon Emission (UPE)
   - Introduction to biophotons and UPE
   - Mitogenetic effect and cell replication

III. Types of Light and Their Biological Effects (20 minutes)

A. Classification of Light
   - Infrared (IR), Visible, and Ultraviolet (UV) Light

B. Beneficial Effects of Infrared Light
   - Mitochondrial function and adenosine triphosphate (ATP) production

C. Visible Light and Its Impact on Health
   - Effects of light on homeostasis, oxidative stress, and chronic diseases

D. UV Light: Risks and Benefits
   - Cancer induction *vs.* beneficial effects of UV light

IV. Collective Behavior and Cellular Interactions (20 minutes)

A. Microtubules and Photodynamics
   - The role of microtubules in the cell
   - Effects of UV and IR light on microtubule dynamics

B. Long-range Electrodynamic Interactions
   - Cellular responses to near-infrared light (NIR)
   - Implications for cancer research and therapy

V. Applications and Future Directions (15 minutes)

A. Medical Diagnostics and Therapeutics
   - UPE as a tool for monitoring cellular health (oxidative stress and cellular functions)
   - Potential applications of THz frequency radiation

B. Biotechnological Innovations
   - Superradiance and hybrid light-matter (polariton) states
   - Potential for manipulating cytoskeleton dynamics using light

VI. Conclusion and Q & A (5 minutes)

   A. Summary of Key Points: The Role of Light in Cell Physiology
   B. Open floor for student questions

Class 9 - Electromagnetic Oscillations in Living Systems

I. Introduction to Microtubules and their Biological Significance (10 minutes)

   A. Overview of Microtubule Structure and Function

      - Structure and composition of microtubules
      - Role in the cytoskeleton: interaction with actin and intermediate filaments

   B. Mechanical and Electrical Properties

      - Unique electrochemical properties of tubulin
      - Microtubule ion transport and charge storage capabilities

II. Dynamic Instability of Microtubules (20 minutes)

   A. Concept of Dynamic Instability

      - Mechanisms of self-assembly and depolymerization
      - Role of guanosine triphosphate (GTP) hydrolysis in dynamic instability

   B. Cellular Functions of Microtubule Dynamics

      - Importance in cell motility, mitosis, and differentiation
      - Impact on tissue homeostasis and disease (*e.g.*, carcinogenesis)

III. Microtubules as Biological Electromagnetic Systems (20 minutes)

   A. Electrical Oscillations in Microtubules

      - Microtubules as biological computing elements and amplifiers
      - Spontaneous electrical oscillations and their implications

   B. Microtubule Interactions with Electromagnetic Fields

      - Effects of oscillating electric fields on microtubule dynamics

- Coupling of microtubule oscillations with cellular electric fields

IV. Quantum Optical Properties of Microtubules (20 minutes)
   A. Cavity Quantum Electrodynamics (cQED) Model of Microtubules
      - Microtubules as quantum optical cavities
      - Sensory-motor integration in the microtubule cytoskeleton
   B. Refractive Index Controversies
      - Discrepancies in the refractive index of tubulin
      - Implications for electromagnetic mode confinement and excitonic interactions

V. Exciton-Polaron Interactions and their Biological Relevance (15 minutes)
   A. Microtubules as biomolecular quantum devices
      - Polariton formation in biological matter and carbon nanotubes
      - Implications for structural dynamics and transport properties
   B. Enhancing Microtubule Behavior Through Electromagnetic Fields
      - Cavity-free light-matter coupling and polariton biochemistry
      - Applications of THz-band oscillating fields in microtubule growth

VI. Conclusion and Q & A (5 minutes)
   A. Summary of Key Concepts: Electromagnetism and the Cytoskeleton
   B. Open floor for student questions

Class 10 – Chemical Control in Biological Systems

I. Microtubules and Reactive Oxygen Species (ROS) in Cellular Processes (10 minutes)
   A. Review of GTP Hydrolysis and Microtubule Dynamics

  - Mechanisms of microtubule self-organization and dynamics
  - Influence of microtubules on calcium signaling and ROS dynamics

B. Cellular Feedback Loops in Signal Transduction

  - Interplay between microtubules, calcium channels, and ROS
  - Interactions between the microtubule cytoskeleton and mitochondria

II. Noise and Signal Amplification in Biological Systems (15 minutes)

A. Role of Noise in Biological Decision-Making

  - Noise as a facilitator for detecting weak signals
  - Cellular ion channels as stochastic resonance amplifiers

B. Microtubules and Noise-Driven Processes

  - Microtubule responses to biochemical and mechanical stress
  - Implications for cellular learning and memory

III. Introduction to Mitochondrial Functions in Cellular Signaling (15 minutes)

A. Overview of Mitochondrial Roles

  - Integration of metabolism, redox chemistry, and apoptosis
  - Mitochondria as critical cell signaling hubs

B. Mitochondrial Dysregulation in Cancer

  - Structural and dynamic changes during carcinogenesis
  - Impact on metabolic regulation and energy transduction

IV. Mitochondrial Dynamics and Reactive Oxygen Species (ROS) (15 minutes)

A. Mitochondrial ROS as Oncogens

  - The role of ROS in cancer metabolism and the Warburg effect
  - Consequences of aberrant mitochondrial morphology

B. ROS and Cellular Stress Responses
   - Effects of external factors (*e.g.*, magnetic fields) on ROS levels
   - Mechanisms of hypoxia and high ROS stress in cancer cells

V. The Interconnectedness of Calcium, ROS, and Cellular Health (15 minutes)

A. Calcium Homeostasis and Mitochondrial Function
   - Importance of calcium in cellular signaling and mitochondrial function
   - Mechanisms of calcium-induced calcium release and implications for health

B. ROS Regulation and Disease Associations
   - Reverse Warburg Effect: extra-cellular matrix stiffening promotes malignancy
   - Connection between ROS dysregulation and various diseases

VI. Therapeutic Implications and Future Directions (15 minutes)

A. Targeting Mitochondrial Function in Cancer Therapy
   - Mitochondria-targeted therapeutic strategies
   - Therapeutic potential of modulating ROS levels

B. Challenges in ROS Amplification for Treatment
   - Balancing efficacy and side effects in therapeutic interventions
   - Towards an understanding of biochemical homeostasis and health

VII. Conclusion and Q & A (5 minutes)

A. Summary of Key Concepts: Cellular Chemodynamics (Noise, Oscillations, & Disease)

B. Open floor for student questions

Class 11 – Magnetic Biomodulation: Biodynamic Control

I. Introduction to Magnetic Fields and Animal Behavior (10 minutes)

A. Overview of Animal Magnetoreception

- Examples of species exhibiting magnetic field perception (e.g., birds, insects, mammals)
- Influence of the Earth's magnetic field on animal migration and behavior

B. Physiological Effects of Geomagnetic Field Deprivation

- Impact on neurogenesis, cognition, and developmental processes
- Influence of magnetic fields on circadian rhythms in humans

II. Mechanisms of Magnetic Field Influence on Biological Systems (15 minutes)

A. Effects of Hypomagnetic Fields (HMFs)

- Impairment of nerve growth and cognition in mammals
- Role of ROS in mediating HMF effects

B. Proposed Mechanisms of Magnetic Field Effects (MFEs)

- RPM, voltage-gated ion channels, and kinase signaling pathways
- Role of pulsed MFEs and flavoproteins in ROS generation

III. Magnetic Fields and Cellular Metabolism (15 minutes)

A. Magnetic Field Effects on Redox Metabolism

- Effects of magnetic fields on cellular redox homeostasis
- Balance between ROS production and metabolic processes

B. Wound Healing and Tissue Repair

- Influence of magnetic fields on ROS and calcium ion concentrations
- Applications of magnetic fields in regenerative medicine

IV. Experimental Findings on Biological Magnetic Field Effects (15 minutes)

A. Studies on Planarians and Fibroblasts

- Effects of static and oscillatory magnetic fields on regeneration and collagen production

  - Autophagy responses in fibroblasts exposed to oscillating magnetic fields
- B. Neutrophil ROS Production and Metabolic Resonance
  - Synchronization of NADPH oscillations with pulsed magnetic fields
  - Regulation of nitric oxide and ROS production in neutrophils

V. Magnetic Field Effects and Photobiology (15 minutes)
- A. Integrating Studies of Cellular Magnetoreception with Photobiology
  - Magnetic field sensitivity of cellular autofluorescence
  - MFEs in flavoproteins and related macromolecules
- B. Magneto-optical Effects in Biological Research
  - Magneto-optical phenomena in plasmonic nanostructures and biosensors
  - Magneto-optical interactions: Implications for ROS modulation and cellular growth

VI. Implications for Health and Therapy (10 minutes)
- A. Potential Applications of Magnetotherapy
  - Overview of current research and therapies involving magnetic fields
  - Magnetic control of cellular processes and functionalized nanoparticles
- B. Considerations for Future Research
  - Combining magnetogenetics with optogenetics: magneto-optical proteins
  - Need for unified theories and mechanistic understanding of biological MFEs

VII. Conclusion and Q & A (10 minutes)
- A. Summary of Key Concepts: Magnetic Physiology from Sensing to Clinical Innovation
- B. Guidance for Mid-Term Examination
- C. Open floor for student questions

Assessment: Mid-Term Exam

Class 12 – Intermolecular Forces: Solvent Effects and Dispersion

I. Introduction to Collective Properties of Biological Systems (15 minutes)
   A. Overview of Collective Behavior in Cells
      - Coordination of molecular activities: differentiation, mobilization, motility, and replication
      - Emergent phenomena and spontaneous synchronization in biology
   B. Structural and Dynamical Properties of Liquid Water
      - Historical Background: Brownian motion and the discovery of atoms
      - Contemporary quantum mechanical studies of liquid water

II. Non-Covalent Interactions and Their Biological Importance (20 minutes)
   A. Types of Non-Covalent Interactions
      - Overview of van der Waals and London dispersion forces, hydrogen bonds, and π–π interactions
      - Influence of non-covalent forces on biomolecular interactions and assembly processes
   B. Solvation Effects and Molecular Dynamics
      - Impact of solvent environments on biomolecular behavior
      - Role of hydrogen-bonding networks in aqueous solutions

III. Non-Covalent Interactions in Biomolecular Systems (20 minutes)
   A. Dispersion Forces and Their Role in Biomolecules
      - Importance of dispersion forces in protein folding and enzyme pathways
      - Dispersion forces and charge transport in DNA and RNA
   B. The Problem of Protein Folding
      - Challenges in understanding protein folding dynamics
      - Role of long-range quantum interactions in life processes

IV. Modeling Many-Body Dispersion in Biological Systems (20 minutes)

   A. Classical Approximations to Dispersive Many-Body Effects
      - Effective Two-Body Approximations (*e.g.*, intrinsically-disordered proteins)
      - Extensions of DFT to account for Many-Body Dispersion Effects (*e.g.*, DFT-MBD)
   B. Advanced Electronic Structure Methods for Liquid Water
      - *GW* Method: Many-Body Green's Function with Screened Coulomb Interaction (Hedin, 1965)
      - Advanced Quantum Many-Body Simulation: Diagrammatic Methods and Dyson Series

V. Conclusion and Q & A (15 minutes)
   A. Summary of Key Concepts: Water and the Hidden Architecture of Life
   B. Open floor for student questions
   C. Review of Midterm Results

Class 13 – Multiscale Modeling of Biomolecular Systems

I. Introduction to Quantum Mechanics in Biological Systems (10 minutes)
   A. Schrödinger's Contributions to Quantum Biology
      - The central dogma of molecular biology and the order-disorder problem
      - The importance of thermodynamic free energy in living systems
   B. From Classical Thermodynamics to Quantum Biochemistry
      - Boltzmann's H-Theorem and Thermodynamic Equilibrium
      - Incorporating quantum effects into classical statistical ensembles

II. Classical and Quantum Approaches to Biological Modeling (20 minutes)
   A. The Role of Classical Molecular Mechanics
      - Overview of classical simulations in studying molecular dynamics (MD)

  - Applications of MD to proteins, nucleic acids, and macromolecular complexes

B. Challenges of Conventional Quantum Chemistry Methods
  - Semiclassical approximations to quantum correlations (i.e., interaction kernels)
  - Limitations of DFT in accounting for strong quantum mechanical correlations

III. Multiscale Modeling Techniques (20 minutes)

A. Quantum Mechanical/Molecular Mechanical (QM/MM) Methods
  - Combining quantum mechanics with classical mechanics
  - Applications in enzymatic reactions and charge transfer

B. Historical Development of QM/MM Techniques
  - Key milestones in the development of multiscale models
  - Recognition of pioneers in the field (Karplus, Levitt, Warshel)

IV. Applications of Multiscale Modeling in Biochemistry (20 minutes)

A. Case Studies in Enzymatic Reactions
  - Mechanistic insights from QM/MM simulations (e.g., lysozyme reaction)
  - Impact of electrostatics on catalytic efficiency

B. Photodynamics and Quantum Processes
  - Investigating vitamin D photosynthesis and phototropin photocycle
  - Role of non-adiabatic quantum processes in biological systems

V. Challenges and Future Directions in Multiscale Modeling (15 minutes)

A. Limitations of Current QM/MM Approaches
  - Issues with strong quantum correlations and non-local effects

- Need for advanced computational methods and dynamical techniques

B. Emerging Techniques and Machine Learning Applications
- Potential of machine learning in enzyme engineering and drug discovery
- Multi-level embedding strategies for enhanced QM/MM simulations

VI. Conclusion and Q & A (5 minutes)

A. Summary of Key Concepts: From Boltzmann's Theorem to Multiscale Modeling

B. Open floor for student questions

Main Assignment #2: Student presentation on quantum biology research topic of interest to be given before final exam study period begins (may require additional instructional time).

Class 14 – Quantum Correlations in Biological Cofactors

I. Role of Metal Ions in Biological Functions (20 minutes)

A. Importance of Iron Cofactors and Iron-Sulfur Clusters
- Functions of iron-sulfur (FeS) clusters in DNA repair and electron transport
- Overview of heme proteins and their versatile roles in biochemistry

B. Quantum Mechanics in Metal-Containing Enzymes
- Influence of quantum correlations on enzyme behavior
- Challenges in studying iron-containing enzymes

II. Advanced Methods for Electronic Structure Calculations (20 minutes)

A. Approaches to Modeling Strongly-Correlated Quantum Effects
- Limitations of local density approximation (LDAs) for simulating quantum correlations
- Importance of correlated electronic effects in metalloproteins

B. Advanced Techniques in Quantum Chemistry
   - Overview of methods like DFT+DMFT and their applications
   - Modeling enzymatic reactions and electron transfer

III. Challenges in Studying Quantum Dynamics (20 minutes)

A. Limitations of Conventional QM/MM Methods
   - Issues with multi-reference methods and quantum correlations (configuration interaction)
   - Need for sophisticated techniques to model complex biochemical systems

B. Emerging Computational Strategies and Applications
   - Hybrid quantum/classical computing approaches
   - Applications in drug discovery and materials science

IV. Quantum Dynamics in Biological Processes (20 minutes)

A. Relevance of Quantum Dynamics in Biochemistry
   - Importance in photosynthesis, light-harvesting, and photocatalysis
   - Non-adiabatic dynamics and their implications for molecular reactions

B. Insights from Recent Quantum Simulations
   - Examples of trapped ion quantum computers in studying molecular dynamics
   - Exploration of photoinduced processes and their significance in metalloproteins

V. Conclusion and Q & A (10 minutes)

A. Summary of Key Concepts: Strongly-Correlated Electronic Systems in Biology

B. Recap of Course Material from Part II (Classes 8-14)

C. Open floor for student questions

Class 15 – Photobiomodulation & Electromagnetic Therapies

I. Introduction to Light-Based Therapy (15 minutes)

A. Historical Context of Photobiomodulation (PBM)
   - Finsen's contributions to light therapy and early applications
   - Rediscovery of red light therapy in the mid-1960s
B. Overview of PBM and Low-Level Light Therapy
   - Definition and significance of PBM in biomedical and health sciences
   - Transition from low-level laser therapy (LLLT) to PBM terminology
C. Overview of Hormesis: *"The dose makes the poison."*
   - Definition hormesis in biology and medicine (e.g., vitamin D_3 metabolism)
   - Comparison of linear and biphasic dose-response models (e.g., in radiation therapies)

II. Mechanisms and Applications of Photobiomodulation (PBM) Therapy (15 minutes)
A. Photophysical Forces and their Effects in the Cell
   - Overview of how light at specific frequencies influences biological processes
   - Action spectra of DNA and ribonucleic acid (RNA) synthesis
B. Mitochondrial Function and Retrograde Signaling
   - Role of mitochondrial cytochrome *c* oxidase in PBM and cellular energy metabolism
   - Mechanisms of nuclear retrograde signaling from mitochondria and other organelles

III. Applications of PBM Therapy for Health (20 minutes)
A. Therapeutic Effects on Musculoskeletal and Neurological Conditions
   - Use of PBM for pain reduction, inflammation, and tissue regeneration
   - Effects on neurological diseases such as stroke and Parkinson's disease

B. PBM Therapy and Cancer Treatment
   - Potential applications of PBM in cancer therapy and chemotherapy-related neuropathies
   - Evidence of PBM effects on cancer cell metabolism and growth

IV. Electromagnetic Therapies and Biological Responses (20 minutes)

A. Overview of Electromagnetic Field (EMF) Therapies
   - Introduction to pulsed electromagnetic field (PEMF) therapy and its applications
   - Therapeutic use of nuclear magnetic resonance (NMR) technologies to treat medical conditions

B. Research on Electromagnetic Fields and Health
   - Effects of extremely low frequency (ELF) EMFs on wound healing and inflammation
   - Potential benefits and risks associated with human exposure to EMFs

V. Future Directions and Challenges in PBM and EMF Research (15 minutes)

A. Need for Standardization of Treatment Parameters
   - Challenges in defining and characterizing optimal PBM protocols
   - Necessity of understanding the physiological mechanisms behind PBM and EMF

B. Emerging Research on Electromagnetic Therapies
   - Exploration of therapeutic applications of electromagnetic fields
   - Integration of PBM and EMF with existing treatment modalities

VI. Conclusion and Q & A (5 minutes)

A. Summary of Key Concepts: Quantum Therapies: Electromagnetic Fields as Medicine

B. Open floor for student questions

Class 16 – Photodynamic Therapy & Nanotheranostics

I. Introduction to Photodynamic Therapy (PDT) (15 minutes)

A. Historical Context: History of Photosensitizers and PDT

- Early photochemical transformations and the role of photosensitizers
- Key developments in PDT from the 20th century to present

B. Photodynamic Mechanism of Action

- Overview of how PDT generates reactive oxygen species (ROS)
- Role of light and photosensitizers in targeting pathological tissues

II. Photosensitizers and Their Applications (15 minutes)

A. Various Types of Photosensitizers

- Flavins, porphyrins, quantum dots and their roles in biological systems
- Development of hematoporphyrin derivatives and their clinical significance

B. Clinical Applications of PDT

- Use in oncology, dermatology, and treatment of infections
- Overview of therapeutic effects and patient tolerance issues

III. Nanomedicine and Its Integration with PDT (25 minutes)

A. Nanomedicine: Concept of Nanoscale Medicine

- Definition and significance of nanoparticles in medical applications
- Role of nanotechnology in enhancing PDT effectiveness

B. Emerging Nanotherapeutic Approaches

- Overview of photothermal therapy (PTT) and its mechanisms
- Overview of chemodynamic therapy (CDT) and its mechanisms
- Overview of electrodynamic therapy (EDT) and its mechanisms

- Overview of radiodynamic therapy (RDT) and its mechanisms
- Overview of sonodynamic therapy (SDT) and its mechanisms

IV. Mechanisms of Action in Nanomedicine (15 minutes)

A. Interaction of Nanoparticles with Biological Systems

- Chemodynamic ROS generation and cellular responses (*i.e.*, the Fenton reaction)
- Importance of targeted delivery and specificity in treatment

B. Examples of Nanomaterials in Therapy

- Functionalized magnetic nanoparticles and their applications
- Combination therapies and their potential benefits

V. Advanced Techniques in PDT and Nanomedicine (15 minutes)

A. Innovations in Photodynamic and Nanotherapeutic Approaches

- Use of optogenetics and light-activated molecules in therapy
- Development of nanotheranostic systems for diagnostics and treatment

B. Challenges and Future Directions

- Limitations in current PDT and nanomedicine practices
- Need for further research into mechanisms and treatment parameters

VI. Conclusion and Q & A (5 minutes)

A. Summary of Key Concepts: Chemical Control in Nanomedicine: Light as Scalpel

B. Open floor for student questions

Class 17 – Regenerative Mechanisms: Cells, Tissues, & Organs

I. Introduction to Circadian Rhythms and Physiological Regulation (10 minutes)

A. Overview of Diurnal Cycles (Daily Cycles)
   - Influence of light and darkness on physiology and behavior
   - Importance of circadian rhythms in biological functions

B. Role of Light in Physiological Processes
   - Effects of light intensity, duration, and color on cellular functions
   - Introduction to chronobiology, the science of biological rhythms

II. Regenerative Mechanisms and Stem Cell Therapy (15 minutes)

A. Overview of Regeneration in Multicellular Organisms
   - Comparison of regenerative capacities across species
   - Focus on planaria and their unique regenerative capabilities

B. Applications of Stem Cell Therapy
   - Mechanisms of stem cell differentiation and proliferation
   - Challenges and risks associated with stem cell treatments

III. Role of Reactive Oxygen Species (ROS) in Regeneration (15 minutes)

A. Mechanistic Insights into ROS Production
   - Role of ROS in tissue growth and healing
   - Evidence from zebrafish and salamander regeneration studies

B. Influence of Electromagnetic Fields on ROS Dynamics
   - Effects of electromagnetic fields on cellular regeneration processes
   - Research findings on electromagnetic radiation and tissue repair

IV. Stem Cell Engineering and Tissue Regeneration (15 minutes)

A. Overview of Stem Cell Therapies
   - Techniques for isolating and directing stem cell differentiation
   - Challenges and risks associated with stem cell treatments

B. Quantum Innovations for Tissue Engineering
   - Enhancing use of biomaterials for tissue repair with PBM and WMFs
   - Advances in creating lab-grown tissues for clinical applications

V. Challenges in Cartilage and Bone Regeneration (15 minutes)

A. Difficulties in Osteochondral Tissue Healing
   - Characteristics of articular cartilage and its regenerative limitations
   - Overview of therapeutic approaches to cartilage regeneration (including PBM)

B. Advances in Osteochondral Tissue Engineering
   - Strategies for safe and effective cartilage regeneration
   - Role of scaffolds and biocompatible materials in cartilage repair

VI. Fibrin Scaffolds and Their Applications (15 minutes)

A. Role of Fibrin Scaffolds in Wound Healing
   - Mechanisms of fibrin formation and its role in tissue repair
   - Applications of fibrin hydrogels in biomedical implants (including PBM)

B. Integration of PBM with Fibrin Scaffolds in Regenerative Medicine
   - Enhancing cell proliferation, differentiation, and function in hydrogel systems with PBM
   - Potential for optimizing outcomes in regenerative therapies and tissue engineering using PBM

VII. Conclusion and Q & A (5 minutes)

A. Summary of Key Concepts: Quantum Effects in Regenerative Medicine from Stem Cells to Tissues
B. Open floor for student questions

Class 18 – Morphogenetic Integration & Immunology

I. Introduction to Cell Behavior and Environmental Integration (10 minutes)

A. Overview of Cellular Responses to Environmental Cues

- Importance of cell behaviors in development, homeostasis, and regeneration
- Role of physico-chemical phenomena in cellular dynamics

B. Mechanotransduction and its Biological Significance

- Impact of biomechanical forces on cellular processes
- Importance of the extracellular matrix (ECM) in cell behavior

II. The Cytoskeleton as a Chemomechanical Processor (20 minutes)

A. Chemomechanical Coupling from Genetic Regulation to Morphogenesis

- Cellular responses to electromagnetic, chemical, and mechanical cues
- Role of the cytoskeleton cell polarization, morphogenesis, and motility

B. Inadequacy of Reaction-Diffusion Mechanisms to Explain Cytoskeletal Organization

- Role of cellular mechanotransduction in embryogenesis, tissue repair, and regeneration
- Influence of ECM on morphological dynamics of cell division, phagocytosis, and cell differentiation

III. Cell Migration and Its Importance in Physiology (20 minutes)

A. Physical Mechanisms of Cell Migration

- Overview of chemotaxis and mechanotaxis
- Role of the cytoskeleton in cell migration and morphology

B. Implications Cell Migration for Growth and Disease

- Relationship between cell migration and cancer metastasis
- Influence of the ECM on cell migration dynamics

IV. Immune Responses and Tissue Regeneration (20 minutes)

A. Role of the Immune System in Tissue Healing

- Overview of immune cell behavior during inflammation and tissue repair
- Interaction between immune cells and tissue microenvironments

B. The Fibroblastic Reticular Cellular Network

- Structure and function of reticular cell networks in lymphoid organs
- Implications for tissue regeneration, bioengineering, and synthetic organ grafts

V. Frontiers in Bioengineering and Synthetic Biology (15 minutes)

A. Current Progress in Developing Synthetic Organ Systems

- Self-organizing organoids and "organ-on-chip" technologies
- 3D-printed tissues and organ manufacturing processes

B. Challenges and Future Prospects for Cell Morphological Science

- Challenge of replicating the complexity of the ECM in synthetic organ systems
- Dynamic control of cellular development for bioengineering and nanobiotechnology

VI. Conclusion and Q & A (5 minutes)

A. Summary of Key Concepts: Cellular Morphogenesis in Synthetic Tissues and Designer Organs

B. Open floor for student questions

Class 19 – Quantum Biotechnology: Universal Applications

I. Chemically-Active Biomolecules as Quantum Mechanical Processing Elements (15 minutes)

A. Biomolecular Cofactors as Molecular Bioelectronics

- Electronic structural basis for functional properties of biomolecules

  - Many-body states of biomolecules: Quantum state preparation by engineered dissipation

  B. Biomolecular Self-Assembly and Organizational Processes

  - Influence of inter-molecular forces on biomolecular form, structure, and organization
  - Role of electromagnetic fields and bioelectric gradients in regulating cellular behavior

II. Quantum Effects in Biological Materials (15 minutes)

  A. Biological Materials: Proteins, Membranes, and Crystals

  - Multi-functional and reconfigurable aspects of biological membranes and materials
  - Optical properties of biological matter: organic crystals and biophotonics

  B. Biological Significance of Metal-Porphyrin Complexes

  - Roles of electron transfer and charge resonance stabilization in enzyme catalysis
  - Applications of natural and synthetic metal–porphyrin complexes in biomedicine and catalysis

III. Advances in Photocatalysis and Nanotechnology (15 minutes)

  A. Light-Driven Redox Catalysis

  - Overview of photocatalytic methods for organic compound synthesis
  - Emerging technologies for sustainable bioenergy production

  B. Single-Atom Photocatalysts and Metal-Organic Systems

  - Benefits of single-atom catalysts in enhancing chemical selectivity and efficiency
  - Comparison with traditional catalytic and nanoparticle systems

IV. Quantum Control in Photoactive Biological Systems (15 minutes)

  A. Advances in Artificial Light-Harvesting Systems

- Overview of developments in artificial photosynthesis and photocatalysis
- Importance of vibronic coupling in enhancing charge transfer efficiency

B. Photo-control of Microtubules and Ion Channels in Biology

- Dynamical photocontrol of microtubule integrity, migration, and mitosis via photoswtiching
- Integrating photochemistry, optomechanics, and optogenetics in ionic signaling cascades

V. Innovations in Synthetic Biology (15 minutes)

A. Overview of Synthetic Biology Approaches

- Comparison of "top-down" and "bottom-up" biosynthetic strategies
- Applications in molecular computing and programmable matter

B. Emerging Potential of DNA-Based Biotechnologies

- Use of DNA for data storage and information processing applications
- Advances in programmable DNA for therapeutic and bioengineering applications

VI. Implications of Quantum Biology for Biotechnology (10 minutes)

A. Technological Innovation and Quantum Biology

- Overview of applications in biomedicine, energy production, and nanotechnology
- Potential for quantum biology to overcome classical limitations on bionic technologies

B. The Future of Quantum Information Processing and DNA Technologies

- Discussion of molecular computing with DNA and its potential applications
- Advances in molecular biosynthesis for programmable quantum matter

VII. Conclusion and Q & A (5 minutes)
   A. Summary of Key Concepts: Quantum Bioengineering and Convergent Biotechnologies
   B. Open floor for student questions

Class 20 – Essential Further Research

I. Importance of Research in Quantum Biology (5 minutes)
   A. Discussion of how quantum effects influence various biological phenomena
      - The challenge of bridging quantum and classical biological models

II. Overview of Quantum Effects in Biological Processes (10 minutes)
   A. Key Quantum Effects in Molecular Biology
      - Overview of electron transfer, photochemical reactions, biomolecular structure, and the role of van der Waals forces in biological systems
   B. Quantum Coherence and Decoherence
      - Review of quantum coherence in biological systems and its implications
      - Discussion of decoherence processes and their significance in living organisms

III. Research Area: Electron and Proton Transport in Living Systems (10 minutes)
   A. Biological Electron Transfer Mechanisms
      - Overview of oxidative metabolism and electron transport chains
      - Importance of chromophores and biomolecular cofactors in electron transfer processes
   B. Proton Transfer and Quantum Mechanics
      - Discussion of how proton transfer accompanies electron transfer in biological reactions
      - Influence of proton tunneling in complex biochemical systems

IV. Research Area: Light-Matter Interactions in Sensing and Physiology (10 minutes)

A. Established Roles of Light-Matter Interactions in Cellular Signaling and Physiology

- Vision, photosynthesis, vitamin D synthesis, and photochemical reactions (e.g., bioluminescence)
- Optically active biomolecules facilitate redox chemistry and catalytic processes

B. Emerging Medical Significance

- Research innovations in photobiomodulation, photodynamic therapy, and optogenetics
- Influence of light on mitochondrial function and retrograde signaling on cellular health

V. Research Area: Photonic and Quantum Optical Effects (10 minutes)

A. Biophotons and Cellular Function

- Review of UPE and their significance in biological oxidative processes
- Links to dynamics of reactive chemical species and photochemical mitochondrial activity

B. Applications in Medical Research

- Association of UPE with weak magnetic fields in regenerative processes require validation
- Potential for non-invasive redox/metabolism monitoring (e.g., cancer diagnostics)

VI. Research Area: Influence of Weak Magnetic Fields on Cell Physiology (10 minutes)

A. Mechanistic Ambiguities in the Radical Pair Model of Magnetoreception

- Radical pair dynamics implicated in animal magnetoreception
- Role of radical scavengers in field amplification remains to be quantified

B. Physiological Impact of Earth-Strength Magnetic Fields
   - Earth-strength fields ($\sim 50\,\mu$T) may influence cell growth and oxygen metabolism
   - Bioelectromagnetic theory needed to unify all observed cellular electromagnetic responses

VII. Research Area: Microtubules Dynamics and Terahertz Radiation (10 minutes)

A. Microtubules as Integrators of Biochemical, Electromagnetic, and Optomechanical Signals
   - Role of oscillating electric fields in assembling microtubules and coordinating cellular activities
   - Tubulin's birefringent and optical traits require thorough ground/excited state characterization

B. Influence Terahertz (THz) Radiation in Biology
   - Influences proliferation, apoptosis, and amino acid release via electrodynamic interactions
   - Combination of THz imaging with biophoton detection for cancer identification

VIII. Research Area: Mechanotransduction and Cellular Regulation (10 minutes)

A. Coupling of Mechanical and Electrochemical Forces in Cells
   - Quantum effects in mechanotransduction (e.g., via molecular mechanotransducers)
   - Involves non-adiabatic phenomena similar to Jahn-Teller effects and intersystem crossings

B. Scaling from molecular interactions to organelles, cells, and tissues
   - Mitochondrial-microtubule interactions central to cellular regulation and morphology
   - Key interactions for understanding photobiomodulation and cellular magnetoreception

IX. Theoretical and Computational Frontiers of Quantum Biology (10 minutes)

A. Outstanding Challenges in the Simulation of Quantum Biochemistry
   - Exchange-correlation potential as a central problem in biomolecular electronic structure modeling
   - Unknown influence of charge and spin transfer dynamics complicate enzyme/cofactor simulations

B. Novel Methods for Approaching the Electronic Structure Problem
   - Quantum computing proposed for simulating intractable systems (e.g., transition metal sites)
   - Advanced quantum electrodynamics (e.g., coupled-cluster and path-integral formulations)

X. Conclusion and Q & A (5 minutes)

A. Summary of Key Concepts: Frontiers and Foundations of Quantum Biology

B. Open Floor for Student Questions

Class 21 – Important Considerations for the Future of Quantum Biology

I. The Quantum-Classical Relationship (10 minutes)

A. Overview of Quantum Theory
   - Definition and significance of quantum biology in understanding living systems
   - Discussion of the historical context of the development of quantum theory

B. The Quantum-Classical Distinction
   - Re-examination of the quantum mechanical measurement problem (Lüders, von Neumann)
   - The Correspondence Principle and Canonical Quantization: The Quantum/Classical Chasm

II. The Need for a Quantum Physical Perspective on Biology (15 minutes)

A. Ubiquity of Non-Trivial Quantum Effects in Quantum Biology

  - Scientific and historical basis for recognizing non-trivial quantum effects in biological systems
  - Making rigorous and unequivocal statements regarding the role of quantum effects in biology

B. Engaging in the Physical Perspective on Quantum Biology
  - Benefits of understanding biological processes from a quantum mechanical perspective
  - Exploration of the hierarchy of quantum approximations from semiclassical to truly quantum

III. Quantum Theoretical Framework for Quantum Biology (25 minutes)

A. The Non-Hermitian Hamiltonian Formalism
  - Exposition of the open quantum systems framework developed by Fano and Feshbach
  - Discussion of environmental influences on open quantum systems (e.g., scattering theory)

B. Incorporating Models of Quantum Dynamics into Molecular Biology
  - Overview of dissipation, external driving, and relaxation in the context of biological systems
  - Implications for understanding the dynamics of life using open quantum systems theory

IV. Interdisciplinary Nature of Quantum Biology (15 minutes)

A. Active Matter Systems in Physics and Biology
  - Review of the concept of active matter and its relevance to quantum biology
  - Discussion of symmetry-breaking effects and their significance in living organisms

B. The Need for Wide Collaboration Across Scientific Disciplines
  - Importance of integrating knowledge from biology, physics, chemistry, computation, and medical science
  - Challenges in developing shared terms, language, and methods for quantum biology research

V. Ethical Considerations and Future Research Directions (10 minutes)

   A. Ethical Implications for Quantum Biology

      - Discussion of the potential societal and ethical challenges associated with emerging technologies
      - Potential impacts of biotechnology on culture, environment, infrastructure, and health
      - Importance of establishing bioethical guidelines for research and applications

   B. Future Research Opportunities

      - Exploration of the potential implications of quantum biology for biomedicine and bioengineering
      - Opportunities for students to engage with emerging biomedical and bioengineering technologies

VI. Final Discussion (15 minutes)

   A. Discussion on the Implications of Quantum Biology

      - Group discussion on how quantum biology is changing our understanding of life and technology

Assessment: Final Exam

Glossary of Terms

ab initio: Latin (trasl. "from the beginning"), a method of approaching a subject or performing calculations from first principles, using basic physical constants without empirical parameters.

biphasic: Refers to a system or process that involves two distinct phases, such as a solid phase and a liquid phase, or a rising phase and a falling phase, often in the context of chemical or physical dynamics.

chemotaxis: The movement of a cell or organism in response to chemical stimuli, often in the study of cell signaling.

chirality: Describes asymmetrical structures that cannot be superimposed onto their mirror images, often in the context of molecular interactions; sometimes known as (left or right) "handedness."

circadian rhythms: Biological cycles that display an endogenous, entrainable oscillation of about $\sim$ 24 hours, influenced by external cues like light, especially in biological and ecological physics.

closed quantum system: A system where quantum states are defined, and interactions with the environment are negligible, allowing for the isolation of quantum phenomena. See also "Hermitian."

coherent quantum effect: Quantum phenomena where particles or states exhibit a definite phase relationship, crucial in the description of quantum interference and superposition.

configuration (electronic): The specification of the quantum state of an electron or electrons, in terms of a specific set of quantum numbers.

configuration interaction: A quantum mechanical method used to describe the electronic state of a molecule or other quantum system by considering interactions between various electron configurations.

conjugated organic molecule: A molecule with alternating single and multiple bonds, resulting in the delocalization of electrons which significantly affects its optical and electronic properties.

decoherent tunneling: A quantum mechanical process wherein a particle moves through an otherwise-insurmountable barrier while losing its quantum coherence, typically due to interactions with the environment.

degeneracy-breaking effect: A phenomenon where a previously degenerate set of quantum states of a system split into distinct energy levels due to an external perturbation or field, to impact the system's behavior.

density functional theory: A computational quantum mechanical modeling method used to investigate the electronic structure of many-body systems by using electron density rather than wave function; typically pertains to Kohn–Sham density functional theory.

dispersion forces: Weak forces arising from fluctuations in transient dipole moments of molecules, significant in understanding molecular interactions.

dynamic instability: A non-equilibrium property of a system that is maintained in metastable state away from thermodynamic equilibrium by continuous energy input; especially in microtubules where it describes polymerization dynamics subject to stochastic switching between growth and collapse.

electron bifurcation: A process in which the pathway of one or more electrons diverges into two separate routes and/or energetic levels, studied in the context of biochemical reactions and energy transduction.

electron delocalization: The distribution of electron density in space, often over several atoms rather than being restricted to a single atom, essential to quantum mechanical models of chemical bonding and reactivity.

electron transfer: The movement of electrons from one molecule or atom to another, fundamental in redox reactions and biological energy conversion; described by Marcus theory as a nonadiabatic quantum mechanical transition predominantly mediated by electron tunneling.

electron transport chain: A series of electronically active molecules in a protein complex that transfer electrons, usually through a membrane; critical in cellular respiration, photosynthesis, and immune function.

electron valence: The number of electrons in the outermost quantum mechanical energy level of an atom that are available for bonding, influencing chemical reactivity and properties.

electrostatic: Relates to stationary arrangements of electric charge, fundamental in understanding forces between charged particles and molecular interactions.

endergonic: Describes a reaction or process that requires energy input to proceed, often used in thermodynamics to describe non-spontaneous reactions.

endogenous: Refers to processes or substances that originate from within an organism, system, or process; significant in biological and medical physics.

endothermic: Pertains to a reaction or process that absorbs heat from its surroundings, impacting thermodynamic behavior and energy transfer.

entanglement: A quantum phenomenon where two or more particles become correlated in such a way that the state of one cannot be described independently of the state of the other, regardless of separation.

exchange-correlation: A term in density functional theory that describes the effective interaction between electrons, taking into account the particle-exchange symmetry and correlated motion of identical pairs of electrons.

extracellular matrix: A complex network of proteins and carbohydrates outside cells that provides structural and biochemical support, studied in biophysics.

Fermi's "Golden Rule": Refers to the second of two formulae popularized by Enrico Fermi, used for calculating the probability of transitions between quantum states, essential in quantum mechanics and statistical physics.

Gibbs free energy: A thermodynamic potential that measures the maximum reversible non-expansion work that can be obtained from a thermodynamic system at constant temperature and pressure; its change determines the spontaneity of chemical and biological processes.

Hartree-Fock theory: A method for approximating the wave function and energy of a quantum many-body system in a stationary state using a single Slater determinant, foundational in quantum chemistry.

Heisenberg cut: A conceptual boundary between quantum and classical systems, typically discussed in the context of quantum measurement.

heliotherapy: The use of sunlight for therapeutic purposes involving the absorption of light by living systems, studied in biophysics and medicine.

Hermitian symmetry: A property of quantum mechanical operators, traditionally used in quantum mechanics to ensure real eigenvalues and orthogonal eigenstates of operators; pertains to the closure of operaters on a Hilbert space with respect to the standard Dirac inner product.

homeostasis: The process by which biological systems maintain stability while adjusting to changing external conditions, used in physiology and medicine.

hormesis: A dose-response phenomenon characterized by low-dose stimulation and high-dose inhibition, showing non-linear, typically biphasic responses in biological systems.

H-theorem: A theorem in statistical mechanics that describes the approach to equilibrium of a thermodynamic system, demonstrating the second law of thermodynamics.

in situ: Latin (transl. "in place" or "on site"), refers to processes or measurements conducted in the original place or context, significant in experimental physics, biology, and medicine.

in vitro: Latin (transl. "in glass"), refers to experiments conducted outside of a living organism, often in controlled environments, important for understanding biological and biochemical processes.

in vivo: Latin (transl. "in life"), refers to experiments or processes that take place within a living organism, crucial for studies in biology and medicine.

incoherent tunneling: See decoherent tunneling.

inelastic: Referring to processes where kinetic energy is not locally conserved, typically involving energy loss during interactions due to heat, light, and noise; important in collision physics.

interference: The phenomenon that occurs when two or more waves superpose to form one resultant wave, significant in optics, electromagnetism, and quantum mechanics.

intersystem crossing: A process in which a molecule undergoes a non-radiative electronic transition between states of different spin multiplicity, important in photochemistry and photophysics.

inverted effect: A phenomenon where a system behaves oppositely to expectations, often discussed in optics and quantum mechanics; especially in Marcus theory when an increase in the force driving a reaction decreases the reaction rate.

ion: An atom or molecule that has gained or lost one or more electrons, resulting in a net electrical charge; fundamental in chemistry and physics.

Jahn-Teller effect: A distortion of molecular or ligand geometry that occurs in certain degenerate electronic states, impacting the properties of the molecule by reducing its symmetry and lowering its energy.

Langevin dynamics: A computational method used to simulate the behavior of systems influenced by both deterministic and stochastic forces, typically used to reproduce the effect of temperature in thermodynamically-realistic simulations.

Lennard-Jones potential: A semiclassical mathematical model that is used to approximately account for interactions between pairs of neutral atoms or molecules, significant in molecular dynamics simulations.

Marcus theory of electron transfer: A theoretical framework that describes the rates of electron transfer using quantum mechanics and the principle of conservation of energy, crucial in physical chemistry and biochemistry.

mean field theory: An approximate method used in statistical mechanics to analyze phase transitions and describe collective behavior of systems by assuming that collective interactions between particles is properly approximated by an averaged effective interaction.

metabolic biophoton emissions: Faint light emissions produced by biological organisms as a result of metabolic processes, studied in biophysics. See also ultraweak photon emissions.

Mitchell's Q cycle: A theoretical model explaining the mechanism of energy conversion in the electron transport chain during cellular respiration; the Q stands for quinone/quinol.

molecular orbital theory: A method for determining molecular structure by considering the distribution of electrons in molecular orbitals (rather than atomic orbitals).

multi-scale models: Approaches that integrate different scales of analysis (from atomic to macroscopic) to understand complex systems in physics and biology, often used in computational physics and numerical modeling.

nanocrystal: A crystalline particle with dimensions on the nanoscale, exhibiting unique physical and chemical properties; studied in materials science and nanotechnology.

nanotheranostic: A field combining diagnostic and therapeutic applications at the nanoscale, significant in medical physics and drug delivery.

nonadiabatic: Referring to processes that occur without the assumption of slow changes, where electronic and nuclear motions may be coupled.

nonlocality: A property of quantum mechanics where particles exhibit correlations that cannot be accounted for by local phenomena; particularly relevant to strongly correlated electron systems including conjugated, aromatic, and organometallic molecular systems where electronic correlations are not adequately described by mean field methods.

non-Hermitian: Referring to operators that do not satisfy Hermitian symmetry, often leading to complex-valued eigenvalues and implications for open quantum systems.

one-particle theory: A theoretical framework that describes a many-body system of interacting particles in terms of individual non-interacting particles, effectively trivializing those interactions.

open quantum systems: Quantum systems that interact with their environment, leading to phenomena like decoherence and loss of information.

optical: Pertaining to light and its properties, often studied in the context of wave phenomena and quantum optics; can refer to wavelengths of light over the infrared, visible, and ultraviolet spectrum.

optoelectronic: Pertains to the combination of optics and electronics, focusing on devices that source, detect, and control light using electronics.

orbital angular momentum: The angular momentum associated with the motion of electrons in their orbitals, significant in atomic, molecular, and optical physics.

photobiomodulation: The use of light to stimulate biological processes, often studied in medical physics for therapeutic applications.

photoelectric effect: The phenomenon where electrons are ejected from a material when it is exposed to light of a sufficiently high frequency, fundamental in quantum mechanics and solid-state physics.

photoentrainment: The process by which light influences biological rhythms, particularly circadian rhythms; significant in chronobiology.

photon: A quantum of electromagnetic radiation, fundamental in the study of light and its interactions with matter using quantum electrodynamics.

photoswitchable: Referring to a molecule that can change its structure or function upon exposure to light, important in materials science and photochemistry.

phototropism: The growth response of plants to light direction, studied in biological physics and investigations of plant behavior.

Planck's constant: A fundamental constant that relates the energy of a photon to its frequency, crucial in quantum mechanics.

polarizable media: Materials that can develop an electric dipole moment in response to an external electric field, significant in electromagnetism.

polaron: A quasiparticle that consists of an electron and its associated polarization field, important in condensed-matter physics.

quanta: Discrete packets of energy, matter, or momentum; fundamental in quantum theory, representing the smallest possible units of a physical quantity.

quantization: The process of constraining a physical quantity to discrete values rather than continuous ones, fundamental in quantum mechanics.

quantum adiabatic theorem: A principle stating that a quantum system remains in its instantaneous eigenstate if a given perturbation is applied slowly enough.

quantum correlations: Non-classical correlations between quantum systems that cannot be explained by classical physics, significant in quantum information theory; can be static or dynamic.

quantum dots: Nanoscale semiconductor particles that have quantum mechanical properties, important in optoelectronics and nanotechnology.

quantum dynamics: The study of time-dependent behavior of quantum systems, focusing on the evolution of quantum states over time.

quantum electrodynamics: A fundamental theory describing how light and matter interact at the quantum level, combining quantum mechanics and electrodynamics.

quantum embedding theory: A framework that allows for the study of a small quantum system within a larger environment, useful in computational chemistry.

quantum field theory: A theoretical framework combining quantum mechanics and special relativity to describe how matter and energy interact.

quantum information processing: The study and application of quantum mechanics to process and transmit information, fundamental in quantum computing.

quantum jump: A sudden transition between quantum states, often observed in the context of atomic or molecular systems; also colloquially referred to as a quantum leap.

quantum spin: A property of electrons, photons, and other quantum mechanical particles that arises as a consequence of intrinsic angular momentum; plays a critical role in chemical and magnetic properties of materials.

quantum tunneling: A quantum phenomenon where particles pass through potential barriers that would be classically impenetrable, significant in nuclear and solid-state physics.

resonance stabilization: The increased stability of a molecular system due to the delocalization of electrons, particularly within a conjugated bonding system.

resonant tunneling: A quantum tunneling process that occurs at specific energy levels, representing the coherent limit of quantum tunneling; often utilized in electronic and optical devices.

self-consistent field: An iterative method used in quantum chemistry to find a solution where the field created by electrons is consistent with their distribution, often used in conjunction with mean field theory.

singlet-triplet interconversion: A process where a singlet state (paired spins) transitions to a triplet state (unpaired spins) and vice versa, relevant in photochemistry and spin dynamics. See also inter-system crossing.

small world network: A type of network characterized by short path lengths and high clustering, often studied in statistical physics and network theory.

solitons: Self-reinforcing wave packets that maintain their shape while traveling at constant speeds, important in nonlinear dynamics.

spacetime: The four-dimensional continuum in which events occur, combining three dimensions of space with one dimension of time; fundamental in relativity.

spatiotemporal: Relating to both space and time, often used in physics to describe phenomena that evolve over both spatial and temporal dimensions.

spin-exchange: A quantum mechanical phenomenon involving the exchange of spin states between particles, significant in magnetic and spin chemical systems.

static magnetic fields: Magnetic fields that do not change over time, important in various applications including imaging techniques and magnetic resonance.

stochastic resonance: A phenomenon where the presence of noise enhances the detection of weak signals in nonlinear systems, studied in statistical mechanics.

subradiant: A state of a quantum mechanical system of light emitters where the emission of radiation is suppressed compared to the normal radiation rate, important in quantum optics.

superexchange: A mechanism for indirect coupling between electronic sites through virtual excitations of an intervening medium; describes both spin coupling in condensed matter physics and electron tunneling through molecular bridges during biological electron transfer. Known as Kramers–Anderson superexchange.

superposition: A fundamental quantum mechanical effect whereby a physical system can exist simultaneously in multiple quantum states.

superradiance: A collective emission phenomenon where a group of excited atoms or molecules emits light more intensely than they can individually, studied in quantum optics.

symmetry-breaking effect: A situation in which a system that is symmetric under certain transformations becomes asymmetric, leading to distinct states or phases; can be spontaneous.

system: A defined set of interacting components, which can be isolated or part of a larger environment.

theranostic: The combination of therapeutic and diagnostic methods, particularly in medicine and nanotechnology.

time-dependent perturbation theory: A mathematical framework in quantum mechanics used to systematically analyze physical systems subject to small time-varying influences.

time-independent perturbation theory: A mathematical framework in quantum mechanics used to systematically analyze physical systems subject to small time-invariant influences.

trivial: Referring to a system or result that is straightforward or without complexity, often used in contrast to non-trivial findings; in physics this typically refers to cases when the correlations, dynamics, or interactions in a system become negligible.

ultraweak photon emissions: Extremely low-level light emissions from biological systems that are produced by oxidative or metabolic processes, often studied in biophysics and medical physics.

uncertainty principle: A fundamental principle in quantum mechanics stating that certain pairs of physical properties cannot be simultaneously known with arbitrary precision.

vacuum field: The lowest energy state of a quantum field, which can still exhibit fluctuations and is important in quantum field theory.

vibronic coupling: The interaction between electronic and vibrational states in molecules, significant in describing molecular spectra and dynamics.

Warburg effect: A metabolic process in which cancer cells break down glucose to produce energy even when oxygen is plentiful, also known as aerobic glycolysis.

weak magnetic field: A magnetic field that is significantly lower in strength compared to typical magnetic fields used in experiments, usually less than one milliTesla; often studied in quantum mechanics and biophysics.

References

[1] M. Bordonaro and V. Ogryzko, Quantum biology at the cellular level–elements of the research program, *BioSystems* **112**, 11–30 (2013).

[2] P. C. Davies, Does quantum mechanics play a non-trivial role in life? *BioSystems* **78**, 69–79 (2004).

[3] W. T. B. Kelvin, Nineteenth century clouds over the dynamical theory of heat and light, in *Baltimore lectures on molecular dynamics and the wave theory of light*. C. J. Clay and Sons, pp. 486–527 (1904).

[4] O. Passon, Kelvin's clouds, *American Journal of Physics* **89**, 1037–1041 (2021).

[5] H. Minkowski, The fundamental equations for electromagnetic processes in moving bodies, *Nachr. Ges. Wiss. Gottingen, Mathematisch Physikalische Klasse*, 53–111 (1908).

[6] J. Lindgren and J. Liukkonen, Maxwell's equations from spacetime geometry and the role of weyl curvature, in *Journal of Physics: Conference Series*, Vol. 1956. IOP Publishing, p. 012017 (2021).

[7] J. Stachel, Why Einstein reinvented the ether, *Physics World* **14**, 55 (2001).

[8] L. Kostro, Einstein and the ether, *Electronics & Wireless World* 238–239 (2000).

[9] R. Laughlin, *A Different Universe: Reinventing Physics From the Bottom Down*. Basic Books (2006).

[10] P. Enders, State, statistics and quantization in einstein's 1907 paper, 'planck's theory of radiation and the theory of specific heat of solids', *ICFAI University Journal of Physics II* **2**, 176–195 (2009).

[11] R. Pathria and D. Haar, *Statistical Mechanics*, International Series of Monographs in Natural Philosophy. Elsevier Science (2017).

[12] F. Mandl, *Statistical Physics*, Vol. 14. John Wiley & Sons (1991).
[13] M. J. Klein, Einstein, specific heats, and the early quantum theory: Einstein's quantum theory of specific heat first showed the power of the new concept of energy quanta. *Science* **148**, 173–180 (1965).
[14] C. Villars, Observables, states and measurements in quantum physics, *European Journal of Physics* **5**, 177 (1984).
[15] I. J. Aitchison, D. A. MacManus and T. M. Snyder, Understanding Heisenberg's "magical" paper of July 1925: A new look at the calculational details, *American Journal of Physics* **72**, 1370–1379 (2004).
[16] P. A. M. Dirac, Bakerian lecture-the physical interpretation of quantum mechanics, *Proceedings of the Royal Society of London. Series A. Mathematical and Physical Sciences* **180**, 1–40 (1942).
[17] P. A. M. Dirac, The fundamental equations of quantum mechanics, *Proceedings of the Royal Society of London. Series A, Containing Papers of a Mathematical and Physical Character* **109**, 642–653 (1925).
[18] M. Born and P. Jordan, Zur quantenmechanik, Zeitschrift für Physik **34**, 858–888 (1925).
[19] E. Schrödinger, *Wave Mechanics.* American Mathematical Society (1926).
[20] J. von Neumann, *Mathematical Foundations of Quantum Mechanics.* Princeton University Press (1955).
[21] S. Ratcliffe, *Oxford Essential Quotations.* Oxford University Press (2017).
[22] X. Fan, T. Myers, B. Sukra and G. Gabrielse, Measurement of the electron magnetic moment, *Physical Review Letters* **130**, 071801 (2023).
[23] M. Arndt, T. Juffmann and V. Vedral, Quantum physics meets biology, *HFSP Journal* **3**, 386–400 (2009).
[24] B. Drummond, Understanding quantum mechanics: A review and synthesis in precise language, *Open Physics* **17**, 390–437 (2019).
[25] W. G. Faris, Martin Schechter, operator methods in quantum mechanics, *Bulletin of the American Mathematical Society* **6**, 105–109 (1982).
[26] A. Einstein and L. Infeld, *Evolution of Physics.* Cambridge University Press (1938).
[27] T. Sigfried, Why quantum mechanics might need an overhaul, (2020), https://www.sciencenews.org/blog/context/why-quantum-mechanics-might-need-overhaul.

[28] P. Dirac, Does renormalization make sense? in *AIP Conference Proceedings*, Vol. 74. American Institute of Physics, 129–130 (1981).

[29] J. G. Cramer, Renormalization: Dodging infinities, *Analog Science Fiction & Fact Magazine* AV-204 (2020).

[30] F. Scholtz, H. Geyer and F. Hahne, Quasi-Hermitian operators in quantum mechanics and the variational principle, *Annals of Physics* **213**, 74–101 (1992).

[31] A. Zeilinger, A foundational principle for quantum mechanics, *Foundations of Physics* **29**, 631–643 (1999).

[32] O. Darrigol, 'shut up and contemplate!': Lucien hardy's reasonable axioms for quantum theory, *Studies in History and Philosophy of Science Part B: Studies in History and Philosophy of Modern Physics* **52**, 328–342 (2015).

[33] C. Marletto and V. Vedral, Gravitationally induced entanglement between two massive particles is sufficient evidence of quantum effects in gravity, *Physical Review Letters* **119**, 240402 (2017).

[34] N. Swanson, A philosopher's guide to the foundations of quantum field theory, *Philosophy Compass* **12**, e12414 (2017).

[35] J. Bernstein, Max born and the quantum theory, *American Journal of Physics* **73**, 999–1008 (2005).

[36] W. Heisenberg, *The Physical Principles of the Quantum Theory*, Dover Books on Physics and Chemistry. Dover Publications (1949).

[37] P. Dirac, *The Principles of Quantum Mechanics*. Clarendon Press (1981).

[38] C. E. Rüter, K. G. Makris, R. El-Ganainy, D. N. Christodoulides, M. Segev and D. Kip, Observation of parity–time symmetry in optics, *Nature Physics* **6**, 192–195 (2010).

[39] T. Lee and G. Wick, Negative metric and the unitarity of the S-matrix, *Nuclear Physics B* **9**, 209–243 (1969).

[40] T. Werlang, M. Matos, F. Brito and D. Valente, Emergence of energyavoiding and energy-seeking behaviors in nonequilibrium dissipative quantum systems, *Communications Physics* **5**, 7 (2022).

[41] J. H. Van Vleck, Quantum mechanics: The key to understanding magnetism, *Reviews of Modern Physics* **50**, 181 (1978).

[42] E. Schrodinger, *What is Life?: With Mind and Matter and Autobiographical Sketches*. Cambridge University Press (2012).

[43] H. Longuet-Higgins, Quantum mechanics and biology, *Biophysical Journal* **2**, 207 (1962).

[44] A. Einstein, B. Podolsky and N. Rosen, Can quantum-mechanical description of physical reality be considered complete? *Physical Review* **47**, 777 (1935).

[45] J. D. Trimmer, The present situation in quantum mechanics: A translation of Schrödinger's "cat paradox" paper, *Proceedings of the American Philosophical Society*, 323–338 (1980).
[46] D. Deutsch, Quantum theory as a universal physical theory, *International Journal of Theoretical Physics* **24**, 1–41 (1985).
[47] J. Bell, Against 'measurement', *Physics World* **3**, 33 (1990).
[48] A. J. Leggett, The quantum measurement problem, *Science* **307**, 871–872 (2005).
[49] M. Gell-Mann, Consciousness, reduction, and emergence: Some remarks, *Annals of the New York Academy of Sciences* **929**, 41–49 (2006).
[50] S. Weinberg, Collapse of the state vector, *Physical Review A* **85**, 062116 (2012).
[51] N. Bohr, Can quantum-mechanical description of physical reality be considered complete? *Physical Review* **48**, 696 (1935).
[52] F. Laudisa, Bohr and von Neumann on the universality of quantum mechanics: Materials for the history of the quantum measurement process, *The European Physical Journal H* **49**, 20 (2024).
[53] C. Chiu, E. Sudarshan and B. Misra, Time evolution of unstable quantum states and a resolution of Zeno's paradox, *Physical Review D* **16**, 520 (1977).
[54] E. Okon and D. Sudarsky, Measurements according to consistent histories, *Studies in History and Philosophy of Science Part B: Studies in History and Philosophy of Modern Physics* **48**, 7–12 (2014).
[55] S. L. Adler and A. Bassi, Is quantum theory exact? *Science* **325**, 275–276 (2009).
[56] A. Bokulich, Open or closed? Dirac, Heisenberg, and the relation between classical and quantum mechanics, *Studies in History and Philosophy of Science Part B: Studies in History and Philosophy of Modern Physics* **35**, 377–396 (2004).
[57] A. Narasimhan, D. Chopra and M. C. Kafatos, The nature of the Heisenberg-von Neumann cut: Enhanced orthodox interpretation of quantum mechanics, *Activitas Nervosa Superior* **61**, 12–17 (2019).
[58] H. D. Zeh, On the interpretation of measurement in quantum theory, *Foundations of Physics* **1**, 69–76 (1970).
[59] N. Hatano, What is the resonant state in open quantum systems? in *Journal of Physics: Conference Series*, Vol. 2038. IOP Publishing, p. 012013 (2021).
[60] D. Hestenes, Quantum mechanics from self-interaction, *Foundations of Physics* **15**, 63–87 (1985).

[61] T. Hübsch, Quantum mechanics is either nonlinear or non-introspective, *Modern Physics Letters A* **13**, 2503–2512 (1998).

[62] I. Basieva, A. Khrennikov and M. Ozawa, Quantum-like modeling in biology with open quantum systems and instruments, *Biosystems* **201**, 104328 (2021).

[63] S. Osat and R. Golestanian, Non-reciprocal multifarious selforganization, *Nature Nanotechnology* **18**, 79–85 (2023).

[64] S. Tzafestas, *Energy, Information, Feedback, Adaptation, and Self-organization.* Springer (2018).

[65] M. Eigen, Self-organization of matter and the evolution of biological macromolecules, *Naturwissenschaften* **58**, 465–523 (1971).

[66] D. Frauchiger and R. Renner, Quantum theory cannot consistently describe the use of itself, *Nature Communications* **9**, 3711 (2018).

[67] A. Khrennikov, I. Basieva, E. M. Pothos and I. Yamato, Quantum probability in decision making from quantum information representation of neuronal states, *Scientific Reports* **8**, 16225 (2018).

[68] I. Reshetnyak, A. Lorin and A. Pasquarello, Many-body screening effects in liquid water, *Nature Communications* **14**, 2705 (2023).

[69] A. Tal, T. Bischoff and A. Pasquarello, Absolute energy levels of liquid water from many-body perturbation theory with effective vertex corrections, *Proceedings of the National Academy of Sciences USA* **121**, e2311472121 (2024).

[70] N. S. Babcock, *Nonadiabatic Control for Quantum Information Processing and Biological Electron Transfer*, Ph.D. thesis, University of Calgary (2015).

[71] S. Rafiq and G. D. Scholes, From fundamental theories to quantum coherences in electron transfer, *Journal of the American Chemical Society* **141**, 708–722 (2018).

[72] A. Kovyrshin, M. Skogh, A. Broo, S. Mensa, E. Sahin, J. Crain and I. Tavernelli, A quantum computing implementation of nuclearelectronic orbital (NEO) theory: Toward an exact pre-born-oppenheimer formulation of molecular quantum systems, *The Journal of Chemical Physics* **158**, 21 (2023).

[73] M. Reiher, N. Wiebe, K. M. Svore, D. Wecker and M. Troyer, Elucidating reaction mechanisms on quantum computers, *Proceedings of the National Academy of Sciences USA* **114**, 7555–7560 (2017).

[74] N. Babcock, G. Montes-Cabrera, K. Oberhofer, M. Chergui, G. Celardo and P. Kurian, Ultraviolet superradiance from mega-networks of tryptophan in biological architectures, *The Journal of Physical Chemistry B* **128**, 4035–4046 (2024).

[75] Y. Meriguet, M. Lechelon, M. Gori, I. Nardecchia, F. Teppe, A. Kudashova, D. Coquillat, L. Varani, M. Pettini and J. Torres,

Collective oscillations of proteins proven by terahertz spectroscopy in aqueous medium, in *2019 44th International Conference on Infrared, Millimeter, and Terahertz Waves (IRMMW-THz)*. IEEE, 1–2 (2019).

[76] A. Nitzan, *Chemical Dynamics in Condensed Phases: Relaxation, Transfer and Reactions in Condensed Molecular Systems*. Oxford University Press (2006).

[77] G. Cerullo and M. Garavelli, A novel spectroscopic window on conical intersections in biomolecules, *Proceedings of the National Academy of Sciences USA* **117**, 26553–26555 (2020).

[78] S. Rafiq, B. Fu, B. Kudisch and G. D. Scholes, Interplay of vibrational wavepackets during an ultrafast electron transfer reaction, *Nature Chemistry* **13**, 70–76 (2021).

[79] K. Mukherjee and P. C. Jana, Controlled optical bistability in parity-time-symmetric coupled micro-cavities: Possibility of all-optical switching, *Physica E: Low-dimensional systems and nanostructures* **117**, 113780 (2020).

[80] H. Eleuch and I. Rotter, Gain and loss in open quantum systems, *Physical Review E* **95**, 062109 (2017).

[81] A. I. Nesterov, G. P. Berman and A. R. Bishop, Non-Hermitian approach for modeling of noise-assisted quantum electron transfer in photosynthetic complexes, *Fortschritte der Physik* **61**, 95–110 (2013).

[82] A. P. Kalra, A. Benny, S. M. Travis, E. A. Zizzi, A. Morales-Sanchez, D. G. Oblinsky, T. J. Craddock, S. R. Hameroff, M. B. MacIver, J. A. Tuszynski *et al.*, Electronic energy migration in microtubules, *ACS Central Science* (2023).

[83] H. Li, Z. Wang, L. Wang, Y. Tan and F. Chen, Enhancing thermal stability of ND:GGG WGM microdisk lasers via silica integration, *Nanophotonics* **13**, 2819–2827 (2024).

[84] C. M. Bender, D. C. Brody and H. F. Jones, Complex extension of quantum mechanics, *Physical Review Letters* **89**, 270401 (2002).

[85] I. Rotter and J. Bird, A review of progress in the physics of open quantum systems: theory and experiment, *Reports on Progress in Physics* **78**, 114001 (2015).

[86] H. Breuer and F. Petruccione, *The Theory of Open Quantum Systems*. Oxford University Press (2002).

[87] H. Eleuch and I. Rotter, Open quantum systems with loss and gain, *International Journal of Theoretical Physics* **54**, 3877–3888 (2015).

[88] D. Valente, F. Brito and T. Werlang, Quantum dissipative adaptation, *Communications Physics* **4**, 11 (2021).

[89] C.-Y. Ju, A. Miranowicz, G.-Y. Chen and F. Nori, Non-Hermitian Hamiltonians and no-go theorems in quantum information, *Physical Review A* **100**, 062118 (2019).

[90] H. Isobe and N. Nagaosa, Anomalous Hall effect from a non-Hermitian viewpoint, *Physical Review B* **107**, L201116 (2023).

[91] H. Zhou, C. Peng, Y. Yoon, C. W. Hsu, K. A. Nelson, L. Fu, J. D. Joannopoulos, M. Soljačić and B. Zhen, Observation of bulk Fermi arc and polarization half charge from paired exceptional points, *Science* **359**, 1009–1012 (2018).

[92] F. Ma, E. Romero, M. R. Jones, V. I. Novoderezhkin and R. Van Grondelle, Vibronic coherence in the charge separation process of the *Rhodobacter sphaeroides* reaction center, *Journal of Physical Chemistry Letters* **9**, 1827–1832 (2018).

[93] L. Wang, M. A. Allodi and G. S. Engel, Quantum coherences reveal excited-state dynamics in biophysical systems, *Nature Reviews Chemistry* **3**, 477–490 (2019).

[94] T. R. Nelson, A. J. White, J. A. Bjorgaard, A. E. Sifain, Y. Zhang, B. Nebgen, S. Fernandez-Alberti, D. Mozyrsky, A. E. Roitberg and S. Tretiak, Non-adiabatic excited-state molecular dynamics: Theory and applications for modeling photophysics in extended molecular materials, *Chemical Reviews* **120**, 2215–2287 (2020).

[95] J. N. Tinsley, M. I. Molodtsov, R. Prevedel, D. Wartmann, J. EspiguléPons, M. Lauwers and A. Vaziri, Direct detection of a single photon by humans, *Nature Communications* **7**, 12172 (2016).

[96] I. Rivalta, A. Nenov and M. Garavelli, Modelling retinal chromophores photoisomerization: from minimal models *in vacuo* to ultimate bidimensional spectroscopy in rhodopsins, *Physical Chemistry Chemical Physics* **16**, 16865–16879 (2014).

[97] M. Ruckenbauer, M. Barbatti, T. Müller and H. Lischka, Nonadiabatic photodynamics of a retinal model in polar and nonpolar environment, *The Journal of Physical Chemistry A* **117**, 2790–2799 (2013).

[98] H. Chung, T. Dai, S. K. Sharma, Y.-Y. Huang, J. D. Carroll and M. R. Hamblin, The nuts and bolts of low-level laser (light) therapy, *Annals of Biomedical Engineering* **40**, 516–533 (2012).

[99] F. Figon, L. F. Deravi and J. Casas, Barriers and promises of the developing pigment organelle field, *Integrative and Comparative Biology* **61**, 1481–1489 (2021).

[100] F. Figon and J. Casas, The integrative biology of pigment organelles, a quantum chemical approach, *Integrative and Comparative Biology* **61**, 1490–1501 (2021).

[101] A. Tamulis and M. Grigalavicius, Quantum entanglement in photoactive prebiotic systems, *Systems and Synthetic Biology* **8**, 117–140 (2014).

[102] E. Tapavicza, A. M. Meyer and F. Furche, Unravelling the details of vitamin D photosynthesis by non-adiabatic molecular dynamics simulations, *Physical Chemistry Chemical Physics* **13**, 20986–20998 (2011).

[103] A. Kohen and J. P. Klinman, Hydrogen tunneling in biology, *Chemistry & Biology* **6**, R191–R198 (1999).

[104] J. P. Klinman and A. Kohen, Hydrogen tunneling links protein dynamics to enzyme catalysis, *Annual Review of Biochemistry* **82**, 471–496 (2013).

[105] T. E. Markland and M. Ceriotti, Nuclear quantum effects enter the mainstream, *Nature Reviews Chemistry* **2**, 0109 (2018).

[106] R. Ono, K. Osawa, Y. Takahashi, Y. Noguchi, N. Kitada, R. Saito-Moriya, T. Hirano, S. A. Maki, K. Shibata, H. Akiyama *et al.*, Quantum yield of near-infrared bioluminescence with firefly luciferin analog: Akalumine, *Journal of Photochemistry and Photobiology A: Chemistry* **434**, 114270 (2023).

[107] R. Sarfati, K. Joshi, O. Martin, J. C. Hayes, S. Iyer-Biswas and O. Peleg, Emergent periodicity in the collective synchronous flashing of fireflies, *eLife* **12**, e78908 (2023).

[108] M. McCrea, B. Ermentrout and J. E. Rubin, A model for the collective synchronization of flashing in photinus carolinus, *Journal of the Royal Society Interface* **19**, 20220439 (2022).

[109] B. A. Trimmer, J. R. Aprille, D. M. Dudzinski, C. J. Lagace, S. M. Lewis, T. Michel, S. Qazi and R. M. Zayas, Nitric oxide and the control of firefly flashing, *Science* **292**, 2486–2488 (2001).

[110] G. N. Lewis, The conservation of photons, *Nature* **118**, 874–875 (1926).

[111] M. O. Scully, The concept of the photon, *Physics Today* **25**, 38 (1972).

[112] K. J. Laidler, The meaning of "adiabatic", *Canadian Journal of Chemistry* **72**, 936–938 (1994).

[113] C. Orzel, How max planck solved the blackbody problem, (2022). https://www.wondriumdaily.com/how-max-planck-solved-the-blackbody-problem/.

[114] W. Heisenberg, The actual content of quantum theoretical kinematics and mechanics, *Zeitschrift für Physik* **43**, 172–98 (1927).

[115] M. Ozawa, Heisenberg's original derivation of the uncertainty principle and its universally valid reformulations, *Current Science* 2006–2016 (2015).

[116] P. W. Milonni, Why spontaneous emission? *American Journal of Physics* **52**, 340–343 (1984).

[117] S. Stierwalt, Einstein's legacy: The photoelectric effect, *Scientific American* **3** August 18 (2015).

[118] P. A. M. Dirac, The quantum theory of the emission and absorption of radiation, *Proceedings of the Royal Society of London. Series A, Containing Papers of a Mathematical and Physical Character* **114**, 243–265 (1927).

[119] R. P. Feynman, *QED: The Strange Theory of Light and Matter.* Princeton University Press (1985).

[120] E. Merzbacher, The early history of quantum tunneling, *Physics Today* **55**, 44–49 (2002).

[121] R. Eisberg and R. Resnick, *Quantum Physics of Atoms, Molecules, Solids, Nuclei, and Particles*, no. v. 1 in Quantum Physics of Atoms, Molecules, Solids, Nuclei, and Particles. Wiley (1985).

[122] P. R. Bueno, Quantum rate theory and electron-transfer dynamics: A theoretical and experimental approach for quantum electrochemistry, *Electrochimica Acta* **466**, 142950 (2023).

[123] V. R. Kaila, Long-range proton-coupled electron transfer in biological energy conversion: towards mechanistic understanding of respiratory complex I, *Journal of The Royal Society Interface* **15**, 20170916 (2018).

[124] R. Dick, Quantum jumps, superpositions, and the continuous evolution of quantum states, *Studies in History and Philosophy of Science Part B: Studies in History and Philosophy of Modern Physics* **57**, 115–125 (2017).

[125] K. Le Gal, E. E. Schmidt and V. I. Sayin, Cellular redox homeostasis, *Antioxidants* **10**, 1377 (2021).

[126] R. Ochoa-Hueso, M. Delgado-Baquerizo, A. C. Risch, M. Schrama, E. Morriën, S. H. Barmentlo, S. Geisen, S. E. Hannula, M. C. Resch, B. L. Snoek *et al.*, Ecosystem coupling: A unifying framework to understand the functioning and recovery of ecosystems, *One Earth* **4**, 951–966 (2021).

[127] L. Orlóci, *Quantum Ecology: Energy Structure and its Analysis.* SCADA Publishing (2015).

[128] J. Halpern and L. Orgel, Oxidation-reduction reactions in ionizing solvents. introductory paper electron-transfer reactions, *Discussions of the Faraday Society* **29**, 7–20 (1960).

[129] D. De Vault and B. Chance, Studies of photosynthesis using a pulsed laser: I. temperature dependence of cytochrome oxidation rate in chromatium. evidence for tunneling, *Biophysical Journal* **6**, 825–847 (1966).

[130] F. Trixler, Quantum tunnelling to the origin and evolution of life, *Current Organic Chemistry* **17**, 1758–1770 (2013).
[131] S. Mostajabi Sarhangi and D. V. Matyushov, Electron tunneling in biology: When does it matter? *ACS Omega* **8**, 27355–27365 (2023).
[132] S. Luryi, Coherent versus incoherent resonant tunneling and implications for fast devices, *Superlattices and Microstructures* **5**, 375–382 (1989).
[133] A. Klimenko, On the effect of decoherence on quantum tunnelling, *SN Applied Sciences* **3**, 710 (2021).
[134] P. J. Dahl, S. M. Yi, Y. Gu, A. Acharya, C. Shipps, J. Neu, J. P. O'Brien, U. N. Morzan, S. Chaudhuri, M. J. Guberman-Pfeffer *et al.*, A 300-fold conductivity increase in microbial cytochrome nanowires due to temperature-induced restructuring of hydrogen bonding networks, *Science Advances* **8**, eabm7193 (2022).
[135] C. Chen, S. Liu, X.-q. Shi, H. Chaté and Y. Wu, Weak synchronization and large-scale collective oscillation in dense bacterial suspensions, *Nature* **542**, 210–214 (2017).
[136] A. de la Lande, N. S. Babcock, J. Řezáč, B. Lévy, B. C. Sanders and D. R. Salahub, Quantum effects in biological electron transfer, *Physical Chemistry Chemical Physics* **14**, 5902–5918 (2012).
[137] D. Dasenbrook, J. Bowles, J. B. Brask, P. P. Hofer, C. Flindt and N. Brunner, Single-electron entanglement and nonlocality, *New Journal of Physics* **18**, 043036 (2016).
[138] E. Schrödinger, Discussion of probability relations between separated systems, in *Mathematical Proceedings of the Cambridge Philosophical Society*, Vol. 31. Cambridge University Press, pp. 555–563 (1935).
[139] A. De la Lande, N. S. Babcock, J. Řezáč, B. C. Sanders and D. R. Salahub, Surface residues dynamically organize water bridges to enhance electron transfer between proteins, *Proceedings of the National Academy of Sciences USA* **107**, 11799–11804 (2010).
[140] S. S. Skourtis, D. N. Beratan and D. H. Waldeck, Coherence in electron transfer pathways, *Procedia Chemistry* **3**, 99–104 (2011).
[141] D. Keilin, On cytochrome, a respiratory pigment, common to animals, yeast, and higher plants, *Proceedings of the Royal Society of London. Series B, Containing Papers of a Biological Character* **98**, 312–339 (1925).
[142] J. Wei, H. Liu, K. Niki, E. Margoliash and D. Waldeck, Probing electron tunneling pathways: Electrochemical study of rat heart cytochrome c and its mutant on pyridine-terminated sams, *The Journal of Physical Chemistry B* **108**, 16912–16917 (2004).

[143] A. Stuchebrukhov, Long-distance electron tunneling in proteins: A new challenge for time-resolved spectroscopy, *Laser Physics* **20**, 125–138 (2010).

[144] Y. Lu, M. Kundu and D. Zhong, Effects of nonequilibrium fluctuations on ultrafast short-range electron transfer dynamics, *Nature Communications* **11**, 2822 (2020).

[145] D. Cahen, I. Pecht and M. Sheves, What can we learn from protein-based electron transport junctions? *The Journal of Physical Chemistry Letters* **12**, 11598–11603 (2021).

[146] J. C. Genereux, S. M. Wuerth and J. K. Barton, Single-step charge transport through DNA over long distances, *Journal of the American Chemical Society* **133**, 3863–3868 (2011).

[147] J. P. Menzel, H. J. de Groot and F. Buda, Photoinduced electron transfer in donor-acceptor complexes: Isotope effect and dynamic symmetry breaking, *The Journal of Physical Chemistry Letters* **10**, 6504–6511 (2019).

[148] S. S. Skourtis, D. H. Waldeck and D. N. Beratan, Inelastic electron tunneling erases coupling-pathway interferences, *The Journal of Physical Chemistry B* **108**, 15511–15518 (2004).

[149] C. Romero-Muñiz, M. Ortega, J. Vilhena, I. Diez-Perez, R. Pérez, J. C. Cuevas and L. A. Zotti, Can electron transport through a blue-copper azurin be coherent? An ab initio study, *The Journal of Physical Chemistry C* **125**, 1693–1702 (2021).

[150] H. Yue, D. Khoshtariya, D. Waldeck, J. Grochol, P. Hildebrandt and D. Murgida, On the electron transfer mechanism between cytochrome c and metal electrodes. Evidence for dynamic control at short distances, *The Journal of Physical Chemistry B* **110**, 19906–19913 (2006).

[151] J. A. Fereiro, X. Yu, I. Pecht, M. Sheves, J. C. Cuevas and D. Cahen, Tunneling explains efficient electron transport via protein junctions, *Proceedings of the National Academy of Sciences USA* **115**, E4577–E4583 (2018).

[152] R. A. Marcus and N. Sutin, Electron transfers in chemistry and biology, *Biochimica et Biophysica Acta (BBA)-Reviews on Bioenergetics* **811**, 265–322 (1985).

[153] R. A. Marcus, Electron transfer reactions in chemistry. theory and experiment, *Reviews of Modern Physics* **65**, 599–610 (1993).

[154] J. R. Winkler and H. B. Gray, Long-range electron tunneling, *Journal of the American Chemical Society* **136**, 2930–2939 (2014).

[155] S. J. Tower, W. J. Hetcher, T. E. Myers, N. J. Kuehl and M. T. Taylor, Selective modification of tryptophan residues in peptides and proteins using a biomimetic electron transfer process, *Journal of the American Chemical Society* **142**, 9112–9118 (2020).

[156] J.-M. Zhang and Y. Liu, Fermi's golden rule: Its derivation and breakdown by an ideal model, *European Journal of Physics* **37**, 065406 (2016).

[157] D. Wang, O. C. Fiebig, D. Harris, H. Toporik, Y. Ji, C. Chuang, M. Nairat, A. L. Tong, J. I. Ogren, S. M. Hart *et al.*, Elucidating interprotein energy transfer dynamics within the antenna network from purple bacteria, *Proceedings of the National Academy of Sciences* **120**, e2220477120 (2023).

[158] A. Klinger, D. Lindorfer, F. Müh and T. Renger, Living on the edge: Light-harvesting efficiency and photoprotection in the core of green sulfur bacteria, *Physical Chemistry Chemical Physics* (2023).

[159] A. Thilagam, Natural light harvesting systems: Unraveling the quantum puzzles, *Journal of Mathematical Chemistry* **53**, 466–494 (2015).

[160] H. Makita and G. Hastings, Inverted-region electron transfer as a mechanism for enhancing photosynthetic solar energy conversion efficiency, *Proceedings of the National Academy of Sciences USA* **114**, 9267–9272 (2017).

[161] R. Marcus, Nonadiabatic processes involving quantum-like and classical-like coordinates with applications to nonadiabatic electron transfers, *The Journal of Chemical Physics* **81**, 4494–4500 (1984).

[162] T. D. Visser, Whose golden rule is it anyway? *American Journal of Physics* **77**, 487–487 (2009).

[163] S. R. Camilo, F. Curtolo, V. V. Galassi and G. M. Arantes, Tunneling and nonadiabatic effects on a proton-coupled electron transfer model for the Q_0, site in cytochrome bc_1, *Journal of Chemical Information and Modeling* **61**, 1840–1849 (2021).

[164] S. Pinna, C. Kunz, A. Halpern, S. A. Harrison, S. F. Jordan, J. Ward, F. Werner and N. Lane, A prebiotic basis for ATP as the universal energy currency, *PLoS Biology* **20**, e3001437 (2022).

[165] M. Karpefors, P. Ädelroth, Y. Zhen, S. Ferguson-Miller and P. Brzezinski, Proton uptake controls electron transfer in cytochrome c oxidase, *Proceedings of the National Academy of Sciences USA* **95**, 1360613611 (1998).

[166] B. C. Westlake, M. K. Brennaman, J. J. Concepcion, J. J. Paul, S. E. Bettis, S. D. Hampton, S. A. Miller, N. V. Lebedeva, M. D. Forbes, A. M. Moran *et al.*, Concerted electron-proton transfer in the optical excitation of hydrogen-bonded dyes, *Proceedings of the National Academy of Sciences* **108**, 8554–8558 (2011).

[167] P. Goyal and S. Hammes-Schiffer, Tuning the ultrafast dynamics of photoinduced proton-coupled electron transfer in energy conversion processes, *ACS Energy Letters* **2**, 512–519 (2017).

[168] Z. Zhou, Z. Chen, X.-W. Kang, Y. Zhou, B. Wang, S. Tang, S. Zou, Y. Zhang, Q. Hu, F. Bai *et al.*, The nature of proton-coupled electron transfer in a blue light using flavin domain, *Proceedings of the National Academy of Sciences* **119**, e2203996119 (2022).

[169] R. I. Cukier and D. G. Nocera, Proton-coupled electron transfer, *Annual Review of Physical Chemistry* **49**, 337–369 (1998).

[170] I. Belevich, M. I. Verkhovsky and M. Wikström, Proton-coupled electron transfer drives the proton pump of cytochrome c oxidase, *Nature* **440**, 829–832 (2006).

[171] A. B. Bandara, J. C. Drake and D. A. Brown, Complex II subunit SDHD is critical for cell growth and metabolism, which can be partially restored with a synthetic ubiquinone analog, *BMC Molecular and Cell Biology* **22**, 1–11 (2021).

[172] A. I. Jonckheere, J. A. Smeitink and R. J. Rodenburg, Mitochondrial ATP synthase: architecture, function and pathology, *Journal of Inherited Metabolic Disease* **35**, 211–225 (2012).

[173] C. C. Moser, T. A. Farid, S. E. Chobot and P. L. Dutton, Electron tunneling chains of mitochondria, *Biochimica et Biophysica Acta (BBA)-Bioenergetics* **1757**, 1096–1109 (2006).

[174] W. A. Cramer, S. S. Hasan and E. Yamashita, The q cycle of cytochrome bc complexes: A structure perspective, *Biochimica et Biophysica Acta (BBA)-Bioenergetics* **1807**, 788–802 (2011).

[175] J. A. Letts, K. Fiedorczuk and L. A. Sazanov, The architecture of respiratory supercomplexes, *Nature* **537**, 644–648 (2016).

[176] X. C. Zhang and B. Li, Towards understanding the mechanisms of proton pumps in complex-I of the respiratory chain, *Biophysics Reports* **5**, 219–234 (2019).

[177] D. Kampjut and L. A. Sazanov, The coupling mechanism of mammalian respiratory complex I, *Science* **370**, eabc4209 (2020).

[178] T. Hayashi and A. A. Stuchebrukhov, Electron tunneling in respiratory complex I, *Proceedings of the National Academy of Sciences USA* **107**, 19157–19162 (2010).

[179] D. Kampjut and L. A. Sazanov, Structure of respiratory complex I-an emerging blueprint for the mechanism, *Current Opinion in Structural Biology* **74**, 102350 (2022).

[180] J. Gu, T. Liu, R. Guo, L. Zhang and M. Yang, The coupling mechanism of mammalian mitochondrial complex I, *Nature Structural & Molecular Biology* **29**, 172–182 (2022).

[181] A. R. Crofts, The modified Q-cycle: a look back at its development and forward to a functional model, *Biochimica et Biophysica Acta (BBA)-Bioenergetics* **1862**, 148417 (2021).

[182] X. Hua, H.-L. Xia, Y.-L. Ying and Y.-T. Long, Proton-coupled electron transfer of coenzyme Q in unbuffered solution by pore confined In Situ liquid ToF-SIMS, *Journal of The Electrochemical Society* **169**, 026525 (2022).

[183] V. Kravchuk, O. Petrova, D. Kampjut, A. Wojciechowska-Bason, Z. Breese and L. Sazanov, A universal coupling mechanism of respiratory complex I, *Nature* **609**, 808–814 (2022).

[184] M. Wikström, A. Djurabekova and V. Sharma, On the role of ubiquinone in the proton translocation mechanism of respiratory complex I, *FEBS Letters* (2022).

[185] J. Gutiérrez-Fernández, K. Kaszuba, G. S. Minhas, R. Baradaran, M. Tambalo, D. T. Gallagher and L. A. Sazanov, Key role of quinone in the mechanism of respiratory complex I, *Nature Communications* **11**, 1–17 (2020).

[186] D. Xia, L. Esser, W.-K. Tang, F. Zhou, Y. Zhou, L. Yu and C.-A. Yu, Structural analysis of cytochrome bc1 complexes: implications to the mechanism of function, *Biochimica et Biophysica Acta (BBA)-Bioenergetics* **1827**, 1278–1294 (2013).

[187] B. L. Trumpower, Cytochrome bc1 complexes of microorganisms, *Microbiological Reviews* **54**, 101–129 (1990).

[188] F. S. Mathews, The structure, function and evolution of cytochromes, *Progress in Biophysics and Molecular Biology* **45**, 1–56 (1985).

[189] J. A. Letts, K. Fiedorczuk, G. Degliesposti, M. Skehel and L. A. Sazanov, Structures of respiratory supercomplex I+III_2 reveal functional and conformational crosstalk, *Molecular Cell* **75**, 1131–1146 (2019).

[190] A. Signes and E. Fernandez-Vizarra, Assembly of mammalian oxidative phosphorylation complexes I–V and supercomplexes, *Essays in Biochemistry* **62**, 255–270 (2018).

[191] D. Milenkovic, J. N. Blaza, N.-G. Larsson and J. Hirst, The enigma of the respiratory chain supercomplex, *Cell Metabolism* **25**, 765–776 (2017).

[192] E. Maranzana, G. Barbero, A. I. Falasca, G. Lenaz and M. L. Genova, Mitochondrial respiratory supercomplex association limits production of reactive oxygen species from complex I, *Antioxidants & Redox Signaling* **19**, 1469–1480 (2013).

[193] I. Vercellino and L. A. Sazanov, The assembly, regulation and function of the mitochondrial respiratory chain, *Nature Reviews Molecular Cell Biology* **23**, 141–161 (2022).

[194] A. Panday, M. K. Sahoo, D. Osorio and S. Batra, Nadph oxidases: an overview from structure to innate immunity-associated pathologies, *Cellular & Molecular Immunology* **12**, 5–23 (2015).

[195] J. D. Hernández-Camacho, M. Bernier, G. López-Lluch and P. Navas, Coenzyme Q_{10} supplementation in aging and disease, *Frontiers in Physiology* **9**, 44 (2018).

[196] Y. Wang and S. Hekimi, The complexity of making ubiquinone, *Trends in Endocrinology & Metabolism* **30**, 929–943 (2019).

[197] M. J. Acosta, L. V. Fonseca, M. A. Desbats, C. Cerqua, R. Zordan, E. Trevisson and L. Salviati, Coenzyme Q biosynthesis in health and disease, *Biochimica et Biophysica Acta (BBA)-Bioenergetics* **1857**, 1079–1085 (2016).

[198] L. N. Laredj, F. Licitra and H. M. Puccio, The molecular genetics of coenzyme Q biosynthesis in health and disease, *Biochimie* **100**, 7887 (2014).

[199] N. J. Watmough and F. E. Frerman, The electron transfer flavoprotein: ubiquinone oxidoreductases, *Biochimica et Biophysica Acta (BBA)-Bioenergetics* **1797**, 1910–1916 (2010).

[200] Y. Wang and S. Hekimi, Mitochondrial respiration without ubiquinone biosynthesis, *Human Molecular Genetics* **22**, 4768–4783 (2013).

[201] S. Deshwal, M. Onishi, T. Tatsuta, T. Bartsch, E. Cors, K. Ried, K. Lemke, H. Nolte, P. Giavalisco and T. Langer, Mitochondria regulate intracellular coenzyme Q transport and ferroptotic resistance via STARD7, *Nature Cell Biology*, 1–12 (2023).

[202] R. Guo, S. Zong, M. Wu, J. Gu and M. Yang, Architecture of human mitochondrial respiratory megacomplex $I_2III_2IV_2$, *Cell* **170**, 1247–1257 (2017).

[203] M. Wikström, K. Krab and V. Sharma, Oxygen activation and energy conservation by cytochrome c oxidase, *Chemical Reviews* **118**, 2469–2490 (2018).

[204] L. Mourokh and M. Vittadello, Physical model of proton-pumping Q-cycle in respiratory and photosynthetic electron transport chains, *Chemical Physics* **530**, 110638 (2020).

[205] F. Tacchino, A. Succurro, O. Ebenhöh and D. Gerace, Optimal efficiency of the Q-cycle mechanism around physiological temperatures from an open quantum systems approach, *Scientific Reports* **9**, 16657 (2019).

[206] A. Taylor, D. J. Heyes and N. S. Scrutton, Catalysis by nature's photoenzymes, *Current Opinion in Structural Biology* **77**, 102491 (2022).

[207] A. Cellini, M. K. Shankar, A. Nimmrich, L. A. Hunt, L. Monrroy, J. Mutisya, A. Furrer, E. V. Beale, M. Carrillo, T. N. Malla *et al.*, Directed ultrafast conformational changes accompany electron

transfer in a photolyase as resolved by serial crystallography, *Nature Chemistry* 1–9 (2024).

[208] D. W. Nebert and D. W. Russell, Clinical importance of the cytochromes P450, *The Lancet* **360**, 1155–1162 (2002).

[209] I. G. Denisov, T. M. Makris, S. G. Sligar and I. Schlichting, Structure and chemistry of cytochrome P450, *Chemical Reviews* **105**, 2253–2278 (2005).

[210] N. Li, S. Yan, P. Wu, J. Li and B. Wang, Local electric fields drives the proton-coupled electron transfer within cytochrome p450 reductase, *ACS Catalysis* **14**, 7893–7900 (2024).

[211] J. J. Goings and S. Hammes-Schiffer, Nonequilibrium dynamics of proton-coupled electron transfer in proton wires: Concerted but asynchronous mechanisms, *ACS Central Science* **6**, 1594–1601 (2020).

[212] S. Hammes-Schiffer, Exploring proton-coupled electron transfer at multiple scales, *Nature Computational Science* **3**, 291–300 (2023).

[213] S. T. Stripp, B. R. Duffus, V. Fourmond, C. Léger, S. Leimkühler, S. Hirota, Y. Hu, A. Jasniewski, H. Ogata and M. W. Ribbe, Second and outer coordination sphere effects in nitrogenase, hydrogenase, formate dehydrogenase, and co dehydrogenase, *Chemical Reviews* **122**, 11900–11973 (2022).

[214] H. Yue and D. H. Waldeck, Understanding interfacial electron transfer to monolayer protein assemblies, *Current Opinion in Solid State and Materials Science* **9**, 28–36 (2005).

[215] K. Michaeli, N. Kantor-Uriel, R. Naaman and D. H. Waldeck, The electron's spin and molecular chirality-how are they related and how do they affect life processes? *Chemical Society Reviews* **45**, 6478–6487 (2016).

[216] G. Likhtenshtein, Basic conceptions: Spin exchange and electron transfer, in *Electron Spin Interactions in Chemistry and Biology.* Springer, 1–26 (2016).

[217] J. Wei, B. P. Bloom, W. A. Dunlap-Shohl, C. B. Clever, J. E. Rivas and D. H. Waldeck, Examining the effects of homochirality for electron transfer in protein assemblies, *The Journal of Physical Chemistry B* (2023).

[218] L. M. Rohas, J. St-Pierre, M. Uldry, S. Jäger, C. Handschin and B. M. Spiegelman, A fundamental system of cellular energy homeostasis regulated by PGC-1α, *Proceedings of the National Academy of Sciences USA* **104**, 7933–7938 (2007).

[219] P. Mitchell, David keilin's respiratory chain concept and its chemiosmotic consequences, *Chemistry,* 1971–1980 **5**, 295 (1992).

[220] R. Guo, J. Gu, S. Zong, M. Wu and M. Yang, Structure and mechanism of mitochondrial electron transport chain, *Biomedical Journal* **41**, 9–20 (2018).

[221] J. Gu, M. Wu, R. Guo, K. Yan, J. Lei, N. Gao and M. Yang, The architecture of the mammalian respirasome, *Nature* **537**, 639–643 (2016).

[222] M. Wu, J. Gu, R. Guo, Y. Huang and M. Yang, Structure of mammalian respiratory supercomplex $I_1III_2IV_1$, *Cell* **167**, 1598–1609 (2016).

[223] C. Biz, M. Fianchini and J. Gracia, Strongly correlated electrons in catalysis: Focus on quantum exchange, *ACS Catalysis* **11**, 14249–14261 (2021).

[224] R. Naaman, Y. Paltiel and D. H. Waldeck, Chiral induced spin selectivity gives a new twist on spin-control in chemistry, *Accounts of Chemical Research* **53**, 2659–2667 (2020a).

[225] R. Naaman, Y. Paltiel and D. H. Waldeck, Chiral molecules and the spin selectivity effect, *The Journal of Physical Chemistry Letters* **11**, 3660–3666 (2020b).

[226] Y. Matsuura, Coherent spin transport in a multi-heme protein molecule, *Chemical Physics* **558**, 111510 (2022).

[227] Y. Matsuura, Coherent spin transport in a natural metalloprotein molecule, *Journal of Applied Physics* **130**, 184301 (2021).

[228] T. P. Fay and D. T. Limmer, Origin of chirality induced spin selectivity in photoinduced electron transfer, *Nano Letters* **21**, 6696–6702 (2021).

[229] D. Waldeck, R. Naaman and Y. Paltiel, The spin selectivity effect in chiral materials, *APL Materials* **9**, 040902 (2021).

[230] R. J. Usselman, C. Chavarriaga, P. R. Castello, M. Procopio, T. Ritz, E. A. Dratz, D. J. Singel and C. F. Martino, The quantum biology of reactive oxygen species partitioning impacts cellular bioenergetics, *Scientific Reports* **6**, 1–6 (2016).

[231] T. P. Fay, Chirality-induced spin coherence in electron transfer reactions, *The Journal of Physical Chemistry Letters* **12**, 1407–1412 (2021).

[232] D. Jaggard, A. Mickelson and C. Papas, On electromagnetic waves in chiral media, *Applied Physics* **18**, 211–216 (1979).

[233] J. Luo and P. Hore, Chiral-induced spin selectivity in the formation and recombination of radical pairs: Cryptochrome magnetoreception and EPR detection, *New Journal of Physics* **23**, 043032 (2021).

[234] H. J. de Groot and F. Buda, Tuning the proton-coupled electron-transfer rate by ligand modification in catalyst-dye supramolecular

complexes for photocatalytic water splitting, *ChemSusChem* **14**, 479–486 (2021).

[235] C. D. Bostick, S. Mukhopadhyay, I. Pecht, M. Sheves, D. Cahen and D. Lederman, Protein bioelectronics: A review of what we do and do not know, *Reports on Progress in Physics* **81**, 026601 (2018).

[236] E. I. Glover, J. Martin, A. Maher, R. E. Thornhill, G. R. Moran and M. A. Tarnopolsky, A randomized trial of coenzyme Q10 in mitochondrial disorders, *Muscle & Nerve* **42**, 739–748 (2010).

[237] C. J. Pastor-Maldonado, J. M. Suárez-Rivero, S. Povea-Cabello, M. Álvarez-Córdoba, I. Villalón-García, M. Munuera-Cabeza, A. Suárez-Carrillo, M. Talaverón-Rey and J. A. Sánchez-Alcázar, Coenzyme Q_{10}: Novel formulations and medical trends, *International Journal of Molecular Sciences* **21**, 8432 (2020).

[238] S. Zhao, W. Wu, J. Liao, X. Zhang, M. Shen, X. Li, Q. Lin and C. Cao, Molecular mechanisms underlying the renal protective effects of coenzyme Q10 in acute kidney injury, *Cellular & Molecular Biology Letters* **27**, 1–19 (2022).

[239] A. Peres, *Quantum Theory: Concepts and Methods*, Fundamental Theories of Physics. Springer Netherlands (2006).

[240] A. Szabo and N. Ostlund, *Modern Quantum Chemistry: Introduction to Advanced Electronic Structure Theory*, Dover Books on Chemistry. Dover Publications (2012).

[241] N. Mardirossian and M. Head-Gordon, Thirty years of density functional theory in computational chemistry: An overview and extensive assessment of 200 density functionals, *Molecular Physics* **115**, 23152372 (2017).

[242] C. Zhou, M. R. Hermes, D. Wu, J. J. Bao, R. Pandharkar, D. S. King, D. Zhang, T. R. Scott, A. O. Lykhin, L. Gagliardi *et al.*, Electronic structure of strongly correlated systems: Recent developments in multiconfiguration pair-density functional theory and multiconfiguration nonclassicalenergy functional theory, *Chemical Science* **13**, 7685–7706 (2022).

[243] Y. Kim, F. Bertagna, E. M. D'Souza, D. J. Heyes, L. O. Johannissen, E. T. Nery, A. Pantelias, A. Sanchez-Pedreño Jimenez, L. Slocombe, M. G. Spencer *et al.*, Quantum biology: An update and perspective, *Quantum Reports* **3**, 80–126 (2021).

[244] G. R. Fleming, G. D. Scholes and Y.-C. Cheng, Quantum effects in biology, *Procedia Chemistry* **3**, 38–57 (2011).

[245] C. D. Aiello, Quantum physics proposes a new way to study biology - and the results could revolutionize our understanding of how life works, (2023), https://theconversation.com/quantum-physics-proposes-a-new-way-to-study-biology-and-the-results-could-revolutionize-our-understanding-of-how-life-works-204995.

[246] H. Wiseman and J. Eisert, Nontrivial quantum effects in biology: A skeptical physicist's view, *Quantum Aspects of Life* 381–402 (2008).

[247] A. Rinaldi, When life gets physical: Quantum effects in selected biological systems have been confirmed experimentally, but how widespread is their role remains unclear, *EMBO Reports* **13**, 24–27 (2012).

[248] N. Lambert, Y.-N. Chen, Y.-C. Cheng, C.-M. Li, G.-Y. Chen and F. Nori, Quantum biology, *Nature Physics* **9**, 10–18 (2013).

[249] Q. Cui, Perspective: Quantum mechanical methods in biochemistry and biophysics, *The Journal of Chemical Physics* **145**, 140901 (2016).

[250] R. E. Amaro and A. J. Mulholland, Multiscale methods in drug design bridge chemical and biological complexity in the search for cures, *Nature Reviews Chemistry* **2**, 0148 (2018).

[251] Q. Cui, T. Pal and L. Xie, Biomolecular QM/MM simulations: What are some of the "burning issues"? *The Journal of Physical Chemistry B* **125**, 689–702 (2021).

[252] K.-S. Csizi and M. Reiher, Universal QM/MM approaches for general nanoscale applications, *Wiley Interdisciplinary Reviews: Computational Molecular Science* **13**, e1656 (2023).

[253] J. Ho, H. Yu, Y. Shao, M. Taylor and J. Chen, How accurate are QM/MM models? *The Journal of Physical Chemistry A* (2024).

[254] R. P. Joshi, A. McNaughton, D. G. Thomas, C. S. Henry, S. R. Canon, L. A. McCue and N. Kumar, Quantum mechanical methods predict accurate thermodynamics of biochemical reactions, *ACS Omega* **6**, 9948–9959 (2021).

[255] T. A. Wesolowski and J. Weber, Applications of density functional theory to biological systems, in *Molecular Orbital Calculations for Biological Systems*. Oxford University Press (1998).

[256] A. I. Nesterov, G. P. Berman, F. A. de la Cruz and J. C. Beas Zepeda, Non-Hermitian quantum annealing and superradiance, in *Non-Hermitian Hamiltonians in Quantum Physics: Selected Contributions from the 15th International Conference on Non-Hermitian Hamiltonians in Quantum Physics, Palermo, Italy, 18–23 May 2015*. Springer, pp. 329–344 (2016).

[257] S. M. Young, H. Häffner and M. Sarovar, Quantum simulation of weak-field light-matter interactions, *Physical Review Research* **5**, 013027 (2023).

[258] H. Tang, X.-W. Shang, Z.-Y. Shi, T.-S. He, Z. Feng, T.-Y. Wang, R. Shi, H.-M. Wang, X. Tan, X.-Y. Xu *et al.*, Simulating photosynthetic energy transport on a photonic network, *npj Quantum Information* **10**, 29 (2024).

[259] D. J. Callaway, Triviality pursuit: Can elementary scalar particles exist? *Physics Reports* **167**, 241–320 (1988).
[260] S. Weinberg, *The Quantum Theory of Fields*. Cambridge University Press (1995).
[261] R. Haag, *Local Quantum Physics: Fields, Particles, Algebras*, Theoretical and Mathematical Physics. Springer Berlin Heidelberg (2012).
[262] L. D. Landau and I. Pomeranchuk, On point interactions in quantum electrodynamics, in *Dokl. Akad. Nauk SSSR*, Vol. 102, 489 (1955).
[263] S. Rivat, Open systems across scales, *Synthese* **205**, 11 (2024).
[264] T. Van Mourik, M. Bühl and M.-P. Gaigeot, Density functional theory across chemistry, physics and biology, *Philosophical Transactions of the Royal Society A: Mathematical, Physical and Engineering Sciences* **372**, 20120488 (2014).
[265] W. Koch and M. C. Holthausen, *A Chemist's Guide to Density Functional Theory*. John Wiley & Sons (2015).
[266] E. Engel and R. Dreizler, *Density Functional Theory: An Advanced Course*, Theoretical and Mathematical Physics. Springer Berlin Heidelberg (2011).
[267] E. Runge and E. K. Gross, Density-functional theory for time-dependent systems, *Physical Review Letters* **52**, 997 (1984).
[268] W. Kohn and L. J. Sham, Self-consistent equations including exchange and correlation effects, *Physical Review* **140**, 4A, A1133 (1965).
[269] J. J. Hopfield, Whatever happened to solid state physics? *Annual Review of Condensed Matter Physics* **5**, 1–13 (2014).
[270] F. Lu, M. Maggioni and S. Tang, Learning interaction kernels in heterogeneous systems of agents from multiple trajectories, *Journal of Machine Learning Research* **22**, 1–67 (2021).
[271] J. Wu, S.-M. Pun, X. Zheng and G. Chen, Construct exchange-correlation functional via machine learning, *The Journal of Chemical Physics* **159**, 9 (2023).
[272] Y. Shi and A. Wasserman, Inverse Kohn-Sham density functional theory: Progress and challenges, *The Journal of Physical Chemistry Letters* **12**, 5308–5318 (2021).
[273] S. Grimme, A. Hansen, J. G. Brandenburg and C. Bannwarth, Dispersion-corrected mean-field electronic structure methods, *Chemical Reviews* **116**, 5105–5154 (2016).
[274] N. Schuch and F. Verstraete, Computational complexity of interacting electrons and fundamental limitations of density functional theory, *Nature Physics* **5**, 732–735 (2009).

[275] J. P. Perdew and K. Schmidt, Jacob's ladder of density functional approximations for the exchange-correlation energy, in *AIP Conference Proceedings*, Vol. 577. *American Institute of Physics*, 1–20 (2001).

[276] J. Schmidt, C. L. Benavides-Riveros and M. A. Marques, Machine learning the physical nonlocal exchange-correlation functional of density functional theory, *The Journal of Physical Chemistry Letters* **10**, 6425–6431 (2019).

[277] C. Fiolhais, F. Nogueira and M. Marques, *A Primer in Density Functional Theory*, Lecture Notes in Physics. Springer Berlin Heidelberg (2008).

[278] S. F. Huelga and M. B. Plenio, Vibrations, quanta and biology, *Contemporary Physics* **54**, 181–207 (2013).

[279] L. Ballentine, *Quantum Mechanics: A Modern Development*, G — Reference, Information and Interdisciplinary Subjects Series. World Scientific (2015).

[280] C. C. Moser, J. R. Anderson and P. L. Dutton, Guidelines for tunneling in enzymes, *Biochimica et Biophysica Acta (BBA)—Bioenergetics* **1797**, 1573–1586 (2010).

[281] G. McLendon and R. Hake, Interprotein electron transfer, *Chemical Reviews* **92**, 481–490 (1992).

[282] G. D. Scholes, G. R. Fleming, L. X. Chen, A. Aspuru-Guzik, A. Buchleitner, D. F. Coker, G. S. Engel, R. Van Grondelle, A. Ishizaki, D. M. Jonas *et al.*, Using coherence to enhance function in chemical and biophysical systems, *Nature* **543**, 647–656 (2017).

[283] J.-K. Ha, I. S. Lee and S. K. Min, Surface hopping dynamics beyond nonadiabatic couplings for quantum coherence, *The Journal of Physical Chemistry Letters* **9**, 1097–1104 (2018).

[284] H.-P. Cheng, E. Deumens, J. K. Freericks, C. Li and B. A. Sanders, Application of quantum computing to biochemical systems: A look to the future, *Frontiers in Chemistry* **8**, 587143 (2020).

[285] A. Extance, Industry adopts quantum computing, qubit by qubit, (2019), https://www.chemistryworld.com/news/industry-adopts-quantum-computing-qubit-by-qubit-/3010591.article.

[286] D. Takahashi, IBM research explains how quantum computing works and why it matters, (2019), https://venturebeat.com/business/ibm-research-explains-how-quantum-computing-works-and-could-be-the-the-supercomputer-of-the-future/.

[287] L. Marchetti, R. Nifosì, P. L. Martelli, E. Da Pozzo, V. Cappello, F. Banterle, M. L. Trincavelli, C. Martini and M. D'Elia, Quantum computing algorithms: getting closer to critical problems in computational biology, *Briefings in Bioinformatics* **23**, bbac437 (2022).

[288] A. Baiardi, M. Christandl and M. Reiher, Quantum computing for molecular biology, *ChemBioChem* **24**, e202300120 (2023).
[289] B. Kanungo, P. M. Zimmerman and V. Gavini, Exact exchange-correlation potentials from ground-state electron densities, *Nature Communications* **10**, 4497 (2019).
[290] S. Tribedi, D.-K. Dang, B. Kanungo, V. Gavini and P. M. Zimmerman, Exchange correlation potentials from full configuration interaction in a slater orbital basis, *The Journal of Chemical Physics* **159**, 5 (2023).
[291] A. Marais, B. Adams, A. K. Ringsmuth, M. Ferretti, J. M. Gruber, R. Hendrikx, M. Schuld, S. L. Smith, I. Sinayskiy, T. P. Krüger *et al.*, The future of quantum biology, *Journal of the Royal Society Interface* **15**, 20180640 (2018).
[292] T. Mirkovic, E. E. Ostroumov, J. M. Anna, R. Van Grondelle, Govindjee and G. D. Scholes, Light absorption and energy transfer in the antenna complexes of photosynthetic organisms, *Chemical Reviews* **117**, 249–293 (2017).
[293] P. Greife, M. Schönborn, M. Capone, R. Assunção, D. Narzi, L. Guidoni and H. Dau, The electron-proton bottleneck of photosynthetic oxygen evolution, *Nature* **617**, 623–628 (2023).
[294] D. Shevela, J. F. Kern, G. Govindjee and J. Messinger, Solar energy conversion by photosystem II: Principles and structures, *Photosynthesis Research* 1–29 (2023).
[295] D. Leister, Enhancing the light reactions of photosynthesis: Strategies, controversies, and perspectives, *Molecular Plant* **16**, 4–22 (2023).
[296] S. Doria, M. Di Donato, R. Borrelli, M. F. Gelin, J. Caram, M. Pagliai, P. Foggi and A. Lapini, Vibronic coherences in light harvesting nanotubes: Unravelling the role of dark states, *Journal of Materials Chemistry C* **10**, 7216–7226 (2022).
[297] K. Wohl, On the mechanism of photosynthesis in purple bacteria and green plants, *The New Phytologist* **40**, 34–55 (1941).
[298] R. K. Clayton, The biophysical problems of photosynthesis, *Science* **149**, 1346–1354 (1965).
[299] E. Thyrhaug, K. Zidek, J. Dostál, D. Bína and D. Zigmantas, Exciton structure and energy transfer in the Fenna-Matthews-Olson complex, *The Journal of Physical Chemistry Letters* **7**, 1653–1660 (2016).
[300] G. D. Scholes, G. R. Fleming, A. Olaya-Castro and R. Van Grondelle, Lessons from nature about solar light harvesting, *Nature Chemistry* **3**, 763–774 (2011).

[301] G. D. Scholes, Quantum-coherent electronic energy transfer: Did nature think of it first? *The Journal of Physical Chemistry Letters* **1**, 2–8 (2010).

[302] H. Dong and G. R. Fleming, Inhomogeneous broadening induced longlived integrated two-color coherence photon echo signal, *The Journal of Physical Chemistry B* **118**, 8956–8961 (2014).

[303] J. S. Higgins, L. T. Lloyd, S. H. Sohail, M. A. Allodi, J. P. Otto, R. G. Saer, R. E. Wood, S. C. Massey, P.-C. Ting, R. E. Blankenship *et al.*, Photosynthesis tunes quantum-mechanical mixing of electronic and vibrational states to steer exciton energy transfer, *Proceedings of the National Academy of Sciences USA* **118**, e2018240118 (2021).

[304] J. Cao, R. J. Cogdell, D. F. Coker, H.-G. Duan, J. Hauer, U. Kleinekathöfer, T. L. Jansen, T. Mančal, R. D. Miller, J. P. Ogilvie *et al.*, Quantum biology revisited, *Science Advances* **6**, eaaz4888 (2020).

[305] S. Savikhin, D. R. Buck and W. S. Struve, Oscillating anisotropies in a bacteriochlorophyll protein: Evidence for quantum beating between exciton levels, *Chemical Physics* **223**, 303–312 (1997).

[306] G. S. Engel, T. R. Calhoun, E. L. Read, T.-K. Ahn, T. Mančal, Y.C. Cheng, R. E. Blankenship and G. R. Fleming, Evidence for wavelike energy transfer through quantum coherence in photosynthetic systems, *Nature* **446**, 782–786 (2007).

[307] G. Panitchayangkoon, D. Hayes, K. A. Fransted, J. R. Caram, E. Harel, J. Wen, R. E. Blankenship and G. S. Engel, Long-lived quantum coherence in photosynthetic complexes at physiological temperature, *Proceedings of the National Academy of Sciences USA* **107**, 12766–12770 (2010).

[308] H. Lee, Y.-C. Cheng and G. R. Fleming, Coherence dynamics in photosynthesis: protein protection of excitonic coherence, *Science* **316**, 1462–1465 (2007).

[309] W. W. Parson, Long live electronic coherence! *Science* **316**, 1438–1439 (2007).

[310] E. Collini, C. Y. Wong, K. E. Wilk, P. M. Curmi, P. Brumer and G. D. Scholes, Coherently wired light-harvesting in photosynthetic marine algae at ambient temperature, *Nature* **463**, 644–647 (2010).

[311] E. Harel and G. S. Engel, Quantum coherence spectroscopy reveals complex dynamics in bacterial light-harvesting complex 2 (LH2), *Proceedings of the National Academy of Sciences USA* **109**, 706–711 (2012).

[312] G. S. Schlau-Cohen, A. Ishizaki, T. R. Calhoun, N. S. Ginsberg, M. Ballottari, R. Bassi and G. R. Fleming, Elucidation of the

timescales and origins of quantum electronic coherence in LHCII, *Nature Chemistry* **4**, 389–395 (2012).

[313] N. Christensson, H. F. Kauffmann, T. Pullerits and T. Mancal, Origin of long-lived coherences in light-harvesting complexes, *The Journal of Physical Chemistry B* **116**, 7449–7454 (2012).

[314] E. Thyrhaug, R. Tempelaar, M. J. Alcocer, K. Žídek, D. Bína, J. Knoester, T. L. Jansen and D. Zigmantas, Identification and characterization of diverse coherences in the Fenna-Matthews-Olson complex, *Nature Chemistry* **10**, 780–786 (2018).

[315] V. Tiwari, W. K. Peters and D. M. Jonas, Electronic resonance with anticorrelated pigment vibrations drives photosynthetic energy transfer outside the adiabatic framework, *Proceedings of the National Academy of Sciences* **110**, 1203–1208 (2013).

[316] V. Butkus, D. Zigmantas, L. Valkunas and D. Abramavicius, Vibrational vs. electronic coherences in 2D spectrum of molecular systems, *Chemical Physics Letters* **545**, 40–43 (2012).

[317] H.-G. Duan, V. I. Prokhorenko, R. J. Cogdell, K. Ashraf, A. L. Stevens, M. Thorwart and R. D. Miller, Nature does not rely on long-lived electronic quantum coherence for photosynthetic energy transfer, *Proceedings of the National Academy of Sciences USA* **114**, 8493–8498 (2017).

[318] R. Tempelaar, T. L. Jansen and J. Knoester, Vibrational beatings conceal evidence of electronic coherence in the FMO light-harvesting complex, *The Journal of Physical Chemistry B* **118**, 12865–12872 (2014).

[319] Y. Fujihashi, G. R. Fleming and A. Ishizaki, Impact of environmentally induced fluctuations on quantum mechanically mixed electronic and vibrational pigment states in photosynthetic energy transfer and 2D electronic spectra, *The Journal of Chemical Physics* **142**, 21 (2015).

[320] A. Ishizaki and G. R. Fleming, Insights into photosynthetic energy transfer gained from free-energy structure: Coherent transport, incoherent hopping, and vibrational assistance revisited, *The Journal of Physical Chemistry B* **125**, 3286–3295 (2021).

[321] D. J. Tibben, G. O. Bonin, I. Cho, G. Lakhwani, J. Hutchison and D. E. Gómez, Molecular energy transfer under the strong light-matter interaction regime, *Chemical Reviews* **123**, 8044–8068 (2023).

[322] Z. Zhang and J. Wang, Origin of long-lived quantum coherence and excitation dynamics in pigment-protein complexes, *Scientific Reports* **6**, 37629 (2016).

[323] F. Caycedo-Soler, A. Mattioni, J. Lim, T. Renger, S. Huelga and M. B. Plenio, Exact simulation of pigment-protein complexes unveils vibronic renormalization of electronic parameters in ultrafast spectroscopy, *Nature Communications* **13**, 2912 (2022).

[324] E. Bukartė, A. Haufe, D. Paleček, C. Büchel and D. Zigmantas, Revealing vibronic coupling in chlorophyll c1 by polarization-controlled 2D electronic spectroscopy, *Chemical Physics* **530**, 110643 (2020).

[325] Z. Kis, J. Janszky, P. Adam, A. V. Vinogradov and T. Kobayashi, Entangled vibrational states in polyatomic molecules, *Physical Review A* **54**, 5110 (1996).

[326] A. G. Dijkstra, C. Wang, J. Cao and G. R. Fleming, Coherent exciton dynamics in the presence of underdamped vibrations, *The Journal of Physical Chemistry Letters* **6**, 627–632 (2015).

[327] M. Mohseni, P. Rebentrost, S. Lloyd and A. Aspuru-Guzik, Environment-assisted quantum walks in photosynthetic energy transfer, *The Journal of Chemical Physics* **129**, 17 (2008).

[328] E. Zerah Harush and Y. Dubi, Do photosynthetic complexes use quantum coherence to increase their efficiency? Probably not, *Science Advances* **7**, eabc4631 (2021).

[329] S. Takahashi and M. R. Badger, Photoprotection in plants: A new light on photosystem II damage, *Trends in Plant Science* **16**, 53–60 (2011).

[330] M. Son, A. Pinnola, S. C. Gordon, R. Bassi and G. S. Schlau-Cohen, Observation of dissipative chlorophyll-to-carotenoid energy transfer in light-harvesting complex II in membrane nanodiscs, *Nature Communications* **11**, 1295 (2020).

[331] D. I. Bennett, G. R. Fleming and K. Amarnath, Energy-dependent quenching adjusts the excitation diffusion length to regulate photosynthetic light harvesting, *Proceedings of the National Academy of Sciences USA* **115**, E9523–E9531 (2018).

[332] E. A. Arsenault, Y. Yoneda, M. Iwai, K. K. Niyogi and G. R. Fleming, Vibronic mixing enables ultrafast energy flow in light-harvesting complex II, *Nature Communications* **11**, 1460 (2020).

[333] N. Friedland, S. Negi, T. Vinogradova-Shah, G. Wu, L. Ma, S. Flynn, T. Kumssa, C.-H. Lee and R. Sayre, Fine-tuning the photosynthetic light harvesting apparatus for improved photosynthetic efficiency and biomass yield, *Scientific Reports* **9**, 13028 (2019).

[334] A. V. Ruban, Nonphotochemical chlorophyll fluorescence quenching: mechanism and effectiveness in protecting plants from photodamage, *Plant Physiology* **170**, 1903–1916 (2016).

[335] S. Doria, T. S. Sinclair, N. D. Klein, D. I. Bennett, C. Chuang, F. S. Freyria, C. P. Steiner, P. Foggi, K. A. Nelson, J. Cao *et al.*, Photochemical control of exciton superradiance in light-harvesting nanotubes, *ACS Nano* **12**, 4556–4564 (2018).

[336] J. Hustings, R. Bonné, R. Cornelissen, F. Morini, R. Valcke, K. Vandewal and J. V. Manca, Charge-transfer states in photosynthesis and organic solar cells, *Frontiers in Photonics* **3**, 1050189 (2022).

[337] F. Häse, L. M. Roch, P. Friederich and A. Aspuru-Guzik, Designing and understanding light-harvesting devices with machine learning, *Nature Communications* **11**, 4587 (2020).

[338] J. E. Runeson, J. E. Lawrence, J. R. Mannouch and J. O. Richardson, Explaining the efficiency of photosynthesis: Quantum uncertainty or classical vibrations? *The Journal of Physical Chemistry Letters* **13**, 3392–3399 (2022).

[339] G. Stock and M. Thoss, Classical description of nonadiabatic quantum dynamics, *Advances in Chemical Physics* **131**, 243–375 (2005).

[340] K. H. Cho and Y. M. Rhee, Computational elucidations on the role of vibrations in energy transfer processes of photosynthetic complexes, *Physical Chemistry Chemical Physics* **23**, 26623–26639 (2021).

[341] M. K. Lee and D. F. Coker, Modeling electronic-nuclear interactions for excitation energy transfer processes in light-harvesting complexes, *The Journal of Physical Chemistry Letters* **7**, 3171–3178 (2016).

[342] I. Layton and J. Oppenheim, The classical-quantum limit, *PRX Quantum* **5**, 020331 (2024).

[343] J. S. Higgins, W. R. Hollingsworth, L. T. Lloyd and G. S. Engel, Quantum coherence in chemical and photobiological systems, in *Emerging Trends in Chemical Applications of Lasers*. ACS Publications, pp. 411–436 (2021).

[344] G. A. Parada, Z. K. Goldsmith, S. Kolmar, B. Pettersson Rimgard, B. Q. Mercado, L. Hammarström, S. Hammes-Schiffer and J. M. Mayer, Concerted proton-electron transfer reactions in the marcus inverted region, *Science* **364**, 471–475 (2019).

[345] E. J. O'Reilly and A. Olaya-Castro, Non-classicality of the molecular vibrations assisting exciton energy transfer at room temperature, *Nature Communications* **5**, 3012 (2014).

[346] E. R. Heller and J. O. Richardson, Semiclassical instanton formulation of Marcus-Levich-Jortner theory, *The Journal of Chemical Physics* **152**, 244117 (2020).

[347] E. Zerah Harush and Y. Dubi, Signature of quantum coherence in the exciton energy pathways of the LH2 photosynthetic complex, *ACS Omega* **8**, 38871–38878 (2023).

[348] S. Tomasi and I. Kassal, Classification of coherent enhancements of light-harvesting processes, *Journal of Physical Chemistry Letters* **11**, 2348–2355 (2020).

[349] A. J. Young, The photoprotective role of carotenoids in higher plants, *Physiologia Plantarum* **83**, 702–708 (1991).

[350] C. Ramanan, J. M. Gruber, P. Malý, M. Negretti, V. Novoderezhkin, T. P. Krüger, T. Mančal, R. Croce and R. van Grondelle, The role of exciton delocalization in the major photosynthetic light-harvesting antenna of plants, *Biophysical Journal* **108**, 1047–1056 (2015).

[351] V. Balevičius Jr, K. F. Fox, W. P. Bricker, S. Jurinovich, I. G. Prandi, B. Mennucci and C. D. Duffy, Fine control of chlorophyll-carotenoid interactions defines the functionality of light-harvesting proteins in plants, *Scientific Reports* **7**, 13956 (2017).

[352] R. Bassi and L. Dall'Osto, Dissipation of light energy absorbed in excess: The molecular mechanisms, *Annual Review of Plant Biology* **72**, 47–76 (2021).

[353] D.-M. Yang, D. Cvetkovic, L. Chen and C. C. Ma, Therapeutic effects of in-vivo radiodynamic therapy (RDT) for lung cancer treatment: a combination of 15MV photons and 5-aminolevulinic acid (5-ALA), *Biomedical Physics & Engineering Express* **8**, 065031 (2022).

[354] A. R. Holzwarth, M. G. Müller, M. Reus, M. Nowaczyk, J. Sander and M. Rögner, Kinetics and mechanism of electron transfer in intact photosystem II and in the isolated reaction center: Pheophytin is the primary electron acceptor, *Proceedings of the National Academy of Sciences* **103**, 6895–6900 (2006).

[355] R. J. Cogdell, A. Gall and J. Köhler, The architecture and function of the light-harvesting apparatus of purple bacteria: From single molecules to in vivo membranes, *Quarterly Reviews of Biophysics* **39**, 227–324 (2006).

[356] X. Hu, A. Damjanović, T. Ritz and K. Schulten, Architecture and mechanism of the light-harvesting apparatus of purple bacteria, *Proceedings of the National Academy of Sciences USA* **95**, 5935–5941 (1998).

[357] P. Krasilnikov, Problems of the theory of electron transfer in biological systems, *Biophysics* **59**, 52–68 (2014).

[358] J. S. Higgins, M. A. Allodi, L. T. Lloyd, J. P. Otto, S. H. Sohail, R. G. Saer, R. E. Wood, S. C. Massey, P.-C. Ting, R. E. Blankenship

and G. S. Engel, Redox conditions correlated with vibronic coupling modulate quantum beats in photosynthetic pigment-protein complexes, *Proceedings of the National Academy of Sciences USA* **118**, e2112817118 (2021).

[359] M. Gross and S. Haroche, Superradiance: An essay on the theory of collective spontaneous emission, *Physics Reports* **93**, 301–396 (1982).

[360] R. Monshouwer, M. Abrahamsson, F. van Mourik and R. van Grondelle, Superradiance and exciton delocalization in bacterial photosynthetic light-harvesting systems, *The Journal of Physical Chemistry B* **101**, 7241–7248 (1997).

[361] T. Malina, R. Koehorst, D. Bína, J. Pšenčík and H. van Amerongen, Superradiance of bacteriochlorophyll c aggregates in chlorosomes of green photosynthetic bacteria, *Scientific Reports* **11**, 8354 (2021).

[362] G. L. Celardo, F. Borgonovi, M. Merkli, V. I. Tsifrinovich and G. P. Berman, Superradiance transition in photosynthetic light-harvesting complexes, *The Journal of Physical Chemistry C* **116**, 22105–22111 (2012).

[363] J. Q. Quach, K. E. McGhee, L. Ganzer, D. M. Rouse, B. W. Lovett, E. M. Gauger, J. Keeling, G. Cerullo, D. G. Lidzey and T. Virgili, Superabsorption in an organic microcavity: Toward a quantum battery, *Science Advances* **8**, eabk3160 (2022).

[364] S. Tomasi, S. Baghbanzadeh, S. Rahimi-Keshari and I. Kassal, Coherent and controllable enhancement of light-harvesting efficiency, *Physical Review A* **100**, 043411 (2019).

[365] R. Zhu, W. Li, Z. Zhen, J. Zou, G. Liao, J. Wang, Z. Wang, H. Chen, S. Qin and Y. Weng, Quantum phase synchronization via exciton-vibrational energy dissipation sustains long-lived coherence in photosynthetic antennas, *Nature Communications* **15**, 3171 (2024).

[366] R. Stones and A. Olaya-Castro, Vibronic coupling as a design principle to optimize photosynthetic energy transfer, *Chem* **1**, 822–824 (2016).

[367] Z. Li, C. Li, G. Xu, W. Chen, Z. Xiong, H. Jing, J. S. Ho and C.-W. Qiu, Synergetic positivity of loss and noise in nonlinear non-hermitian resonators, *Science Advances* **9**, eadi0562 (2023).

[368] P. M. Harrington, E. J. Mueller and K. W. Murch, Engineered dissipation for quantum information science, *Nature Reviews Physics* **4**, 660–671 (2022).

[369] L. M. Sieberer, M. Buchhold, J. Marino and S. Diehl, Universality in driven open quantum matter, *Reviews of Modern Physics* **97**, 025004 (2025).

[370] R. Mayne and A. Adamatzky, Slime mould foraging behaviour as optically coupled logical operations, *International Journal of General Systems* **44**, 305–313 (2015).

[371] H. Takala, A. Björling, O. Berntsson, H. Lehtivuori, S. Niebling, M. Hoernke, I. Kosheleva, R. Henning, A. Menzel, J. A. Thalainen *et al.*, Signal amplification and transduction in phytochrome photosensors, *Nature* **509**, 245–248 (2014).

[372] K. Moazed, *Quantum Biology of the Eye: Understanding the Essentials.* Springer International Publishing (2023).

[373] M. A. Crawford, C. L. Broadhurst, M. Guest, A. Nagar, Y. Wang, K. Ghebremeskel and W. F. Schmidt, A quantum theory for the irreplaceable role of docosahexaenoic acid in neural cell signalling throughout evolution, *Prostaglandins, Leukotrienes and Essential Fatty Acids* **88**, 5–13 (2013).

[374] Y. Imamoto and Y. Shichida, Cone visual pigments, *Biochimica et Biophysica Acta (BBA)-Bioenergetics* **1837**, 664–673 (2014).

[375] V. J. Kefalov, Rod and cone visual pigments and phototransduction through pharmacological, genetic, and physiological approaches, *Journal of Biological Chemistry* **287**, 1635–1641 (2012).

[376] I. Provencio, G. Jiang, W. J. De Grip, W. P. Hayes and M. D. Rollag, Melanopsin: An opsin in melanophores, brain, and eye, *Proceedings of the National Academy of Sciences USA* **95**, 340–345 (1998).

[377] J. Rosenthal, Circadian and mitochondrial effects of light, in *Nutrition and Integrative Medicine for Clinicians.* CRC Press, 339–372 (2023).

[378] A. Terakita, The opsins, *Genome Biology* **6**, 213 (2005).

[379] T. Matsuyama, T. Yamashita, H. Imai and Y. Shichida, Covalent bond between ligand and receptor required for efficient activation in rhodopsin, *Journal of Biological Chemistry* **285**, 8114–8121 (2010).

[380] B. Honig, A. D. Greenberg, U. Dinur and T. G. Ebrey, Visual-pigment spectra: Implications of the protonation of the retinal Schiff base, *Biochemistry* **15**, 4593–4599 (1976).

[381] J. H. Park, T. Morizumi, Y. Li, J. E. Hong, E. F. Pai, K. P. Hofmann, H.-W. Choe and O. P. Ernst, Opsin, a structural model for olfactory receptors? *Angewandte Chemie* **125**, 11227–11230 (2013).

[382] R. L. Brown and P. R. Robinson, Melanopsin-shedding light on the elusive circadian photopigment, *Chronobiology International* **21**, 189–204 (2004).

[383] O. I. Krzysztynska-Kuleta, M. M. Olchawa and T. J. Sarna, Melanopsin signaling pathway in HEK293 cell line with stable

expression of human melanopsin: Possible participation of phospholipase C beta 4 and diacylglycerol, *Photochemistry and Photobiology* **97**, 1136–1144 (2021).

[384] L. Sardon Puig, M. Valera-Alberni, C. Cantó and N. J. Pillon, Circadian rhythms and mitochondria: Connecting the dots, *Frontiers in Genetics* **9**, 452 (2018).

[385] A. Neufeld-Cohen, M. S. Robles, R. Aviram, G. Manella, Y. Adamovich, B. Ladeuix, D. Nir, L. Rousso-Noori, Y. Kuperman, M. Golik *et al.*, Circadian control of oscillations in mitochondrial rate-limiting enzymes and nutrient utilization by period proteins, *Proceedings of the National Academy of Sciences USA* **113**, E1673–E1682 (2016).

[386] N. F. Ruby, T. J. Brennan, X. Xie, V. Cao, P. Franken, H. C. Heller and B. F. O'Hara, Role of melanopsin in circadian responses to light, *Science* **298**, 2211–2213 (2002).

[387] C. L. Thompson, C. B. Rickman, S. J. Shaw, J. N. Ebright, U. Kelly, A. Sancar and D. W. Rickman, Expression of the blue-light receptor cryptochrome in the human retina, *Investigative Ophthalmology & Visual Science* **44**, 4515–4521 (2003).

[388] J. Vanderstraeten, P. Gailly and E. P. Malkemper, Light entrainment of retinal biorhythms: cryptochrome 2 as candidate photoreceptor in mammals, *Cellular and Molecular Life Sciences* **77**, 875–884 (2020).

[389] A. Sancar, Structure and function of DNA photolyase and cryptochrome blue-light photoreceptors, *Chemical Reviews* **103**, 2203–2238 (2003).

[390] N. Ozturk, Phylogenetic and functional classification of the photolyase/cryptochrome family, *Photochemistry and Photobiology* **93**, 104–111 (2017).

[391] B. D. Zoltowski, Resolving cryptic aspects of cryptochrome signaling, *Proceedings of the National Academy of Sciences* **112**, 8811–8812 (2015).

[392] G. C. G. Parico, I. Perez, J. L. Fribourgh, B. N. Hernandez, H.-W. Lee and C. L. Partch, The human CRY1 tail controls circadian timing by regulating its association with CLOCK: BMAL1, *Proceedings of the National Academy of Sciences USA* **117**, 27971–27979 (2020).

[393] Y. L. Vechtomova, T. Telegina and M. Kritsky, Evolution of proteins of the DNA photolyase/cryptochrome family, *Biochemistry (Moscow)* **85**, 131–153 (2020).

[394] M. H. Vitaterna, K. Shimomura and P. Jiang, Genetics of circadian rhythms, *Neurologic clinics* **37**, 487–504 (2019).

[395] Z. Yang, B. Liu, J. Su, J. Liao, C. Lin and Y. Oka, Cryptochromes orchestrate transcription regulation of diverse blue light responses in plants, *Photochemistry and Photobiology* **93**, 112–127 (2017).

[396] I. Chaves, R. Pokorny, M. Byrdin, N. Hoang, T. Ritz, K. Brettel, L.O. Essen, G. T. J. van der Horst, A. Batschauer and M. Ahmad, The cryptochromes: Blue light photoreceptors in plants and animals, *Annual Review of Plant Biology* **62**, 335–364 (2011).

[397] C. Darwin, *The Power of Movement in Plants.* John Murray, London (1880).

[398] A. W. Galston, Riboflavin, light, and the growth of plants, *Science* **111**, 619–624 (1950).

[399] M. Ahmad and A. R. Cashmore, Seeing blue: The discovery of cryptochrome, *Plant Molecular Biology* **30**, 851–861 (1996).

[400] K. L. Poff and W. L. Butler, Absorbance changes induced by blue light in *Phycomyces blakesleeanus* and *Dictyostelium discoideum*, *Nature* **248**, 799–801 (1974).

[401] J. Gressel, Blue light photoreception, *Photochemistry and Photobiology* **30**, 749–754 (1979).

[402] M. Koornneef, E. Rolff and C. J. P Spruit, Genetic control of light inhibited hypocotyl elongation in *Arabidopsis thaliana*, in *The Genetics of Some Plant Hormones and Photoreceptors in Aradopsis Thaliana*, November 5, pp. 147–160 (1982).

[403] M. Ahmad and A. R. Cashmore, HY4 gene of *A. thaliana* encodes a protein with characteristics of a blue-light photoreceptor, *Nature Letters* **366**, 162–166 (1993).

[404] C. Lin, M. Ahmad, P. Dutton, D. Robertson, M. Jorns, A. Cashmore and A. Raibekas, Association of flavin adenine dinucleotide with the Arabidopsis blue light receptor CRY1, *Science* **269**, 968–970 (1995).

[405] G. Whitelam, Plant Photomorphogenesis: A green light for cryptochrome research, *Current Biology* **5**, 1351–1353 (1995).

[406] Similarity among the Drosophila (6-4)Photolyase, a human photolyase homolog, and the DNA Photolyase-Blue-Light photoreceptor family, *Science* **272**, 109–112 (1996).

[407] D. S. Hsu, X. Zhao, S. Zhao, A. Kazantsev, R. P. Wang, T. Todo, Y. F. Wei and A. Sancar, Putative human blue-light photoreceptors hCRY1 and hCRY2 are flavoproteins, *Biochemistry* **35**, 13871–13877 (1996).

[408] R. Stanewsky, M. Kaneko, P. Emery, B. Beretta, K. Wager-smith, S. A. Kay, M. Rosbash and J. C. Hall, The cry_b mutation identifies cryptochrome as a circadian photoreceptor in *Drosophila*, *Cell* **95**, 681–692 (1998).

[409] P. Emery, W. V. So, M. Kaneko, J. C. Hall and M. Rosbash, CRY, a Drosophila clock and light-regulated cryptochrome, is a major contributor to circadian rhythm resetting and photosensitivity *Cell* **95**, 669–679 (1998).

[410] Y. Miyamoto and A. Sancar, Vitamin B2-based blue-light photoreceptors in the retinohypothalamic tract as the photoactive pigments, *Proceedings of the National Academy of Sciences* **95**, 6097–6102 (1998).

[411] R. J. Thresher, M. H. Vitaterna, Y. Miyamoto, A. Kazantsev, David S. Hsu, C. Petit, C. P. Selby, L. Dawut, O. Smithies, J. S. Takahashi, A. Sancar, T. R. A, V. MH;, M. Y;, K. A;, H. DS; and P. C. S. C. D. L. S. O. T. J. Sancar, Role of mouse cryptochrome blue-light photoreceptor in circadian photoresponses. *Science* **282**, 1490–1494 (1998).

[412] A. R. Cashmore, J. A. Jarillo, Y.-J. Wu and D. Liu, Cryptochromes: Blue light receptors for plants and animals, *Science* **284**, 760–765 (1999).

[413] A. Sancar, Cryptochrome: The second photoactive pigment in the eye and its role in circadian photoreception, *Annual Review of Biochemistry* **69**, 31–67 (2002).

[414] A. Jagannath, L. Taylor, Z. Wakaf, S. R. Vasudevan and R. G. Foster, The genetics of circadian rhythms, sleep and health, *Human Molecular Genetics* **26**, R2, R128–R138 (2017).

[415] I. N. Karatsoreos, Links between circadian rhythms and psychiatric disease, *Frontiers in Behavioral Neuroscience* **8**, 162 (2014).

[416] A. Charrier, B. Olliac, P. Roubertoux and S. Tordjman, Clock genes and altered sleep-wake rhythms: Their role in the development of psychiatric disorders, *International Journal of Molecular Sciences* **18**, 1–22 (2017).

[417] R. Wever, Human circadian rhythms under the influence of weak electric fields and the different aspects of these studies, *International Journal of Biometeorology* **17**, 227–232 (1973).

[418] M. C. Moore-Ede, S. S. Campbell and R. J. Reiter, *Electromagnetic Fields and Circadian Rhythmicity*. Springer (1992).

[419] L. Lundberg, Z. Sienkiewicz, D. C. Anthony and K. A. Broom, Effects of 50 Hz magnetic fields on circadian rhythm control in mice, *Bioelectromagnetics* **40**, 250–259 (2019).

[420] L. Contalbrigo, C. Stelletta, L. Falcioni, S. Casella, G. Piccione, M. Soffritti and M. Morgante, Effects of different electromagnetic fields on circadian rhythms of some haematochemical parameters in rats, *Biomedical and Environmental Sciences* **22**, 348–353 (2009).

[421] T. Yoshii, M. Ahmad and C. Helfrich-Förster, Cryptochrome mediates light-dependent magnetosensitivity of Drosophila's circadian clock, *PLoS Biology* **7**, 0813–0819 (2009).
[422] M. Liedvogel and H. Mouritsen, Cryptochromes–a potential magnetoreceptor: What do we know and what do we want to know? *Journal of the Royal Society Interface* **7**, S147–S162 (2010).
[423] V. Balland, M. Byrdin, A. P. Eker, M. Ahmad and K. Brettel, What makes the difference between a cryptochrome and DNA photolyase? A spectroelectrochemical comparison of the flavin redox transitions, *Journal of the American Chemical Society* **131**, 426–427 (2009).
[424] A. Sancar, Structure and function of DNA photolyase, *Biochemistry* **33**, 2–9 (1994).
[425] T. Ritz, S. Adem and K. Schulten, A model for photoreceptor-based magnetoreception in birds, *Biophysical Journal* **78**, 707–718 (2000).
[426] P. Müller, J. Yamamoto, R. Martin, S. Iwai and K. Brettel, Discovery and functional analysis of a 4th electron-transferring tryptophan conserved exclusively in animal cryptochromes and (6–4) photolyases, *Chemical Communications* **51**, 15502–15505 (2015).
[427] F. Cintolesi, T. Ritz, C. W. Kay, C. R. Timmel and P. J. Hore, Anisotropic recombination of an immobilized photoinduced radical pair in a 50-μT magnetic field: A model avian photomagnetoreceptor, *Chemical Physics* **294**, 385–399 (2003).
[428] C. Lin, D. Top, C. C. Manahan, M. W. Young and B. R. Crane, Circadian clock activity of cryptochrome relies on tryptophan-mediated photoreduction, *Proceedings of the National Academy of Sciences USA* **115**, 3822–3827 (2018).
[429] A. Günther, A. Einwich, E. Sjulstok, R. Feederle, P. Bolte, K. W. Koch, I. A. Solov'yov and H. Mouritsen, Double-cone localization and seasonal expression pattern suggest a role in magnetoreception for European robin cryptochrome 4. *Current Biology* **28**, 211–223.e4 (2018).
[430] K. Schulten, C. E. Swenberg and A. Weller, A biomagnetic sensory mechanism based on magnetic field modulated coherent electron spin motion, *Zeitschrift für Physikalische Chemie* **111**, 1–5 (1978).
[431] K. Schulten and A. Windemuth, Model for a physiological magnetic compass, *Springer Proceedings in Physics* **11**, 99–106 (1986).
[432] P. J. Hore and H. Mouritsen, The radical-pair mechanism of magnetoreception, *Annual Review of Biophysics* **45**, 299–344 (2016).

[433] W. Wiltschko, J. Traudt, O. Güntürkün, H. Prior and R. Wiltschko, Lateralization of magnetic compass orientation in a migratory bird, *Nature* **419**, 467–470 (2002).
[434] T. Ritz, P. Thalau, J. B. Phillips, R. Wiltschko and W. Wiltschko, Resonance effects indicate a radical-pair mechanism for avian magnetic compass, *Nature* **429**, 177–180 (2004).
[435] P. Thalau, T. Ritz, K. Stapput, R. Wiltschko and W. Wiltschko, Magnetic compass orientation of migratory birds in the presence of a 1.315 MHz oscillating field, *Naturwissenschaften* **92**, 86–90 (2005).
[436] T. Ritz, R. Wiltschko, P. J. Hore, C. T. Rodgers, K. Stapput, P. Thalau, C. R. Timmel and W. Wiltschko, Magnetic compass of birds is based on a molecule with optimal directional sensitivity, *Biophysical Journal* **96**, 3451–3457 (2009).
[437] T. Ritz, Quantum effects in biology: Bird navigation, *Procedia Chemistry* **3**, 262–275 (2011).
[438] W. Wiltschko, J. Traudt, O. Güntürkün, H. Prior and R. Wiltschko, Wiltschko *et al.* reply, *Nature* **471**, 5–7 (2011).
[439] W. Wiltschko and R. Wiltschko, Magnetoreception in birds: Two receptors for two different tasks, *Journal of Ornithology* **148** (2007).
[440] S. Engels, N.-L. Schneider, N. Lefeldt, C. M. Hein, M. Zapka, A. Michalik, D. Elbers, A. Kittel, P. Hore and H. Mouritsen, Anthropogenic electromagnetic noise disrupts magnetic compass orientation in a migratory bird, *Nature* **509**, 353—356 (2014).
[441] K. Kavokin, N. Chernetsov, A. Pakhomov, J. Bojarinova, D. Kobylkov and B. Namozov, Magnetic orientation of garden warblers (Sylvia borin) under 1.4 MHz radiofrequency magnetic field, *Journal of the Royal Society Interface* **11**, 20140451 (2014).
[442] A. Pakhomov, J. Bojarinova, R. Cherbunin, R. Chetverikova, P. S. Grigoryev, K. Kavokin, D. Kobylkov, R. Lubkovskaja and N. Chernetsov, Very weak oscillating magnetic field disrupts the magnetic compass of songbird migrants, *Journal of the Royal Society Interface* **14**, 20170364 (2017).
[443] H. Zadeh-Haghighi and C. Simon, Magnetic field effects in biology from the perspective of the radical pair mechanism, *Journal of the Royal Society Interface* **19**, 20220325 (2022).
[444] R. J. Gegear, A. Casselman, S. Waddell and S. M. Reppert, Cryptochrome mediates light-dependent magnetosensitivity in Drosophila, *Nature* **454**, 1014–1018 (2008).
[445] G. Fedele, E. W. Green, E. Rosato and C. P. Kyriacou, An electromagnetic field disrupts negative geotaxis in Drosophila via a CRY-dependent pathway, *Nature Communications* **5**, 1–6 (2014a).

[446] G. Fedele, M. D. Edwards, S. Bhutani, J. M. Hares, M. Murbach, E. W. Green, S. Dissel, M. H. Hastings, E. Rosato and C. P. Kyriacou, Genetic analysis of circadian responses to low frequency electromagnetic fields in Drosophila melanogaster, *PLoS Genetics* **10**, e1004804 (2014).

[447] R. Marley, C. N. Giachello, N. S. Scrutton, R. A. Baines and A. R. Jones, Cryptochrome-dependent magnetic field effect on seizure response in Drosophila larvae, *Scientific Reports* **4**, 1–4 (2014).

[448] C. L. Wu, T. F. Fu, M. H. Chiang, Y. W. Chang, J. L. Her and T. Wu, Magnetoreception regulates male courtship activity in Drosophila, *PLoS One* **11**, 1–11 (2016).

[449] O. Bazalova, M. Kvicalova, T. Valkova, P. Slaby, P. Bartos, R. Netusil, K. Tomanova, P. Braeunig, H.-J. Lee, I. Sauman, M. Damulewicz, J. Provaznik, R. Pokorny, D. Dolezel and M. Vacha, Cryptochrome 2 mediates directional magnetoreception in cockroaches, *Proceedings of the National Academy of Sciences USA* **113**, 1660–1665 (2016).

[450] M. Ahmad, P. Galland, T. Ritz, R. Wiltschko and W. Wiltschko, Magnetic intensity affects cryptochrome-dependent responses in *Arabidopsis thaliana*, *Planta* **225**, 615–624 (2007).

[451] C. Xu, X. Yin, Y. Lv, C. Wu, Y. Zhang and T. Song, A near-null magnetic field affects cryptochrome-related hypocotyl growth and flowering in Arabidopsis, *Advances in Space Research* **49**, 834–840 (2012).

[452] C. Xu, Y. Lv, C. Chen, Y. Zhang and S. Wei, Blue light-dependent phosphorylations of cryptochromes are affected by magnetic fields in Arabidopsis, *Advances in Space Research* **53**, 1118–1124 (2014).

[453] C. Xu, Y. Li, Y. Yu, Y. Zhang and S. Wei, Suppression of Arabidopsis flowering by near-null magnetic field is affected by light, *Bioelectromagnetics* **36**, 476–479 (2015).

[454] T. Thoradit, K. Thongyoo, K. Kamoltheptawin, L. Tunprasert, M. A. El-Esawi, B. Aguida, N. Jourdan, K. Buddhachat and M. Pooam, Cryptochrome and quantum biology: Unraveling the mysteries of plant magnetoreception, *Frontiers in Plant Science* **14**, 1266357 (2023).

[455] W. Wiltschko, Über den Einfluß statischer Magnetfelder auf die Zugorientierung der Rotkehlchen (Erithacus rubecula), *Zeitschrift für Tierpsychologie* **25**, 537–558 (1968).

[456] W. Wiltschko and R. Wiltschko, Magnetic compass of European robins, *Science* **176**, 62–64 (1972).

[457] C. Walcott and R. P. Green, Orientation of homing pigeons altered by a change in the direction of an applied magnetic field, *Science* **184**, 180–182 (1974).

[458] C. Walcott, Magnetic fields and the orientation of homing pigeons under sun, *Journal of Experimental Biology* **70**, 105–123 (1977).
[459] W. Wiltschko and R. Wiltschko, Disorientation of inexperienced young pigeons after transportation in total darkness, *Nature* **291**, 433–434 (1981).
[460] C. Walcott, Pigeons at magnetic anomalies: The effects of loft location, *Journal of Experimental Biology* **170**, 127–141 (1992).
[461] W. Wiltschko, U. Munro, H. Ford and R. Wiltschko, Red light disrupts magnetic orientation of migratory birds, *Letters to Nature* **364**, 525–527 (1993).
[462] W. Wiltschko and R. Wiltschko, Migratory orientation of European Robins is affected by the wavelength of light as well as by a magnetic pulse, *Journal of Comparative Physiology A* **177**, 363–369 (1995).
[463] W. Wiltschko and R. Wiltschko, The effect of yellow and blue light on magnetic compass orientation in European robins, Erithacus rubecula, *Journal of Comparative Physiology — A Sensory, Neural, and Behavioral Physiology* **184**, 295–299 (1999).
[464] R. Rappl, R. Wiltschko, P. Weindler, P. Berthold and W. Wiltschko, Orientation behavior of garden warblers (Sylvia borin) under monochromatic light of various wavelengths, *The Auk* **117**, 256–260 (2000).
[465] W. Wiltschko and R. Wiltschko, Light-dependent magnetoreception in birds, *The Journal of Experimental Biology* **204**, 3295–3302 (2001).
[466] W. Wiltschko, M. Gesson and R. Wiltschko, Magnetic compass orientation of European robins under 565 nm green light, *Naturwissenschaften* **88**, 387–390 (2001).
[467] R. Muheim, J. Bäckman and S. Åkesson, Magnetic compass orientation in European robins is dependent on both wavelength and intensity of light, *Journal of Experimental Biology* **205**, 3845–3856 (2002).
[468] K. Stapput, P. Thalau, R. Wiltschko and W. Wiltschko, Orientation of birds in total darkness, *Current Biology* **18**, 602–606 (2008).
[469] C. M. Hein, S. Engels, D. Kishkinev and H. Mouritsen, Robins have a magnetic compass in both eyes, *Nature* **471**, E1–E1 (2011).
[470] R. Wiltschko, D. Gehring, S. Denzau, C. Niessner and W. Wiltschko, Magnetoreception in birds: II. Behavioural experiments concerning the cryptochrome cycle, *Journal of Experimental Biology* **217**, 4225–4228 (2014).
[471] R. Wiltschko and W. Wiltschko, Sensing magnetic directions in birds: Radical pair processes involving cryptochrome, *Biosensors* **4**, 221–242 (2014).

[472] R. Wiltschko, U. Munro, H. Ford, K. Stapput, P. Thalau and W. Wiltschko, Orientation of migratory birds under ultraviolet light, *Journal of Comparative Physiology A: Neuroethology, Sensory, Neural, and Behavioral Physiology* **200**, 399–407 (2014).

[473] R. Wiltschko, P. Thalau, D. Gehring, C. Nießner, T. Ritz and W. Wiltschko, Magnetoreception in birds: The effect of radio-frequency fields, *Journal of the Royal Society Interface* **12**, 1–6 (2015).

[474] R. Wiltschko, M. Ahmad, C. Nießner, D. Gehring and W. Wiltschko, Light-dependent magnetoreception in birds: The crucial step occurs in the dark, *Journal of The Royal Society Interface* **13**, 20151010 (2016).

[475] R. Wiltschko, C. Nießner and W. Wiltschko, The magnetic compass of birds: The role of cryptochrome, *Frontiers in Physiology* **12**, 667000 (2021).

[476] S. Schwarze, N.-L. Schneider, T. Reichl, D. Dreyer, N. Lefeldt, S. Engels, N. Baker, P. J. Hore and H. Mouritsen, Weak Broadband Electromagnetic Fields are More Disruptive to Magnetic Compass Orientation in a Night-Migratory Songbird (Erithacus rubecula) than Strong Narrow Band Fields, *Frontiers in Behavioral Neuroscience* **10**, 1–13 (2016).

[477] D. Kobylkov, J. Wynn, M. Winklhofer, R. Chetverikova, J. Xu, H. Hiscock, P. Hore and H. Mouritsen, Electromagnetic 0.1–100 kHz noise does not disrupt orientation in a night-migrating songbird implying a spin coherence lifetime of less than 10 μs, *Journal of the Royal Society Interface* **16**, 20190716 (2019).

[478] A. R. Jones, Magnetic field effects in proteins, *Molecular Physics* **114**, 1691–1702 (2016).

[479] C. T. Rodgers, Magnetic field effects in chemical systems, *Pure and Applied Chemistry* **81**, 19–43 (2009).

[480] U. E. Steiner and T. Ulrich, Magnetic field effects in chemical kinetics and related phenomena, *Chemical Reviews* **89**, 51–147 (1989).

[481] P. Hore, K. L. Ivanov and M. R. Wasielewski, Spin chemistry, *The Journal of Chemical Physics* **152**, 12 (2020).

[482] N. S. Babcock and D. R. Kattnig, Electron-electron dipolar interaction poses a challenge to the radical pair mechanism of magnetoreception, *The Journal of Physical Chemistry Letters* **11**, 2414–2421 (2020).

[483] G. L. Closs, Mechanism explaining nuclear spin polarizations in radical combination reactions, *Journal of the American Chemical Society* **91**, 4552–4554 (1969).

[484] R. Kaptein and J. Oosterhoff, Chemically induced dynamic nuclear polarization II: (relation with anomalous ESR spectra), *Chemical Physics Letters* **4**, 195–197 (1969).

[485] H. G. Hiscock, H. Mouritsen, D. E. Manolopoulos and P. Hore, Disruption of magnetic compass orientation in migratory birds by radiofrequency electromagnetic fields, *Biophysical Journal* **113**, 1475–1484 (2017).

[486] B. Leberecht, D. Kobylkov, T. Karwinkel, S. Döge, L. Burnus, S. Y. Wong, S. Apte, K. Haase, I. Musielak, R. Chetverikova *et al.*, Broadband 75–85 MHz radiofrequency fields disrupt magnetic compass orientation in night-migratory songbirds consistent with a flavin-based radical pair magnetoreceptor, *Journal of Comparative Physiology A* **208**, 97106 (2022).

[487] B. Leberecht, S. Y. Wong, B. Satish, S. Döge, J. Hindman, L. Venkatraman, S. Apte, K. Haase, I. Musielak, G. Dautaj *et al.*, Upper bound for broadband radiofrequency field disruption of magnetic compass orientation in night-migratory songbirds, *Proceedings of the National Academy of Sciences* **120**, e2301153120 (2023).

[488] J. C. Weaver, T. E. Vaughan and R. D. Astumlan, Biological sensing of small field differences by magnetically sensitive chemical reactions, *Nature* **405**, 707–709 (2000).

[489] C. P. Kyriacou and E. Rosato, Genetic analysis of cryptochrome in insect magnetosensitivity, *Frontiers in Physiology*, 1522 (2022).

[490] J. Deviers, F. Cailliez, B. Z. Gutiérrez, D. R. Kattnig and A. de la Lande, *Ab initio* derivation of flavin hyperfine interactions for the protein magnetosensor cryptochrome, *Physical Chemistry Chemical Physics* **24**, 16784–16798 (2022).

[491] V. Bezchastnov and T. Domratcheva, Quantum-mechanical insights into the anisotropic response of the cryptochrome radical pair to a weak magnetic field, *The Journal of Chemical Physics* (2022).

[492] N. Lambert, S. De Liberato, C. Emary and F. Nori, Radical-pair model of magnetoreception with spin-orbit coupling, *New Journal of Physics* **15**, 083024 (2013).

[493] T. P. Fay, L. P. Lindoy, D. E. Manolopoulos and P. Hore, How quantum is radical pair magnetoreception? *Faraday Discussions* **221**, 77–91 (2020).

[494] Z. Schulten and K. Schulten, The generation, diffusion, spin motion, and recombination of radical pairs in solution in the nanosecond time domain, *The Journal of Chemical Physics* **66**, 4616–4634 (1977).

[495] A. M. Lewis, T. P. Fay, D. E. Manolopoulos, C. Kerpal, S. Richert and C. R. Timmel, On the low magnetic field effect in radical

pair reactions, *The Journal of Chemical Physics* **149**, 034103 (2018).
[496] U. Till, C. Timmel, B. Brocklehurst and P. Hore, The influence of very small magnetic fields on radical recombination reactions in the limit of slow recombination, *Chemical Physics Letters* **298**, 7–14 (1998).
[497] V. Bagryansky, V. Borovkov and Y. N. Molin, Singlet–triplet oscillations of spin-correlated radical pairs due to the larmor precession in low magnetic fields, *Molecular Physics* **100**, 1071–1078 (2002).
[498] J. Woodward, Radical pairs in solution, *Progress in Reaction Kinetics and Mechanism* **27**, 3, 165–207(2002).
[499] E. Condon and J. Mack, An interpretation of Pauli's exclusion principle, *Physical Review* **35**, 579 (1930).
[500] J. Deviers, F. Cailliez, A. de la Lande and D. R. Kattnig, Anisotropic magnetic field effects in the re-oxidation of cryptochrome in the presence of scavenger radicals, *The Journal of Chemical Physics* **156**, 025101 (2022).
[501] D. R. Kattnig, Radical-pair-based magnetoreception amplified by radical scavenging: Resilience to spin relaxation, *The Journal of Physical Chemistry B* **121**, 10215–10227 (2017).
[502] G. C. Nordmann, T. Hochstoeger and D. A. Keays, Magnetoreception — a sense without a receptor, *PLoS Biology* **15**, e2003234 (2017).
[503] N. S. Babcock and D. R. Kattnig, Radical scavenging could answer the challenge posed by electron-electron dipolar interactions in the cryptochrome compass model, *JACS Au* **1**, 2033–2046 (2021).
[504] B. Zhang, X. Yuan, H. Lv, J. Che, S. Wang and P. Shang, Biophysical mechanisms underlying the effects of static magnetic fields on biological systems, *Progress in Biophysics and Molecular Biology* **177**, 14–23 (2023).
[505] L. Gerhards, A. Deser, D. R. Kattnig, J. Matysik and I. A. Solov'yov, Weak radiofrequency field effects on biological systems mediated through the radical pair mechanism, *Chemical Reviews* **125**, 8051–8088 (2025).
[506] J. Cukras and J. Sadlej, Towards quantum-chemical modeling of the activity of anesthetic compounds, *International Journal of Molecular Sciences* **22**, 9272 (2021).
[507] A. Gupta, Y. Sang, C. Fontanesi, L. Turin and R. Naaman, A possible unitary mechanism for general anesthesia, *bioRxiv* 2022–11 (2022).
[508] H. Zadeh-Haghighi and C. Simon, Entangled radicals may explain lithium effects on hyperactivity, *Scientific Reports* **11**, 12121 (2021).

[509] H. Zadeh-Haghighi and C. Simon, Radical pairs can explain magnetic field and lithium effects on the circadian clock, *Scientific Reports* **12**, 269 (2022).
[510] J. D. Livingstone, M. J. Gingras, Z. Leonenko and M. A. Beazely, Search for lithium isotope effects in neuronal ht 22 cells, *Biochemistry and Biophysics Reports* **34**, 101461 (2023).
[511] I. Bukhteeva, F. A. Rahman, B. Kendall, R. E. Duncan, J. Quadrilatero, E. V. Pavlov, M. J. Gingras and Z. Leonenko, Effects of lithium isotopes on sodium/lithium co-transport and calcium efflux through the sodium/calcium/lithium exchanger in mitochondria, *Frontiers in Physiology* **15**, 1354091 (2024).
[512] S. M. Reppert, Magnetic field effects on behaviour in *Drosophila*, *Nature* **629**, E1–E2 (2024).
[513] C. P. Kyriacou, Magnetic field responses in *Drosophila*, *Nature* **629**, E3–E5 (2024).
[514] M. Bassetto, T. Reichl, D. Kobylkov, D. R. Kattnig, M. Winklhofer, P. Hore and H. Mouritsen, Bassetto *et al.* reply, *Nature* **629**, E6–E7 (2024).
[515] N. Ikeya and J. R. Woodward, Cellular autofluorescence is magnetic field sensitive, *Proceedings of the National Academy of Sciences USA* **118**, e2018043118 (2021).
[516] K. Maeda, A. J. Robinson, K. B. Henbest, H. J. Hogben, T. Biskup, M. Ahmad, E. Schleichere, S. Webere, C. R. Timmel and P. J. Hore, Magnetically sensitive light-induced reactions in cryptochrome are consistent with its proposed role as a magnetoreceptor, *Proceedings of the National Academy of Sciences USA* **109**, 4774–4779 (2012).
[517] D. M. Sheppard, J. Li, K. B. Henbest, S. R. Neil, K. Maeda, J. Storey, E. Schleicher, T. Biskup, R. Rodriguez, S. Weber, P. J. Hore, C. R. Timmel and S. R. MacKenzie, Millitesla magnetic field effects on the photocycle of an animal cryptochrome, *Scientific Reports* **7**, 1–7 (2017).
[518] J. Xu, L. E. Jarocha, T. Zollitsch, M. Konowalczyk, K. B. Henbest, S. Richert, M. J. Golesworthy, J. Schmidt, V. Déjean, D. J. Sowood *et al.*, Magnetic sensitivity of cryptochrome 4 from a migratory songbird, *Nature* **594**, 535–540 (2021).
[519] D. R. Kattnig, E. W. Evans, V. Déjean, C. A. Dodson, M. I. Wallace, S. R. Mackenzie, C. R. Timmel and P. Hore, Chemical amplification of magnetic field effects relevant to avian magnetoreception, *Nature Chemistry* **8**, 384–391 (2016).
[520] Z. Zeng, J. Wei, Y. Liu, W. Zhang and T. Mabe, Magnetoreception of Photoactivated Cryptochrome 1 in Electrochemistry and Electron Transfer, *ACS Omega* **3**, 4752–4759 (2018).

[521] J. L. Kirschvink, M. Winklhofer and M. M. Walker, Biophysics of magnetic orientation: Strengthening the interface between theory and experimental design, *Journal of the Royal Society Interface* **7**, S179–S191 (2010).

[522] T. Hochstoeger, T. Al Said, D. Maestre, F. Walter, A. Vilceanu, M. Pedron, T. D. Cushion, W. Snider, S. Nimpf, G. C. Nordmann *et al.*, The biophysical, molecular, and anatomical landscape of pigeon CRY4: A candidate light-based quantal magnetosensor, *Science Advances* **6**, eabb9110 (2020).

[523] R. Kavet and J. Brain, Cryptochromes in mammals and birds: Clock or magnetic compass? *Physiology* (2021).

[524] C. N. Giachello, N. S. Scrutton, A. R. Jones and R. A. Baines, Magnetic fields modulate blue-light-dependent regulation of neuronal firing by cryptochrome, *The Journal of Neuroscience* **36**, 10742–10749 (2016).

[525] G. C. G. Parico and C. L. Partch, The tail of cryptochromes: an intrinsically disordered cog within the mammalian circadian clock, *Cell Communication and Signaling* **18**, 1–9 (2020).

[526] P. Li, H. Cheng, V. Kumar, C. S. Lupala, X. Li, Y. Shi, C. Ma, K. Joo, J. Lee, H. Liu *et al.*, Direct experimental observation of bluelightinduced conformational change and intermolecular interactions of cryptochrome, *Communications Biology* **5**, 1103 (2022).

[527] A. A. Lee, J. C. Lau, H. J. Hogben, T. Biskup, D. R. Kattnig and P. Hore, Alternative radical pairs for cryptochrome-based magnetoreception, *Journal of The Royal Society Interface* **11**, 20131063 (2014).

[528] I. A. Solov'yov and K. Schulten, Magnetoreception through cryptochrome may involve superoxide, *Biophysical Journal* **96**, 4804–4813 (2009).

[529] P. Müller and M. Ahmad, Light-activated cryptochrome reacts with molecular oxygen to form a flavin-superoxide radical pair consistent with magnetoreception, *Journal of Biological Chemistry* **286**, 21033–21040 (2011).

[530] T. C. Player and P. Hore, Viability of superoxide-containing radical pairs as magnetoreceptors, *The Journal of chemical physics* **151**, 22 (2019).

[531] K. M. Salerno, J. Domenico, N. Q. Le, K. Balakrishnan, R. J. McQuillen, C. D. Stiles, I. A. Solov'yov and C. F. Martino, Longtime oxygen and superoxide localization in arabidopsis thaliana cryptochrome, *Journal of Chemical Information and Modeling* **63**, 6756–6767 (2023).

[532] C. Nielsen, D. R. Kattnig, E. Sjulstok, P. Hore and I. A. Solov'yov, Ascorbic acid may not be involved in cryptochrome-based magnetoreception, *Journal of the Royal Society Interface* **14**, 20170657 (2017).

[533] A. Sotoodehfar, H. Zadeh-Haghighi, C. Simon *et al.*, Quantum theory of a potential biological magnetic field sensor: Radical pair mechanism in flavin adenine dinucleotide biradicals, *Computational and Structural Biotechnology Journal* **26**, 70–77 (2024).

[534] L. Landler and D. A. Keays, Cryptochrome: The magnetosensor with a sinister side? *PLoS Biology* **16**, e3000018 (2018).

[535] P. J. Hore, Spin chemistry in living systems, *National Science Review*, nwae126 (2024).

[536] V. Binhi, Nonspecific magnetic biological effects: A model assuming the spin-orbit coupling, *The Journal of Chemical Physics* **151**, 204101 (2019).

[537] T. J. Penfold, E. Gindensperger, C. Daniel and C. M. Marian, Spinvibronic mechanism for intersystem crossing, *Chemical Reviews* **118**, 6975–7025 (2018).

[538] R. H. Keens, S. Bedkihal and D. R. Kattnig, Magnetosensitivity in dipolarly coupled three-spin systems, *Physical Review Letters* **121**, 096001 (2018).

[539] M. C. Denton, L. D. Smith, W. Xu, J. Pugsley, A. Toghill and D. R. Kattnig, Magnetosensitivity of tightly bound radical pairs in cryptochrome is enabled by the quantum Zeno effect, *Nature Communications* **15**, 10823 (2024).

[540] S. Qin, H. Yin, C. Yang, Y. Dou, Z. Liu, P. Zhang, H. Yu, Y. Huang, J. Feng, J. Hao *et al.*, A magnetic protein biocompass, *Nature Materials* **15**, 217–226 (2016).

[541] A. I. Krylov, The quantum chemistry of open-shell species, *Reviews in Computational Chemistry* **30**, 151–224 (2017).

[542] R. Haberkorn, Density matrix description of spin-selective radical pair reactions, *Molecular Physics* **32**, 1491–1493 (1976).

[543] T. P. Fay, L. P. Lindoy and D. E. Manolopoulos, Spin-selective electron transfer reactions of radical pairs: Beyond the haberkorn master equation, *The Journal of Chemical Physics* **149**, 6 (2018).

[544] B. Adams, I. Sinayskiy and F. Petruccione, An open quantum system approach to the radical pair mechanism, *Scientific Reports* **8**, 15719 (2018).

[545] K. B. Henbest, K. Maeda, P. J. Hore, M. Joshi, A. Bacher, R. Bittl, S. Weber, C. R. Timmel and E. Schleicher, Magnetic-field effect on the photoactivation reaction of *Escherichia coli* DNA photolyase, *Proceedings of the National Academy of Sciences USA* **105**, 14395–14399 (2008).

[546] T. J. Zwang, E. C. Tse, D. Zhong and J. K. Barton, A compass at weak magnetic fields using thymine dimer repair, *ACS Central Science* **4**, 405–412 (2018).

[547] K. Maeda, K. B. Henbest, F. Cintolesi, I. Kuprov, C. T. Rodgers, P. A. Liddell, D. Gust, C. R. Timmel and P. J. Hore, Chemical compass model of avian magnetoreception, *Nature* **453**, 387–390 (2008).

[548] A. V. Van Huizen, J. M. Morton, L. J. Kinsey, D. G. Von Kannon, M. A. Saad, T. R. Birkholz, J. M. Czajka, J. Cyrus, F. S. Barnes and W. S. Beane, Weak magnetic fields alter stem cell-mediated growth, *Science Advances* **5**, eaau7201 (2019).

[549] M. M. Walker, T. E. Dennis and J. L. Kirschvink, The magnetic sense and its use in long-distance navigation by animals, *Current Opinion in Neurobiology* **12**, 735–744 (2002).

[550] R. K. Adair, Effects of very weak magnetic fields on radical pair reformation, Bioelectromagnetics: Journal of the Bioelectromagnetics Society, The Society for Physical Regulation in Biology and Medicine, *The European Bioelectromagnetics Association* **20**, 255–263 (1999).

[551] J. C. Weaver, T. E. Vaughan and G. T. Martin, Biological effects due to weak electric and magnetic fields: The temperature variation threshold, *Biophysical Journal* **76**, 3026–3030 (1999).

[552] D. R. Kattnig and P. Hore, The sensitivity of a radical pair compass magnetoreceptor can be significantly amplified by radical scavengers, *Scientific Reports* **7**, 11640 (2017).

[553] C. C. DeOliveira and B. R. Crane, A structural decryption of cryptochromes, *Frontiers in Chemistry* **12**, 1436322 (2024).

[554] B. Aguida, J. Babo, S. Baouz, N. Jourdan, M. Procopio, M. A. El-Esawi, D. Engle, S. Mills, S. Wenkel, A. Huck *et al.*, 'seeing' the electromagnetic spectrum: spotlight on the cryptochrome photocycle, *Frontiers in Plant Science* **15**, 1340304 (2024).

[555] Q. Mei and V. Dvornyk, Evolutionary history of the photolyase/cryptochrome superfamily in eukaryotes, *PloS One* **10**, e0135940 (2015).

[556] I. H. Kavakli, I. Baris, M. Tardu, Ş. Gül, H. Öner, S. Cal, S. Bulut, D. Yarparvar, C. Berkel, P. Ustaoğlu *et al.*, The photolyase/cryptochrome family of proteins as DNA repair enzymes and transcriptional repressors, *Photochemistry and Photobiology* **93**, 93103 (2017).

[557] Y. Hosokawa, H. Morita, M. Nakamura and J. Yamamoto, A deazariboflavin chromophore kinetically stabilizes reduced FAD state in a bifunctional cryptochrome, *Scientific Reports* **13**, 16682 (2023).

[558] L. Yan, X. Cao, L. Wang, J. Chen, A. Sancar and D. Zhong, Dynamics and mechanism of DNA repair by a bifunctional cryptochrome, *Proceedings of the National Academy of Sciences* **121**, e2417633121 (2024).

[559] J. McFadden and J. Al-Khalili, The origins of quantum biology, *Proceedings of the Royal Society A* **474**, 20180674 (2018).

[560] J. A. Tuszynski, From quantum chemistry to quantum biology: A path toward consciousness, *Journal of Integrative Neuroscience* **19**, 687700 (2020).

[561] C. Tan, Ultrafast Dynamics of Flavin Cofactor in Photolyase/Cryptochrome Family, Ph.D. thesis, *The Ohio State University* (2013).

[562] X. Cao, L. Yan, C. Yang, L. Wang, M. Zhang and D. Zhong, Dynamics of DNA repair by class-II photolyases via a unified electron-transfer bifurcating mechanism, *Journal of the American Chemical Society* **147**, 11291–11300 (2025).

[563] S. Zou, X.-W. Kang, Y. Lu, Z. Chen, Z. Zhou, S. Tang, D. Zhong and B. Ding, Alternative photoreduction pathway involving electron tunneling and proton transfer in a bifunctional cryptochrome, *Journal of Physical Chemistry Letters* **16**, 6615–6621 (2025).

[564] M. J. Golesworthy, T. Zollitsch, J. Luo, D. Selby, L. E. Jarocha, K. B. Henbest, O. Paré-Labrosse, R. Bartölke, J. Schmidt, J. Xu *et al.*, Singlettriplet dephasing in radical pairs in avian cryptochromes leads to timedependent magnetic field effects, *The Journal of Chemical Physics* **159**, 10 (2023).

[565] A. Hernando, F. Galvez, M. A. García, V. Soto-León, C. Alonso-Bonilla, J. Aguilar and A. Oliviero, Effects of moderate static magnetic field on neural systems is a non-invasive mechanical stimulation of the brain possible theoretically? *Frontiers in Neuroscience* **14**, 419 (2020).

[566] C. Sampson, R. H. Keens and D. R. Kattnig, On the magnetosensitivity of lipid peroxidation: Two-versus three-radical dynamics, *Physical Chemistry Chemical Physics* **21**, 13526–13538 (2019).

[567] P. S. Nair, H. Zadeh-Haghighi and C. Simon, Radical pair model for magnetic field effects on NMDA receptor activity, *Scientific Reports* **14**, 3628 (2024).

[568] X. Wang, Y. Ye, H. Zuo and Y. Li, Neurobiological effects and mechanisms of magnetic fields: A review from 2000 to 2023, *BMC Public Health* **24**, 3094 (2024).

[569] H. Wang, W. Zou and Y. Cao, Bioelectromagnetic fields as signaling currents of life, *Radiation Medicine and Protection* **6**, 112–118 (2025).

[570] C. P. Selby and A. Sancar, A cryptochrome/photolyase class of enzymes with single-stranded DNA-specific photolyase activity, *Proceedings of the National Academy of Sciences* **103**, 17696–17700 (2006).

[571] J. Wang, X. Du, W. Pan, X. Wang and W. Wu, Photoactivation of the cryptochrome/photolyase superfamily, *Journal of Photochemistry and Photobiology C: Photochemistry Reviews* **22**, 84–102 (2015).

[572] W. R. Briggs and E. Huala, Blue-light photoreceptors in higher plants, *Annual Review of Cellular Developmental Biology* **15**, 33–62 (1999).

[573] V. G. Tagua, M. Pausch, M. Eckel, G. Gutiérrez, A. Miralles-Durán, C. Sanz, A. P. Eslava, R. Pokorny, L. M. Corrochano and A. Batschauer, Fungal cryptochrome with DNA repair activity reveals an early stage in cryptochrome evolution, *Proceedings of the National Academy of Sciences* **112**, 15130–15135 (2015).

[574] V. Kruusvee, A. M. Toft, B. Aguida, M. Ahmad and S. Wenkel, Stop crying! inhibition of cryptochrome function by small proteins, *Biochemical Society Transactions* **50**, 773–782 (2022).

[575] S. Weber, T. Biskup, A. Okafuji, A. Marino, T. Berthold, L. G., H. K., G. E.D. and S. E., Origin of light-induced spin-correlated radical pairs in cryptochrome, *Journal of Physical Chemistry B* **114**, 14745–14754 (2010).

[576] Z. Liu, M. Zhang, X. Guo, C. Tan, J. Li, L. Wang, A. Sancar and D. Zhong, Dynamic determination of the functional state in photolyase and the implication for cryptochrome, *Proceedings of the National Academy of Sciences* **110**, 12972–12977 (2013).

[577] Y. Terai, R. Sato, R. Matsumura, S. Iwai and J. Yamamoto, Enhanced DNA repair by DNA photolyase bearing an artificial light-harvesting chromophore, *Nucleic Acids Research* **48**, 10076–10086 (2020).

[578] C. F. Menck, Shining a light on photolyases, *Nature Genetics* **32**, 338–339 (2002).

[579] R. Hakem, DNA-damage repair; the good, the bad, and the ugly, *The EMBO Journal* **27**, 589–605 (2008).

[580] C. Yi and C. He, Dna repair by reversal of DNA damage, *Cold Spring Harbor Perspectives in Biology* **5**, a012575 (2013).

[581] D. Zhong, Electron transfer mechanisms of DNA repair by photolyase, *Annual Review of Physical Chemistry* **66**, 691–715 (2015).

[582] M. Zhang, L. Wang and D. Zhong, Photolyase: Dynamics and electron-transfer mechanisms of DNA repair, *Archives of Biochemistry and Biophysics* **632**, 158–174 (2017).

[583] C. Tan, Z. Liu, J. Li, X. Guo, L. Wang, A. Sancar and D. Zhong, The molecular origin of high DNA-repair efficiency by photolyase, *Nature Communications* **6**, 7302 (2015).
[584] G. B. Schuster, Long-range charge transfer in DNA: Transient structural distortions control the distance dependence, *Accounts of Chemical Research* **33**, 253–260 (2000).
[585] G. B. Sancar, F. W. Smith, R. Reid, G. Payne, M. Levy and A. Sancar, Action mechanism of escherichia coli DNA photolyase. i. Formation of the enzyme-substrate complex. *Journal of Biological Chemistry* **262**, 478–485 (1987).
[586] D. Ramírez-Gamboa, A. L. Díaz-Zamorano, E. R. Meléndez-Sánchez, H. Reyes-Pardo, K. R. Villaseñor-Zepeda, M. E. López-Arellanes, J. E. Sosa-Hernández, K. G. Coronado-Apodaca, A. Gámez-Méndez, S. Afewerki *et al.*, Photolyase production and current applications: A review, *Molecules* **27**, 5998 (2022).
[587] Y. Hosokawa, P. Müller, H. Kitoh-Nishioka, S. Iwai and J. Yamamoto, Limited solvation of an electron donating tryptophan stabilizes a photoinduced charge-separated state in plant (6-4) photolyase, *Scientific Reports* **12**, 1–11 (2022).
[588] R. P. Rastogi, A. Kumar, M. B. Tyagi, R. P. Sinha *et al.*, Molecular mechanisms of ultraviolet radiation-induced DNA damage and repair, *Journal of Nucleic Acids* **2010**, 592980 (2010).
[589] N. Chatterjee and G. C. Walker, Mechanisms of DNA damage, repair, and mutagenesis, *Environmental and Molecular Mutagenesis* **58**, 235–263 (2017).
[590] L. Gao and Y. Bu, Molecular dynamics insights into electron-catalyzed dissociation repair of cyclobutane pyrimidine dimer, *Chinese Journal of Chemical Physics* **34**, 850–860 (2021).
[591] D. Nohr, S. Franz, R. Rodriguez, B. Paulus, L. O. Essen, S. Weber and E. Schleicher, Extended electron-transfer in animal cryptochromes mediated by a tetrad of aromatic amino acids, *Biophysical Journal* **111**, 301–311 (2016).
[592] D. Timmer, A. Frederiksen, D. C. Lünemann, A. R. Thomas, J. Xu, R. Bartölke, J. Schmidt, K. Tomáš, A. De Sio, I. A. Solov'yov *et al.*, Tracking the electron transfer cascade in european robin cryptochrome 4 mutants, *Journal of the American Chemical Society* **145**, 11566–11578 (2023).
[593] A. Cellini, M. K. Shankar, W. Y. Wahlgren, A. Nimmrich, A. Furrer, D. James, M. Wranik, S. Aumonier, E. V. Beale, F. Dworkowski *et al.*, Structural basis of the radical pair state in photolyases and cryptochromes, *Chemical Communications* **58**, 4889–4892 (2022).

[594] S. Weber, Light-driven enzymatic catalysis of DNA repair: A review of recent biophysical studies on photolyase, *Biochimica et Biophysica Acta (BBA)-Bioenergetics* **1707**, 1–23 (2005).

[595] F. Lacombat, A. Espagne, N. Dozova, P. Plaza, P. Müller, H.-J. Emmerich, M. Saft and L.-O. Essen, Ultrafast photoreduction dynamics of a new class of CPD photolyases, *Photochemical & Photobiological Sciences* **20**, 733–746 (2021).

[596] P. Mathis, E. Sage and M. Byrdin, Pushing the limits of flash photolysis to unravel the secrets of biological electron and proton transfer, *Photochemical & Photobiological Sciences*, 1–12 (2022).

[597] A. A. Zieba, C. Richardson, C. Lucero, S. D. Dieng, Y. M. Gindt and J. P. Schelvis, Evidence for concerted electron proton transfer in charge recombination between $FADH^-$ and $^{306}TRP^{\bullet}$ in Escherichia coli photolyase, *Journal of the American Chemical Society* **133**, 7824–7836 (2011).

[598] Y. Ando, K. Niwa, N. Yamada, T. Enomoto, T. Irie, H. Kubota, Y. Ohmiya and H. Akiyama, Firefly bioluminescence quantum yield and colour change by ph-sensitive green emission, *Nature Photonics* **2**, 44–47 (2008).

[599] M. Mereis, R. J. Wanders, M. Schoonen, M. Dercksen, I. Smuts and F. H. Van der Westhuizen, Disorders of flavin adenine dinucleotide metabolism: Madd and related deficiencies, *The International Journal of Biochemistry & Cell Biology* **132**, 105899 (2021).

[600] J. C. Molano-Arevalo, D. R. Hernandez, W. G. Gonzalez, J. Miksovska, M. E. Ridgeway, M. A. Park and F. Fernandez-Lima, Flavin adenine dinucleotide structural motifs: From solution to gas phase, *Analytical Chemistry* **86**, 10223–10230 (2014).

[601] H. W. Park, S. T. Kim, A. Sancar and J. Deisenhofer, Crystal structure of DNA photolyase from *Escherichia coli*, *Science* **268**, 18661872 (1995).

[602] A. R. Cashmore, J. A. Jarillo, Y.-J. Wu and D. Liu, Cryptochromes: Blue light receptors for plants and animals, *Science* **284**, 760–765 (1999).

[603] T. Zhang, K. Papson, R. Ochran and D. P. Ridge, Stability of flavin semiquinones in the gas phase: The electron affinity, proton affinity, and hydrogen atom affinity of lumiflavin, *The Journal of Physical Chemistry A* **117**, 11136–11141 (2013).

[604] M. S. Afanasyeva, M. B. Taraban, P. A. Purtov, T. V. Leshina and C. B. Grissom, Magnetic spin effects in enzymatic reactions: Radical oxidation of NADH by horseradish peroxidase, *Journal of the American Chemical Society* **128**, 8651–8658 (2006).

[605] L. Xu, S. Chen, B. Wen, H. Shi, C. Chi, C. Liu, K. Wang, X. Tao, M. Wang, J. Lv *et al.*, Identification of a novel class of photolyases as possible ancestors of their family, *Molecular Biology and Evolution* **38**, 4505–4519 (2021).

[606] Y. Huang, B. S. Smith, C. L. Colbert, R. Baxter, C. L. Partch and J. Deisenhofer, Crystal structure of cryptochrome 3 from Arabidopsis thaliana and its implications for photolyase activity, *Proceedings of the National Academy of Sciences USA* **103**, 17701–17706 (2006).

[607] T. Klar, R. Pokorny, J. Moldt, A. Batschauer and L. O. Essen, Cryptochrome 3 from Arabidopsis thaliana: Structural and functional analysis of its complex with a folate light antenna, *Journal of Molecular Biology* **366**, 954–964 (2007).

[608] R. Pokorny, T. Klar, U. Hennecke, T. Carell, A. Batschauer and L.O. Essen, Recognition and repair of UV lesions in loop structures of duplex DNA by DASH-type cryptochrome, *Proceedings of the National Academy of Sciences USA* **105**, 21023–21027 (2008).

[609] M. Byrdin, A. P. Eker, M. H. Vos and K. Brettel, Dissection of the triple tryptophan electron transfer chain in escherichia coli DNA photolyase: Trp382 is the primary donor in photoactivation, *Proceedings of the National Academy of Sciences USA* **100**, 8676–8681 (2003).

[610] S. Y. Wong, Y. Wei, H. Mouritsen, I. A. Solov'yov and P. Hore, Cryptochrome magnetoreception: Four tryptophans could be better than three, *Journal of the Royal Society Interface* **18**, 20210601 (2021).

[611] N. Karki, S. Vergish and B. D. Zoltowski, Cryptochromes: Photochemical and structural insight into magnetoreception, *Protein Science* **30**, 1521–1534 (2021).

[612] F. Schuhmann, J. L. Ramsay, D. R. Kattnig and I. A. Solov'yov, Structural rearrangements of pigeon cryptochrome 4 undergoing a complete redox cycle, *Journal of Physical Chemistry B* **128**, 3844–3855 (2024).

[613] P. J. Kabacinski, *Ultrafast UV Spectroscopy of Primary Photoinduced Processes in Biomolecules*, Ph.D. thesis, Politecnico Milano (2022).

[614] J. T. Vivian and P. R. Callis, Mechanisms of tryptophan fluorescence shifts in proteins, *Biophysical Journal* **80**, 2093–2109 (2001).

[615] D. Wu, Q. Hu, Z. Yan, W. Chen, C. Yan, X. Huang, J. Zhang, P. Yang, H. Deng, J. Wang *et al.*, Structural basis of ultraviolet-B perception by UVR8, *Nature* **484**, 214–219 (2012).

[616] X. Li, H. Ren, M. Kundu, Z. Liu, F. W. Zhong, L. Wang, J. Gao and D. Zhong, A leap in quantum efficiency through light harvesting in photoreceptor UVR8, *Nature Communications* **11**, 1–9 (2020).

[617] H. Patwa, N. S. Babcock and P. Kurian, Quantum-enhanced photoprotection in neuroprotein architectures emerges from collective light-matter interactions, *Frontiers in Physics* **12**, 1387271 (2024).

[618] É. Boulais, N. P. Sawaya, R. Veneziano, A. Andreoni, J. L. Banal, T. Kondo, S. Mandal, S. Lin, G. S. Schlau-Cohen, N. W. Woodbury *et al.*, Programmed coherent coupling in a synthetic DNA-based excitonic circuit, *Nature Materials* **17**, 159–166 (2018).

[619] I. Y. Makeeva and T. Puzina, The functional state of the photosynthetic apparatus of potato plants upon destruction of the tubulin cytoskeleton, *Biophysics* **65**, 769–774 (2020).

[620] A. Valzelli, A. Boschetti, F. Mattiotti, A. Kargol, C. Green, F. Borgonovi and G. L. Celardo, Large scale simulations of photosynthetic antenna systems: Interplay of cooperativity and disorder, *The Journal of Physical Chemistry B* **128**, 9643–9655 (2024).

[621] A. Alachkar, Aromatic patterns: Tryptophan aromaticity as a catalyst for the emergence of life and rise of consciousness, *Physics of Life Reviews* (2022).

[622] C. Juhász, Z. Nahleh, I. Zitron, D. C. Chugani, M. Z. Janabi, S. Bandyopadhyay, R. Ali-Fehmi, T. J. Mangner, P. K. Chakraborty, S. Mittal *et al.*, Tryptophan metabolism in breast cancers: molecular imaging and immunohistochemistry studies, *Nuclear Medicine and Biology* **39**, 926–932 (2012).

[623] M. Platten, E. A. Nollen, U. F. Röhrig, F. Fallarino and C. A. Opitz, Tryptophan metabolism as a common therapeutic target in cancer, neurodegeneration and beyond, *Nature Reviews Drug Discovery* **18**, 379–401 (2019).

[624] Y. Liu, T. Wang, J. Yang, T. Yu and Q. Liao, Modulations of superradiance and quantum entanglement of quantum emitters in terahertz frequency, *Physica B: Condensed Matter* **684**, 415972 (2024).

[625] G. Celardo, M. Angeli, T. Craddock and P. Kurian, On the existence of superradiant excitonic states in microtubules, *New Journal of Physics* **21**, 023005 (2019).

[626] D. Ferrari, G. Celardo, G. P. Berman, R. Sayre and F. Borgonovi, Quantum biological switch based on superradiance transitions, *The Journal of Physical Chemistry C* **118**, 20–26 (2014).

[627] F. Mattiotti, M. Sarovar, G. G. Giusteri, F. Borgonovi and G. L. Celardo, Efficient light harvesting and photon sensing via engineered cooperative effects, *New Journal of Physics* **24**, 013027 (2022).

[628] M.-O. Pleinert, J. von Zanthier and G. S. Agarwal, Hyperradiance from collective behavior of coherently driven atoms, *Optica* **4**, 779–785 (2017).

[629] A. I. Nesterov, J. C. B. Zepeda and G. P. Berman, Non-Hermitian quantum annealing in the ferromagnetic ising model, *Physical Review A* **87**, 042332 (2013).
[630] A. Ichiki and M. Ohzeki, Violation of detailed balance accelerates relaxation, *Physical Review E-Statistical, Nonlinear, and Soft Matter Physics* **88**, 020101 (2013).
[631] F. A. Quinton, P. A. S. Myhr, M. Barani, P. Crespo del Granado and H. Zhang, Quantum annealing applications, challenges and limitations for optimisation problems compared to classical solvers, *Scientific Reports* **15**, 12733 (2025).
[632] R. Y. Li, R. Di Felice, R. Rohs and D. A. Lidar, Quantum annealing versus classical machine learning applied to a simplified computational biology problem, *npj Quantum Information* **4**, 14 (2018).
[633] X. Chu and J. Wang, Specificity and affinity quantification of flexible recognition from underlying energy landscape topography, *PLoS Computational Biology* **10**, e1003782 (2014).
[634] A. Frutiger, A. Tanno, S. Hwu, R. F. Tiefenauer, J. Voros and N. Nakatsuka, Nonspecific binding-fundamental concepts and consequences for biosensing applications, *Chemical Reviews* **121**, 8095–8160 (2021).
[635] Z. Liu, X. Guo, C. Tan, J. Li, Y.-T. Kao, L. Wang, A. Sancar and D. Zhong, Electron tunneling pathways and role of adenine in repair of cyclobutane pyrimidine dimer by DNA photolyase, *Journal of the American Chemical Society* **134**, 8104–8114 (2012).
[636] J. Fadanni, A. Acocella and F. Zerbetto, White and colored noises as driving forces of electron transfer: The photolyase repair mechanism as a test case, *The Journal of Physical Chemistry Letters* **10**, 4511–4516 (2019).
[637] M. Zhang, L. Wang, S. Shu, A. Sancar and D. Zhong, Bifurcating electron-transfer pathways in DNA photolyases determine the repair quantum yield, *Science* **354**, 209–213 (2016).
[638] Y. Hosokawa and M. Maestre-Reyna, Light-driven DNA repair at atomic and picosecond resolution revealed via time-resolved serial femtosecond crystallography, *Clinical and Translational Medicine* **14**, e1566 (2024).
[639] Y. He, M. Barone, S. R. Meech, A. Lukacs and P. J. Tonge, Light-driven enzyme catalysis: Ultrafast mechanisms and biochemical implications, *Biochemistry* **64**, 2491–2505 (2025).
[640] P. L. Benjamin, L. Gerhards, I. A. Solov'yov and P. Hore, Magnetosensitivity of model flavin-tryptophan radical pairs in a dynamic protein environment, *Journal of Physical Chemistry B* **129**, 5937–5947 (2025).

[641] M. Rossini, O. Ammerpohl, R. Siebert and J. Ankerhold, Effects of environmental noise on quantum charge diffusion in DNA sequences, *Scientific Reports* **15**, 17543 (2025).

[642] J. C. Genereux, A. K. Boal and J. K. Barton, DNA-mediated charge transport in redox sensing and signaling, *Journal of the American Chemical Society* **132**, 891–905 (2010).

[643] J.-C. Lin, R. R. Singh and D. L. Cox, Theoretical study of DNA damage recognition via electron transfer from the [4fe-4s] complex of MutY, *Biophysical Journal* **95**, 3259–3268 (2008).

[644] J. S. Trimble, R. Crawshaw, F. J. Hardy, C. W. Levy, M. J. Brown, D. E. Fuerst, D. J. Heyes, R. Obexer and A. P. Green, A designed photoenzyme for enantioselective [2+2] cycloadditions, *Nature* **611**, 709–714 (2022).

[645] P. K. Agarwal, C. Schultz, A. Kalivretenos, B. Ghosh and S. Broedel, Engineering hyper-catalytic enzyme by photo-activated conformation modulation, *Biophysical Journal* **102**, 273a (2012).

[646] A. C. Kneuttinger, A guide to designing photocontrol in proteins: Methods, strategies and applications, *Biological Chemistry* **403**, 573–613 (2022).

[647] M. Maestre-Reyna, P.-H. Wang, E. Nango, Y. Hosokawa, M. Saft, A. Furrer, C.-H. Yang, E. P. Gusti Ngurah Putu, W.-J. Wu, H.-J. Emmerich *et al.*, Visualizing the DNA repair process by a photolyase at atomic resolution, *Science* **382**, eadd7795 (2023).

[648] Z. Liu, C. Tan, X. Guo, Y.-T. Kao, J. Li, L. Wang, A. Sancar and D. Zhong, Dynamics and mechanism of cyclobutane pyrimidine dimer repair by DNA photolyase, *Proceedings of the National Academy of Sciences USA* **108**, 14831–14836 (2011).

[649] R. Huang and P.-K. Zhou, DNA damage repair: Historical perspectives, mechanistic pathways and clinical translation for targeted cancer therapy, *Signal Transduction and Targeted Therapy* **6**, 1–35 (2021).

[650] G. Cooper, *The Cell: A Molecular Approach*, ASM Press (2000).

[651] T. C. Bruice and S. J. Benkovic, Chemical basis for enzyme catalysis, *Biochemistry* **39**, 6267–6274 (2000).

[652] R. Wolfenden and M. J. Snider, The depth of chemical time and the power of enzymes as catalysts, *Accounts of Chemical Research* **34**, 938–945 (2001).

[653] A. Warshel, P. K. Sharma, M. Kato, Y. Xiang, H. Liu and M. H. Olsson, Electrostatic basis for enzyme catalysis, *Chemical Reviews* **106**, 3210–3235 (2006).

[654] A. Warshel, Electrostatic origin of the catalytic power of enzymes and the role of preorganized active sites, *Journal of Biological Chemistry* **273**, 27035–27038 (1998).

[655] A. Warshel, Electrostatic basis of structure-function correlation in proteins, *Accounts of Chemical Research* **14**, 284–290 (1981).
[656] N. G. Léonard, R. Dhaoui, T. Chantarojsiri and J. Y. Yang, Electric fields in catalysis: From enzymes to molecular catalysts, *ACS Catalysis* **11**, 10923–10932 (2021).
[657] S. D. Fried and S. G. Boxer, Electric fields and enzyme catalysis, *Annual Review of Biochemistry* **86**, 387–415 (2017).
[658] C. Zheng, Z. Ji, I. I. Mathews and S. G. Boxer, A unifying electrostatic basis for designing enzymes faster than natural ones, *Biophysical Journal* **122**, 483a (2023).
[659] A. Prah, E. Frančišković, J. Mavri and J. Stare, Electrostatics as the driving force behind the catalytic function of the monoamine oxidase a enzyme confirmed by quantum computations, *ACS Catalysis* **9**, 1231–1240 (2019).
[660] D. Burschowsky, A. van Eerde, M. Ökvist, A. Kienhöfer, P. Kast, D. Hilvert and U. Krengel, Electrostatic transition state stabilization rather than reactant destabilization provides the chemical basis for efficient chorismate mutase catalysis, *Proceedings of the National Academy of Sciences* **111**, 17516–17521 (2014).
[661] S. C. Kamerlin and A. Warshel, At the dawn of the 21st century: Is dynamics the missing link for understanding enzyme catalysis? *Proteins: Structure, Function, and Bioinformatics* **78**, 1339–1375 (2010).
[662] A. Warshel and M. Levitt, Theoretical studies of enzymic reactions: Dielectric, electrostatic and steric stabilization of the carbonium ion in the reaction of lysozyme, *Journal of Molecular Biology* **103**, 227–249 (1976).
[663] A. Warshel, Multiscale modeling of biological functions: From enzymes to molecular machines (Nobel lecture), *Angewandte Chemie International Edition* **53**, 10020–10031 (2014).
[664] S. O. Mansoorabadi, C. J. Thibodeaux and H.-w. Liu, The diverse roles of flavin coenzymes nature's most versatile thespians, *The Journal of Organic Chemistry* **72**, 6329–6342 (2007).
[665] C. T. Walsh and T. A. Wencewicz, Flavoenzymes: Versatile catalysts in biosynthetic pathways, *Natural Product Reports* **30**, 175–200 (2013).
[666] V. Piano, B. A. Palfey and A. Mattevi, Flavins as covalent catalysts: New mechanisms emerge, *Trends in Biochemical Sciences* **42**, 457–469 (2017).
[667] C. E. Paul, D. Eggerichs, A. H. Westphal, D. Tischler and W. J. van Berkel, Flavoprotein monooxygenases: Versatile biocatalysts, *Biotechnology Advances* **51**, 107712 (2021).

[668] R. K. Kar, S. Chasen, M.-A. Mroginski and A.-F. Miller, Tuning the quantum chemical properties of flavins via modification at C_8, *The Journal of Physical Chemistry B* **125**, 12654–12669 (2021).

[669] R. K. Kar, A.-F. Miller and M.-A. Mroginski, Understanding flavin electronic structure and spectra, *Wiley Interdisciplinary Reviews: Computational Molecular Science* **12**, e1541 (2022).

[670] E. W. Evans, C. A. Dodson, K. Maeda, T. Biskup, C. Wedge and C. R. Timmel, Magnetic field effects in flavoproteins and related systems, *Interface Focus* **3**, 20130037 (2013).

[671] J. C. Patterson, B. A. Joughin, B. van de Kooij, D. C. Lim, D. A. Lauffenburger and M. B. Yaffe, ROS and oxidative stress are elevated in mitosis during asynchronous cell cycle progression and are exacerbated by mitotic arrest, *Cell Systems* **8**, 163–167 (2019).

[672] D. C. Lim, V. Joukov and M. B. Yaffe, Are redox changes a critical switch for mitotic progression? *Molecular & Cellular Oncology* **7**, 1832419 (2020).

[673] A. Warshel and R. P. Bora, Perspective: Defining and quantifying the role of dynamics in enzyme catalysis, *The Journal of Chemical Physics* **144**, 180901 (2016).

[674] B. D. Dratch, Y. Orozco-Gonzalez, G. Gadda and S. Gozem, Ionic atmosphere effect on the absorption spectrum of a flavoprotein: A reminder to consider solution ions, *The Journal of Physical Chemistry Letters* **12**, 8384–8396 (2021).

[675] S. Foroutannejad, L. L. Good, C. Lin, Z. I. Carter, M. G. Tadesse, A. L. Lucius, B. R. Crane and R. A. Maillard, The cofactor-dependent folding mechanism of drosophila cryptochrome revealed by single-molecule pulling experiments, *Nature Communications* **14**, 1057 (2023).

[676] M. P. Kabir, P. Ghosh and S. Gozem, Electronic structure methods for simulating flavin's spectroscopy and photophysics: Comparison of multireference, TD-DFT, and single-reference wave function methods, *The Journal of Physical Chemistry B* **128**, 7545–7557 (2024).

[677] M. Toplak, A. Matthews and R. Teufel, The devil is in the details: The chemical basis and mechanistic versatility of flavoprotein monooxygenases, *Archives of Biochemistry and Biophysics* **698**, 108732 (2021).

[678] J. Stare, Oxidation of flavin by molecular oxygen: Computational insights into a possible radical mechanism, *ACS Omega* **9**, 23431–23441 (2024).

[679] A. Mattevi, To be or not to be an oxidase: Challenging the oxygen reactivity of flavoenzymes, *Trends in Biochemical Sciences* **31**, 276–283 (2006).
[680] L. Reuter and A. Lüchow, Real space electron delocalization, resonance, and aromaticity in chemistry, *Nature Communications* **12**, 4820 (2021).
[681] S. Nordholm, Analysis of bonding by quantum chemistry–resolving delocalization stabilization in a mechanistic basis and new Hückel model, *The Journal of Physical Chemistry A* **127**, 3449–3471 (2023).
[682] A. Gobbi and G. Frenking, Resonance stabilization in allyl cation, radical, and anion, *Journal of the American Chemical Society* **116**, 9275–9286 (1994).
[683] D. G. Truhlar, The concept of resonance, *Journal of Chemical Education* **84**, 781 (2007).
[684] R. W. Góra, M. Maj and S. J. Grabowski, Resonance-assisted hydrogen bonds revisited. resonance stabilization vs. charge delocalization, *Physical Chemistry Chemical Physics* **15**, 2514–2522 (2013).
[685] H. Tamura, K. Saito and H. Ishikita, Acquirement of water-splitting ability and alteration of the charge-separation mechanism in photosynthetic reaction centers, *Proceedings of the National Academy of Sciences* **117**, 16373–16382 (2020).
[686] P. Kanchanawong, M. G. Dahlbom, T. P. Treynor, J. R. Reimers, N. S. Hush and S. G. Boxer, Charge delocalization in the special-pair radical cation of mutant reaction centers of rhodobacter sphaeroides from stark spectra and nonadiabatic spectral simulations, *The Journal of Physical Chemistry B* **110**, 18688–18702 (2006).
[687] H. Zhou and S. G. Boxer, Charge resonance effects on electronic absorption line shapes: application to the heterodimer absorption of bacterial photosynthetic reaction centers, *The Journal of Physical Chemistry B* **101**, 5759–5766 (1997).
[688] H. Zhou and S. G. Boxer, Probing excited-state electron transfer by resonance stark spectroscopy. 1. Experimental results for photosynthetic reaction centers, *The Journal of Physical Chemistry B* **102**, 9139–9147 (1998).
[689] H. Zhou and S. G. Boxer, Probing excited-state electron transfer by resonance stark spectroscopy. 2. Theory and application, *The Journal of Physical Chemistry B* **102**, 9148–9160 (1998).
[690] S. Volpert, Z. Hashemi, J. M. Foerster, M. R. Marques, I. Schelter, S. Kümmel and L. Leppert, Delocalized electronic excitations and their role in directional charge transfer in the reaction center of rhodobacter sphaeroides, *The Journal of Chemical Physics* **158**, 195102 (2023).

[691] J. Geng, K. Dornevil, V. L. Davidson and A. Liu, Tryptophan-mediated charge-resonance stabilization in the bis-Fe(IV) redox state of MauG, *Proceedings of the National Academy of Sciences* **110**, 9639–9644 (2013).

[692] J. Geng, I. Davis and A. Liu, Probing Bis-Fe^{IV} MauG: Experimental evidence for the long-range charge-resonance model, *Angewandte Chemie* **127**, 3763–3767 (2015).

[693] S. Shin and V. L. Davidson, Maug, a diheme enzyme that catalyzes tryptophan tryptophylquinone biosynthesis by remote catalysis, *Archives of Biochemistry and Biophysics* **544**, 112–118 (2014).

[694] I. Davis, T. Koto and A. Liu, Radical trapping study of the relaxation of bis-Fe(IV) MauG, *Reactive Oxygen Species (Apex)* **5**, 46 (2018).

[695] Z. Ma, H. R. Williamson and V. L. Davidson, Roles of multiple-proton transfer pathways and proton-coupled electron transfer in the reactivity of the bis-Fe^{IV} state of MauG, *Proceedings of the National Academy of Sciences* **112**, 10896–10901 (2015).

[696] H. Kwon, J. Basran, C. M. Casadei, A. J. Fielding, T. E. Schrader, A. Ostermann, J. M. Devos, P. Aller, M. P. Blakeley, P. C. Moody *et al.*, Direct visualization of a Fe(IV)–OH intermediate in a heme enzyme, *Nature Communications* **7**, 13445 (2016).

[697] X. Huang and J. T. Groves, Oxygen activation and radical transformations in heme proteins and metalloporphyrins, *Chemical Reviews* **118**, 2491–2553 (2017).

[698] G. E. Kenney, L. M. Dassama, A. C. Manesis, M. O. Ross, S. Chen, B. M. Hoffman and A. C. Rosenzweig, MbnH is a diheme MauG-like protein associated with microbial copper homeostasis, *Journal of Biological Chemistry* **294**, 16141–16151 (2019).

[699] A. M. Garcia Costas, S. Poudel, A.-F. Miller, G. J. Schut, R. N. Ledbetter, K. R. Fixen, L. C. Seefeldt, M. W. Adams, C. S. Harwood, E. S. Boyd *et al.*, Defining electron bifurcation in the electron-transferring flavoprotein family, *Journal of Bacteriology* **199**, e00440–17 (2017).

[700] G. J. Schut, D. K. Haja, X. Feng, F. L. Poole, H. Li and M. W. Adams, An abundant and diverse new family of electron bifurcating enzymes with a non-canonical catalytic mechanism, *Frontiers in Microbiology* **13**, 946711 (2022).

[701] C. E. Wise, A. E. Ledinina, J. L. Yuly, J. H. Artz and C. E. Lubner, The role of thermodynamic features on the functional activity of electron bifurcating enzymes, *Biochimica et Biophysica Acta (BBA)-Bioenergetics* **1862**, 148377 (2021).

[702] P. Zhang, J. L. Yuly, C. E. Lubner, D. W. Mulder, P. W. King, J. W. Peters and D. N. Beratan, Electron bifurcation: Thermodynamics and kinetics of two-electron brokering in biological redox chemistry, *Accounts of Chemical Research* **50**, 2410–2417 (2017).

[703] D. Xia, L. Esser, L. Yu and C.-A. Yu, Structural basis for the mechanism of electron bifurcation at the quinol oxidation site of the cytochrome bc 1 complex, *Photosynthesis Research* **92**, 17–34 (2007).

[704] L. A. Malone, M. S. Proctor, A. Hitchcock, C. N. Hunter and M. P. Johnson, Cytochrome b_6f-orchestrator of photosynthetic electron transfer, *Biochimica et Biophysica Acta (BBA)-Bioenergetics* **1862**, 148380 (2021).

[705] G. J. Schut, N. Mohamed-Raseek, M. Tokmina-Lukaszewska, D. W. Mulder, D. M. Nguyen, G. L. Lipscomb, J. P. Hoben, A. Patterson, C. E. Lubner, P. W. King *et al.*, The catalytic mechanism of electronbifurcating electron transfer flavoproteins (ETFs) involves an intermediary complex with NAD^+, *Journal of Biological Chemistry* **294**, 32713283 (2019).

[706] D. Das and A.-F. Miller, A single hydrogen bond that tunes flavin redox reactivity and activates it for modification, *Chemical Science* **15**, 7610–7622 (2024).

[707] J. L. Yuly, P. Zhang, C. E. Lubner, J. W. Peters and D. N. Beratan, Universal free-energy landscape produces efficient and reversible electron bifurcation, *Proceedings of the National Academy of Sciences* **117**, 21045–21051 (2020).

[708] J. W. Peters, D. N. Beratan, B. Bothner, R. B. Dyer, C. S. Harwood, Z. M. Heiden, R. Hille, A. K. Jones, P. W. King, Y. Lu *et al.*, A new era for electron bifurcation, *Current Opinion in Chemical Biology* **47**, 32–38 (2018).

[709] J. L. Yuly, P. Zhang, X. Ru, K. Terai, N. Singh and D. N. Beratan, Efficient and reversible electron bifurcation with either normal or inverted potentials at the bifurcating cofactor, *Chem* **7**, 1870–1886 (2021).

[710] A. N. Tikhonov, The cytochrome b_6f complex: Biophysical aspects of its functioning in chloroplasts, *Membrane Protein Complexes: Structure and Function*, 287–328 (2018).

[711] S. Ransac, N. Parisey and J.-P. Mazat, The loneliness of the electrons in the bc1 complex, *Biochimica et Biophysica Acta (BBA)-Bioenergetics* **1777**, 1053–1059 (2008).

[712] J. L. Yuly, P. Zhang and D. N. Beratan, Energy transduction by reversible electron bifurcation, *Current Opinion in Electrochemistry* **29**, 100767 (2021).

[713] K. Terai, J. L. Yuly, P. Zhang and D. N. Beratan, Correlated particle transport enables biological free energy transduction, *Biophysical Journal* **122**, 1762–1771 (2023).

[714] C. E. Lubner, D. P. Jennings, D. W. Mulder, G. J. Schut, O. A. Zadvornyy, J. P. Hoben, M. Tokmina-Lukaszewska, L. Berry, D. M. Nguyen, G. L. Lipscomb *et al.*, Mechanistic insights into energy conservation by flavin-based electron bifurcation, *Nature Chemical Biology* **13**, 655–659 (2017).

[715] X. Feng, G. J. Schut, D. K. Haja, M. W. Adams and H. Li, Structure and electron transfer pathways of an electron-bifurcating NiFe-hydrogenase, *Science Advances* **8**, eabm7546 (2022).

[716] D. T. Murray, X. Ge, G. J. Schut, D. J. Rosenberg, M. Hammel, J. C. Bierma, R. Hille, M. W. Adams and G. L. Hura, Correlating conformational equilibria with catalysis in the electron bifurcating EtfABCX of Thermotoga maritima, *Biochemistry* **63**, 128–140 (2023).

[717] W. Buckel and R. K. Thauer, Flavin-based electron bifurcation, a new mechanism of biological energy coupling, *Chemical Reviews* **118**, 3862–3886 (2018).

[718] J. S. Rieske, Composition, structure, and function of complex III of the respiratory chain, *Biochimica et Biophysica Acta (BBA)-Reviews on Bioenergetics* **456**, 195–247 (1976).

[719] P. Mitchell, The protonmotive q cycle: A general formulation, *FEBS Letters* **59**, 137–139 (1975).

[720] P. Mitchell, Possible molecular mechanisms of the protonmotive function of cytochrome systems, *Journal of Theoretical Biology* **62**, 327–367 (1976).

[721] J. P. Hoben, C. E. Lubner, M. W. Ratzloff, G. J. Schut, D. M. Nguyen, K. W. Hempel, M. W. Adams, P. W. King and A.-F. Miller, Equilibrium and ultrafast kinetic studies manipulating electron transfer: A short-lived flavin semiquinone is not sufficient for electron bifurcation, *Journal of Biological Chemistry* **292**, 14039–14049 (2017).

[722] B. J. Henriques, R. K. J. Olsen, C. M. Gomes and P. Bross, Electron transfer flavoprotein and its role in mitochondrial energy metabolism in health and disease, *Gene* **776**, 145407 (2021).

[723] K. M. Salerno, J. Domenico, N. Q. Le, C. D. Stiles, I. A. Solov'yov and C. F. Martino, Long-time oxygen localization in electron transfer flavoprotein, *Journal of Chemical Information and Modeling* **62**, 4191–4199 (2022).

[724] J. Sucharitakul, W. Buckel and P. Chaiyen, Rapid kinetics reveal surprising flavin chemistry in bifurcating electron transfer flavoprotein from acidaminococcus fermentans, *Journal of Biological Chemistry* **296**, 100124 (2021).

[725] B. K. Carpenter, Kinetic isotope effects: Unearthing the unconventional, *Nature Chemistry* **2**, 80–82 (2010).

[726] J. P. Roth, R. Wincek, G. Nodet, D. E. Edmondson, W. S. McIntire and J. P. Klinman, Oxygen isotope effects on electron transfer to o_2 probed using chemically modified flavins bound to glucose oxidase, *Journal of the American Chemical Society* **126**, 15120–15131 (2004).

[727] W. A. Francisco, G. Tian, P. F. Fitzpatrick and J. P. Klinman, Oxygen18 kinetic isotope effect studies of the tyrosine hydroxylase reaction: Evidence of rate limiting oxygen activation, *Journal of the American Chemical Society* **120**, 4057–4062 (1998).

[728] L. M. Mirica, K. P. McCusker, J. W. Munos, H.-w. Liu and J. P. Klinman, ^{18}O kinetic isotope effects in non-heme iron enzymes: Probing the nature of Fe/O_2 intermediates, *Journal of the American Chemical Society* **130**, 8122–8123 (2008).

[729] M. P. Meyer and J. P. Klinman, Investigating inner-sphere reorganization via secondary kinetic isotope effects in the C–H cleavage reaction catalyzed by soybean lipoxygenase: Tunneling in the substrate backbone as well as the transferred hydrogen, *Journal of the American Chemical Society* **133**, 430–439 (2011).

[730] H. D. Duan, C. E. Lubner, M. Tokmina-Lukaszewska, G. H. Gauss, B. Bothner, P. W. King, J. W. Peters and A.-F. Miller, Distinct properties underlie flavin-based electron bifurcation in a novel electron transfer flavoprotein FixAB from rhodopseudomonas palustris, *Journal of Biological Chemistry* **293**, 4688–4701 (2018).

[731] D. Balzer and I. Kassal, Even a little delocalization produces large kinetic enhancements of charge-separation efficiency in organic photovoltaics, *Science Advances* **8**, eabl9692 (2022).

[732] A. Gluchowski, K. L. Gray, S. N. Hood and I. Kassal, Increases in the charge separation barrier in organic solar cells due to delocalization, *The Journal of Physical Chemistry Letters* **9**, 1359–1364 (2018).

[733] N. Mohamed-Raseek, C. van Galen, R. Stanley and A.-F. Miller, Unusual reactivity of a flavin in a bifurcating electron-transferring flavoprotein leads to flavin modification and a charge-transfer complex, *Journal of Biological Chemistry* **298**, 12 (2022).

[734] N. Mohamed-Raseek, Flavin modification and redox tuning in the bifurcating electron transfer flavoprotein from rhodopseudomonas palustris: Two arginines with different roles, *Journal of Biological Chemistry* **298**, 102606 (2021).

[735] P. K. Agarwal, A biophysical perspective on enzyme catalysis, *Biochemistry* **58**, 438–449 (2018).

[736] R. Hille, D. Niks, W. Vigil Jr, J. Tran, S. Ortiz, K. Menjivar and D. Nguyen, Kinetic studies of bifurcating flavoproteins, *Archives of Biochemistry and Biophysics* **764**, 110278 (2025).

[737] W. Vigil Jr, J. Tran, D. Niks, G. J. Schut, X. Ge, M. W. Adams and R. Hille, The reductive half-reaction of two bifurcating electrontransferring flavoproteins: Evidence for changes in flavin reduction potentials mediated by specific conformational changes, *Journal of Biological Chemistry* **298**, 101927 (2022).

[738] K. Kayastha, S. Vitt, W. Buckel and U. Ermler, Flavins in the electron bifurcation process, *Archives of Biochemistry and Biophysics* **701**, 108796 (2021).

[739] A. Robinson, Did Einstein really say that? *Nature* **557**, 30–31 (2018).

[740] C. W. Saleeby, The advance of heliotherapy, *Nature* **109**, 663 (1922).

[741] F. P. Ellinger, *Medical Radiation Biology.* Charles C. Thomas (1957).

[742] M. Planck, On the theory of the energy distribution law of the normal spectrum, *Verh. Deut. Phys. Ges.* **2**, 237–245 (1900).

[743] M. Nauenberg, Max Planck and the birth of the quantum hypothesis, *American Journal of Physics* **84**, 709–720 (2016).

[744] R. Loudon, The Quantum Theory of Light. *Oxford University Press* (2000).

[745] J. Shen and J. Tower, Effects of light on aging and longevity, *Ageing Research Reviews* **53**, 100913 (2019).

[746] S.-R. Tsai and M. R. Hamblin, Biological effects and medical applications of infrared radiation, *Journal of Photochemistry and Photobiology B: Biology* **170**, 197–207 (2017).

[747] M. C. Giménez, M. Luxwolda, E. G. Van Stipriaan, P. P. Bollen, R. L. Hoekman, M. A. Koopmans, P. R. Arany, M. R. Krames, A. C. Berends, R. A. Hut *et al.*, Effects of near-infrared light on well-being and health in human subjects with mild sleep-related complaints: A double-blind, randomized, placebo-controlled study, *Biology (Basel)* **12**, 60 (2022).

[748] A. G. Gurwitsch, Die natur des spezifischen erregers der zellteilung, *Archiv für mikroskopische Anatomie und Entwicklungsmechanik* **100**, 11–40 (1923).

[749] A. A. Gurwitsch, A historical review of the problem of mitogenetic radiation, *Experientia* **44**, 545–550 (1988).

[750] F.-A. Popp, Biophotons–background, experimental results, theoretical approach and applications, in *Integrative Biophysics: Biophotonics.* Springer, 387–438 (2003).

[751] J. Du, T. Deng, B. Cao, Z. Wang, M. Yang and J. Han, The application and trend of ultra-weak photon emission in biology and medicine, *Frontiers in Chemistry* **11**, 1140128 (2023).
[752] R. R. Mould, A. M. Mackenzie, I. Kalampouka, A. V. Nunn, E. L. Thomas, J. D. Bell and S. W. Botchway, Ultra weak photon emission–a brief review, *Frontiers in Physiology* **15**, 1348915 (2024).
[753] P. Pospíšil, A. Prasad and M. Rác, Role of reactive oxygen species in ultra-weak photon emission in biological systems, *Journal of Photochemistry and Photobiology B: Biology* **139**, 11–23 (2014).
[754] M. Cifra and P. Pospíšil, Ultra-weak photon emission from biological samples: definition, mechanisms, properties, detection and applications, *Journal of Photochemistry and Photobiology B: Biology* **139**, 2–10 (2014).
[755] A. Gurwitsch and L. Gurwitsch, *Das Problem der Zellteilung Physiologisch Betrachtet*, Vol. 11. Verlag von Julius Springer (1926).
[756] A. Gurwitsch and L. Gurwitsch, *Die Mitogenetische Strahlung*, Monographien aus dem Gesamtgebiet der Physiologie der Pflanzen und der Tiere. Springer-Verlag Berlin Heidelberg GmbH (1932).
[757] O. Rahn, *Invisible Radiations Of Organisms*. Verlag von Gebrüder Borntraeger (1936).
[758] N. S. Babcock, Open quantum systems theory of ultraweak ultraviolet photon emissions: Revisiting gurwitsch's onion experiment as a prototype for quantum biology, *Computational and Structural Biotechnology Journal* **26**, 78–91 (2024).
[759] I. Volodyaev and L. V. Beloussov, Revisiting the mitogenetic effect of ultra-weak photon emission, *Frontiers in Physiology* **6**, 241 (2015).
[760] E. Naumova, A. Naumova, D. Isaev and I. Volodyaev, Historical review of early researches on mitogenetic radiation: From discovery to cancer diagnostics, *Journal of Biomedical Photonics & Engineering* **4**, 040201 (2018).
[761] G. Cilento and W. Adam, Photochemistry and photobiology without light, *Photochemistry and Photobiology* **48**, 361–368 (1988).
[762] F. A. Armstrong, Photons in biology, *Interface Focus* **3**, 20130039 (2013).
[763] A. Prasad and P. Pospíšil, Towards the two-dimensional imaging of spontaneous ultra-weak photon emission from microbial, plant and animal cells, *Scientific Reports* **3**, 1211 (2013).
[764] W. J. Baader, C. V. Stevani and E. J. Bechara, "Photo" chemistry without light? *Journal of the Brazilian Chemical Society* **26**, 2430–2447 (2015).

[765] F. Ortega-Ojeda, M. Calcerrada, A. Ferrero, J. Campos and C. GarciaRuiz, Measuring the human ultra-weak photon emission distribution using an electron-multiplying, charge-coupled device as a sensor, *Sensors* **18**, 1152 (2018).

[766] W. M. Bayliss, Heliotherapy, *Nature* **112**, 197–198 (1923).

[767] A. Rollier and A. Rosselet, *Heliotherapy*, Oxford Medical Publications. H. Frowde; Hodder & Stoughton (1923).

[768] G. S. Stent, Light and life: Niels bohr's legacy to contemporary biology, in *Niels Bohr: Physics and the World*. Routledge, 231–244 (2014).

[769] H. Nussenzveig, Bohr's 'light and life' revisited, *Physica Scripta* **90**, 118001 (2015).

[770] P. Gunter and H. Bergson, *Bergson and the Evolution of Physics*. University of Tennessee Press (1969).

[771] M. Sinclair, *Bergson*. Routledge (2019).

[772] M. Delbrück, Light and life III, *Carlsberg Research Communications* **41**, 299–309 (1976).

[773] N. Wiener, *Cybernetics or Control and Communication in the Animal and the Machine*. MIT press (2019).

[774] A. Szent-Györgyi, Towards a new biochemistry? *Science* **93**, 609–611 (1941).

[775] J. B. S. Haldane, Quantum mechanics as a basis for philosophy, *Philosophy of Science* **1**, 78–98 (1934).

[776] H. Fröhlich, Further evidence for coherent excitations in biological systems, *Physics Letters A* **110**, 480–481 (1985).

[777] H. Fröhlich, *Biological Coherence and Response to External Stimuli*. Springer-Verlag (1988).

[778] N. W. Timoféef-Ressovsky, K. G. Zimmer and M. Delbrück, *Über die Natur der Genmutation und der Genstruktur*, Nachrichten von der Gesellschaft der Wissenschaften zu Göttingen, mathematisch-physikalische Klasse. Weidmannasche Buchhandlung (1935).

[779] M. Cobb and N. Comfort, What Rosalind Franklin truly contributed to the discovery of DNA's structure, *Nature* **616**, 657–660 (2023).

[780] N. Bohr, Essays 1958–1962 on atomic physics and human knowledge, John Wiley & Sons Ltd. (1964).

[781] A. T. Domondon, Bringing physics to bear on the phenomenon of life: the divergent positions of Bohr, Delbrück, and Schrödinger, *Studies in History and Philosophy of Science Part C: Studies in History and Philosophy of Biological and Biomedical Sciences* **37**, 433–458 (2006).

[782] S. Cohen and F. Popp, Biophoton emission of the human body, *Journal of Photochemistry and Photobiology B: Biology* **40**, 187–189 (1997).

[783] L. Colli, U. Facchini, G. Guidotti, R. D. Lonati, M. Orsenigo and O. Sommariva, Further measurements on the bioluminescence of the seedlings, *Experientia* **11**, 479–481 (1955).

[784] T. Quickenden and R. Tilbury, Luminescence spectra of exponential and stationary phase cultures of respiratory deficient saccharomyces cerevisiae, *Journal of Photochemistry and Photobiology B: Biology* **8**, 169–174 (1991).

[785] G. Albrecht-Buehler, Rudimentary form of cellular "vision", *Proceedings of the National Academy of Sciences USA* **89**, 8288–8292 (1992).

[786] G. Albrecht-Buehler, Cellular infrared detector appears to be contained in the centrosome, *Cell Motility and the Cytoskeleton* **27**, 262–271 (1994).

[787] G. Albrecht-Buehler, Changes of cell behavior by near-infrared signals, *Cell Motility and the Cytoskeleton* **32**, 299–304 (1995).

[788] G. Albrecht-Buehler, Altered drug resistance of microtubules in cells exposed to infrared light pulses: Are microtubules the "nerves" of cells? *Cell Motility and the Cytoskeleton* **40**, 183–192 (1998).

[789] G. Albrecht-Buehler, A long-range attraction between aggregating 3T3 cells mediated by near-infrared light scattering, *Proceedings of the National Academy of Sciences USA* **102**, 5050–5055 (2005).

[790] N. Medina-Villalobos, R. Avila, M. Marsal, J. Andilla, P. Loza-Álvarez, M. M. Ojeda-Ramírez and E. Tamariz, Infrared laser effects on cell projection depend on irradiation intermittence and cell activity, *Cells* **12**, 540 (2023).

[791] M. Lechelon, Y. Meriguet, M. Gori, S. Ruffenach, I. Nardecchia, E. Floriani, D. Coquillat, F. Teppe, S. Mailfert, D. Marguet *et al.*, Experimental evidence for long-distance electrodynamic intermolecular forces, *Science Advances* **8**, eabl5855 (2022).

[792] S. Ruffenach, M. Lechelon, Y. Meriguet, M. Gori, I. Nardecchia, E. Floriani, D. Coquillat, F. Teppe, S. Mailfert, P. Ferrier *et al.*, Long-range electrodynamic interactions between proteins assessed by thz spectroscopy, in *Terahertz Emitters, Receivers, and Applications XIII*, Vol. 12230. SPIE, 74–79 (2022).

[793] I. Nardecchia, J. Torres, M. Lechelon, V. Giliberti, M. Ortolani, P. Nouvel, M. Gori, Y. Meriguet, I. Donato, J. Preto *et al.*, Out-of-equilibrium collective oscillation as phonon condensation in a model protein, *Physical Review X* **8**, 031061 (2018).

[794] Z. Zhang, G. S. Agarwal and M. O. Scully, Quantum fluctuations in the Fröhlich condensate of molecular vibrations driven far from equilibrium, *Physical Review Letters* **122**, 158101 (2019).

[795] Y. Chen, H. I. Okur, N. Gomopoulos, C. Macias-Romero, P. S. Cremer, P. B. Petersen, G. Tocci, D. M. Wilkins, C. Liang,

M. Ceriotti *et al.*, Electrolytes induce long-range orientational order and free energy changes in the H-bond network of bulk water, *Science Advances* **2**, e1501891 (2016).

[796] R. Van Wijk, E. P. Van Wijk, J. Pang, M. Yang, Y. Yan and J. Han, Integrating ultra-weak photon emission analysis in mitochondrial research, *Frontiers in Physiology* **11**, 717 (2020).

[797] M. Pietruszka and M. Marzec, Ultra-weak photon emission from DNA, *Scientific Reports* **14**, 28915 (2024).

[798] X. Shen, W. Mei and X. Xu, Activation of neutrophils by a chemically separated but optically coupled neutrophil population undergoing respiratory burst, *Experientia* **50**, 963–968 (1994).

[799] F. Zapata, V. Pastor-Ruiz, F. Ortega-Ojeda, G. Montalvo, A. V. Ruiz-Zolle and C. García-Ruiz, Human ultra-weak photon emission as non-invasive spectroscopic tool for diagnosis of internal states – A review, *Journal of Photochemistry and Photobiology B: Biology* **216**, 112141 (2021).

[800] E. Downs, A. D. Bottrell and K. Naylor, Identifying the effects of reactive oxygen species on mitochondrial dynamics and cytoskeleton stability in dictyostelium discoideum, *Cells* **10**, 2147 (2021).

[801] J. Blumberger, Recent advances in the theory and molecular simulation of biological electron transfer reactions, *Chemical Reviews* **115**, 11191–11238 (2015).

[802] N. J. Murugan, M. A. Persinger, L. M. Karbowski and B. T. Dotta, Ultraweak photon emissions as a non-invasive, early-malignancy detection tool: An in vitro and in vivo study, *Cancers* **12**, 1001 (2020).

[803] A. Ritter and N.-N. Kreis, Microtubule dynamics and cancer, *Cancers* **14**, 4368 (2022).

[804] J. Pokorný, J. Pokorný, A. Foletti, J. Kobilková, J. Vrba and J. Vrba Jr, Mitochondrial dysfunction and disturbed coherence: Gate to cancer, *Pharmaceuticals* **8**, 675–695 (2015).

[805] B. A. Weaver, How taxol/paclitaxel kills cancer cells, *Molecular Biology of the Cell* **25**, 2677–2681 (2014).

[806] B. T. Castle and D. J. Odde, Optical control of microtubule dynamics in time and space, *Cell* **162**, 243–245 (2015).

[807] T. J. A. Craddock, D. Friesen, J. Mane, S. Hameroff and J. A. Tuszynski, The feasibility of coherent energy transfer in microtubules, *Journal of the Royal Society Interface* **11**, 20140677 (2014).

[808] Y. A. Krasylenko, A. I. Yemets and Y. B. Blume, Plant microtubules reorganization under the indirect UV-B exposure and during UV-B-induced programmed cell death, *Plant Signaling & Behavior* **8**, e24031 (2013).

[809] A. Müller-Deku, J. Meiring, K. Loy, Y. Kraus, C. Heise, R. Bingham, K. I. Jansen, X. Qu, F. Bartolini, L. C. Kapitein *et al.*, Photoswitchable paclitaxel-based microtubule stabilisers allow optical control over the microtubule cytoskeleton, *Nature Communications* **11**, 4640 (2020).

[810] N. Ankenbruck, T. Courtney, Y. Naro and A. Deiters, Optochemical control of biological processes in cells and animals, *Angewandte Chemie International Edition* **57**, 2768–2798 (2018).

[811] M. S. Islam, A. M. R. Kabir, D. Inoue, K. Sada and A. Kakugo, Enhanced dynamic instability of microtubules in a ROS free inert environment, *Biophysical Chemistry* **211**, 1–8 (2016).

[812] P. Livanos, B. Galatis and P. Apostolakos, The interplay between ROS and tubulin cytoskeleton in plants, *Plant Signaling & Behavior* **9**, e28069 (2014).

[813] D. Stojkov, P. Amini, K. Oberson, C. Sokollik, A. Duppenthaler, H.-U. Simon and S. Yousefi, Ros and glutathionylation balance cytoskeletal dynamics in neutrophil extracellular trap formation, *Journal of Cell Biology* **216**, 4073–4090 (2017).

[814] N. Nambiar, P. K. R. Nagireddy, R. Pedapati, S. Kantevari and M. Lopus, Tubulin-and ROS-dependent antiproliferative mechanism of a potent analogue of noscapine, N-propargyl noscapine, *Life Sciences* **258**, 118238 (2020).

[815] P. Nasehi, N. Omidkhah and R. Ghodsi, A review of recent advances in quinoline/isoquinoline based hybrids as microtubule targeted cancer therapeutics: Synthesis, binding mode, QSAR and docking studies, *Journal of Molecular Structure* **1295**, 136720 (2024).

[816] M. Gullì, A. Valzelli, F. Mattiotti, M. Angeli, F. Borgonovi and G. L. Celardo, Macroscopic coherence as an emergent property in molecular nanotubes, *New Journal of Physics* **21**, 013019 (2019).

[817] K. Higgins, S. Benjamin, T. Stace, G. Milburn, B. W. Lovett and E. Gauger, Superabsorption of light via quantum engineering, *Nature Communications* **5**, 4705 (2014).

[818] N. M. Ennist, S. Wang, M. A. Kennedy, M. Curti, G. A. Sutherland, C. Vasilev, R. L. Redler, V. Maffeis, S. Shareef, A. V. Sica *et al.*, De novo design of proteins housing excitonically coupled chlorophyll special pairs, *Nature Chemical Biology* **20**, 906–915 (2024).

[819] M. Reitz, C. Sommer and C. Genes, Cooperative quantum phenomena in light-matter platforms, *PRX Quantum* **3**, 010201 (2022).

[820] V. A. Shahnazaryan, V. A. Saroka, I. A. Shelykh, W. L. Barnes and M. E. Portnoi, Strong light–matter coupling in carbon nanotubes as a route to exciton brightening, *ACS Photonics* **6**, 904–914 (2019).

[821] V. Schroeder, S. Savagatrup, M. He, S. Lin and T. M. Swager, Carbon nanotube chemical sensors, *Chemical Reviews* **119**, 599–663 (2018).

[822] P. Nick, Microtubules, signalling and abiotic stress, *The Plant Journal* **75**, 309–323 (2013).

[823] A. L. Parker, M. Kavallaris and J. A. McCarroll, Microtubules and their role in cellular stress in cancer, *Frontiers in Oncology* **4**, 153 (2014).

[824] Y. Guo, Y. Liang, Y. Li, B. Tian, X. Fan, Y. He, M. Liu, L. Peng, N. Tang, T. Tan *et al.*, Optical microcavities empowered biochemical sensing: status and prospects, *Advanced Devices & Instrumentation* **5**, 0041 (2024).

[825] A. Fontcuberta i Morral and F. Stellacci, Ultrastrong routes to new chemistry, *Nature Materials* **11**, 272–273 (2012).

[826] J. A. Hutchison, T. Schwartz, C. Genet, E. Devaux and T. W. Ebbesen, Modifying chemical landscapes by coupling to vacuum fields, *Angewandte Chemie International Edition* **51**, 1592–1596 (2012).

[827] P. A. Thomas, K. S. Menghrajani and W. L. Barnes, Cavity-free ultrastrong light-matter coupling, *Journal of Physical Chemistry Letters* **12**, 6914–6918 (2021).

[828] S. Pannir-Sivajothi, J. A. Campos-Gonzalez-Angulo, L. A. Martínez-Martínez, S. Sinha and J. Yuen-Zhou, Driving chemical reactions with polariton condensates, *Nature Communications* **13**, 1645 (2022).

[829] R. Bhuyan, J. Mony, O. Kotov, G. W. Castellanos, J. Gómez Rivas, T. O. Shegai and K. Borjesson, The rise and current status of polaritonic photochemistry and photophysics, *Chemical Reviews* **123**, 10877–10919 (2023).

[830] B. Xiang and W. Xiong, Molecular polaritons for chemistry, photonics and quantum technologies, *Chemical Reviews* **124**, 2512–2552 (2024).

[831] T. Chervy, P. Knüppel, H. Abbaspour, M. Lupatini, S. Fält, W. Wegscheider, M. Kroner and A. Imamoğlu, Accelerating polaritons with external electric and magnetic fields, *Physical Review X* **10**, 011040 (2020).

[832] D. Basov, A. Asenjo-Garcia, P. J. Schuck, X. Zhu, A. Rubio, A. Cavalleri, M. Delor, M. M. Fogler and M. Liu, Polaritonic quantum matter, *Nanophotonics* (2025).

[833] S. Kumar, K. Boone, J. Tuszyński, P. Barclay and C. Simon, Possible existence of optical communication channels in the brain, *Scientific Reports* **6**, 36508 (2016).

[834] P. Zarkeshian, S. Kumar, J. Tuszynski, P. Barclay and C. Simon, Are there optical communication channels in the brain? *Frontiers in Bioscience* **23**, 407–1421 (2018).

[835] J. R. Reimers, L. K. McKemmish, R. H. McKenzie, A. E. Mark and N. S. Hush, Weak, strong, and coherent regimes of Fröhlich condensation and their applications to terahertz medicine and quantum consciousness, *Proceedings of the National Academy of Sciences* **106**, 4219–4224 (2009).

[836] H. J. Geesink and M. Schmieke, Organizing and disorganizing resonances of microtubules, stem cells, and proteins calculated by a quantum equation of coherence, *Journal of Modern Physics* **13**, 1530–1580 (2022).

[837] M. Cavaglià, M. A. Deriu and J. A. Tuszynski, Toward a holographic brain paradigm: a lipid-centric model of brain functioning, *Frontiers in Neuroscience* **17**, 1302519 (2023).

[838] M. A. Schroer, S. Schewa, A. Y. Gruzinov, C. Rönnau, J. M. Lahey-Rudolph, C. E. Blanchet, T. Zickmantel, Y.-H. Song, D. I. Svergun and M. Roessle, Probing the existence of non-thermal terahertz radiation induced changes of the protein solution structure, *Scientific Reports* **11**, 22311 (2021).

[839] E. Perez-Martin, T. Beranger, L. Bonnet, F. Teppe, A. Lisauskas, K. Ikamas, E. Vrouwe, E. Floriani, G. Katona, D. Marguet *et al.*, Unveiling long-range forces in light-harvesting proteins: Pivotal roles of temperature and light, *Science Advances* **11**, eadv0346 (2025).

[840] S. Ruffenach, M. Lechelon, Y. Meriguet, M. Gori, I. Nardecchia, E. Floriani, D. Coquillat, F. Teppe, S. Mailfert, D. Marguet *et al.*, Thz spectroscopy to probe long-range electrodynamic interactions between proteins, in *2022 47th International Conference on Infrared, Millimeter and Terahertz Waves (IRMMW-THz)*. IEEE, pp. 1–2 (2022).

[841] M. Rahnama, J. A. Tuszynski, I. Bokkon, M. Cifra, P. Sardar and V. Salari, Emission of mitochondrial biophotons and their effect on electrical activity of membrane via microtubules, *Journal of Integrative Neuroscience* **10**, 65–88 (2011).

[842] Y. Kataoka, Y. Cui, A. Yamagata, M. Niigaki, T. Hirohata, N. Oishi and Y. Watanabe, Activity-dependent neural tissue oxidation emits intrinsic ultraweak photons, *Biochemical and Biophysical Research Communications* **285**, 1007–1011 (2001).

[843] M. Kobayashi, M. Takeda, T. Sato, Y. Yamazaki, K. Kaneko, K.-I. Ito, H. Kato and H. Inaba, In vivo imaging of spontaneous ultraweak photon emission from a rat's brain correlated with cerebral

energy metabolism and oxidative stress, *Neuroscience Research* **34**, 103–113 (1999).

[844] P. A. Thomas, K. S. Menghrajani and W. L. Barnes, All-optical control of phase singularities using strong light-matter coupling, *Nature Communications* **13**, 1809 (2022).

[845] C. P. Dietrich, A. Steude, L. Tropf, M. Schubert, N. M. Kronenberg, K. Ostermann, S. Höfling and M. C. Gather, An exciton-polariton laser based on biologically produced fluorescent protein, *Science Advances* **2**, e1600666 (2016).

[846] P. Bhattacharya, T. Frost, S. Deshpande, M. Z. Baten, A. Hazari and A. Das, Room temperature electrically injected polariton laser, *Physical Review Letters* **112**, 236802 (2014).

[847] E. Kokin, H. J. An, D. Koo, S. Han, K. Whang, T. Kang, I. Choi and L. P. Lee, Quantum electrodynamic behavior of chlorophyll in a plasmonic nanocavity, *Nano Letters* **22**, 9861–9868 (2022).

[848] A. Debnath and A. Rubio, Entangled biphoton enhanced double quantum coherence signal as a probe for cavity polariton correlations in presence of phonon induced dephasing, *Frontiers in Physics* **10**, 879113 (2022).

[849] T. G. Zaremba, T. R. LeBon, D. B. Millar, R. M. Smejkal and R. J. Hawley, Effects of ultraviolet light on the in vitro assembly of microtubules, *Biochemistry* **23**, 1073–1080 (1984).

[850] I. Staxén, C. Bergounioux and J. F. Bornman, Effect of ultraviolet radiation on cell division and microtubule organization in petunia hybrida protoplasts, *Protoplasma* **173**, 70–76 (1993).

[851] Y. Zhao and Q. Zhan, Electric fields generated by synchronized oscillations of microtubules, centrosomes and chromosomes regulate the dynamics of mitosis and meiosis, *Theoretical Biology and Medical Modelling* **9**, 1–10 (2012).

[852] L. Gao, J. C. Meiring, C. Heise, A. Rai, A. Müller-Deku, A. Akhmanova, J. Thorn-Seshold and O. Thorn-Seshold, Photoswitchable epothilonebased microtubule stabilisers allow GFP-imaging-compatible, optical control over the microtubule cytoskeleton, *Angewandte Chemie International Edition* **61**, e202114614 (2022).

[853] S. Watari, H. Inaba, T. Tamura, A. M. R. Kabir, A. Kakugo, K. Sada, I. Hamachi and K. Matsuura, Light-induced stabilization of microtubules by photo-crosslinking of a Tau-derived peptide, *Chemical Communications* **58**, 9190–9193 (2022).

[854] E. Dubas, A. Castillo, I. Żur, M. Krzewska and M. Vallés, Microtubule organization changes severely after mannitol and n-butanol treatments inducing microspore embryogenesis in bread wheat, *BMC Plant Biology* **21**, 1–19 (2021).

[855] H. Y. G. Lim and N. Plachta, Cytoskeletal control of early mammalian development, *Nature Reviews Molecular Cell Biology* **22**, 548–562 (2021).

[856] P. Tolar, Cytoskeletal control of B cell responses to antigens, *Nature Reviews Immunology* **17**, 621–634 (2017).

[857] F. C. Ramaekers and F. T. Bosman, The cytoskeleton and disease, *The Journal of Pathology: A Journal of the Pathological Society of Great Britain and Ireland* **204**, 351–354 (2004).

[858] C. Janke and M. M. Magiera, The tubulin code and its role in controlling microtubule properties and functions, *Nature Reviews Molecular Cell Biology* **21**, 307–326 (2020).

[859] E. Frixione, Recurring views on the structure and function of the cytoskeleton: A 300-year epic, *Cell Motility and the Cytoskeleton* **46**, 73–94 (2000).

[860] L. Cirillo, M. Gotta and P. Meraldi, The elephant in the room: The role of microtubules in cancer, *Cell Division Machinery and Disease*, 93–124 (2017).

[861] M. Keller, K. Reis, A. Hjerpe, K. Dobra and P. Aspenström, Cytoskeletal organization correlates to motility and invasiveness of malignant mesothelioma cells, *Cancers* **13**, 685 (2021).

[862] M. T. Kelliher, H. A. Saunders and J. Wildonger, Microtubule control of functional architecture in neurons, *Current Opinion in Neurobiology* **57**, 39–45 (2019).

[863] T. J. Craddock, J. A. Tuszynski and S. Hameroff, Cytoskeletal signaling: is memory encoded in microtubule lattices by camkii phosphorylation? *PLoS Computational Biology* **8**, e1002421 (2012).

[864] A. Priel, J. A. Tuszynski and N. J. Woolf, Neural cytoskeleton capabilities for learning and memory, *Journal of Biological Physics* **36**, 3–21 (2010).

[865] M. Lasser, J. Tiber and L. A. Lowery, The role of the microtubule cytoskeleton in neurodevelopmental disorders, *Frontiers in Cellular Neuroscience* **12**, 165 (2018).

[866] B. J. Gentil, M. Tibshirani and H. D. Durham, Neurofilament dynamics and involvement in neurological disorders, *Cell and tissue research* **360**, 609–620 (2015).

[867] H. Xiao, P. Verdier-Pinard, N. Fernandez-Fuentes, B. Burd, R. Angeletti, A. Fiser, S. B. Horwitz and G. A. Orr, Insights into the mechanism of microtubule stabilization by taxol, *Proceedings of the National Academy of Sciences* **103**, 10166–10173 (2006).

[868] N. B. Gudimchuk and J. R. McIntosh, Regulation of microtubule dynamics, mechanics and function through the growing tip, *Nature Reviews Molecular Cell Biology* **22**, 777–795 (2021).

[869] K. S. Burbank and T. J. Mitchison, Microtubule dynamic instability, *Current Biology* **16**, R516-R517 (2006).

[870] T. Mitchison and M. Kirschner, Dynamic instability of microtubule growth, *Nature* **312**, 237–242 (1984).

[871] A. P. Kalra, B. B. Eakins, S. D. Patel, G. Ciniero, V. Rezania, K. Shankar and J. A. Tuszynski, All wired up: An exploration of the electrical properties of microtubules and tubulin, *ACS Nano* **14**, 16301–16320 (2020).

[872] K. Liew, P. Xiang and L. Zhang, Mechanical properties and characteristics of microtubules: A review, *Composite Structures* **123**, 98–108 (2015).

[873] D. L. Sekulić, B. M. Satarić, J. A. Tuszynski and M. V. Satarić, Nonlinear ionic pulses along microtubules, *The European Physical Journal E* **34**, 49 (2011).

[874] A. P. Kalra, B. B. Eakins, S. I. Vagin, H. Wang, S. D. Patel, P. Winter, M. Aminpour, J. D. Lewis, V. Rezania, K. Shankar *et al.*, A nanometric probe of the local proton concentration in microtubule-based biophysical systems, *Nano Letters* **22**, 517–523 (2022).

[875] A. P. Kalra, S. D. Patel, B. B. Eakins, S. Riddell, P. Kumar, P. Winter, J. Preto, K. W. Carlson, J. D. Lewis, V. Rezania *et al.*, Revealing and attenuating the electrostatic properties of tubulin and its polymers, *Small* **17**, 2003560 (2021).

[876] M. d. R. Cantero, P. L. Perez, N. Scarinci and H. F. Cantiello, Two-dimensional brain microtubule structures behave as memristive devices, *Scientific Reports* **9**, 1–10 (2019).

[877] A. Priel, A. J. Ramos, J. A. Tuszynski and H. F. Cantiello, A biopolymer transistor: electrical amplification by microtubules, *Biophysical Journal* **90**, 4639–4643 (2006).

[878] J. Pokorný, J. Pokorný and J. Vrba, Generation of electromagnetic field by microtubules, *International Journal of Molecular Sciences* **22**, 8215 (2021).

[879] N. E. Mavromatos, Quantum mechanical aspects of cell microtubules: science fiction or realistic possibility? in *Journal of Physics: Conference Series*, Vol. 306. IOP Publishing, 012008 (2011).

[880] J. A. Tuszynski, D. Friesen, H. Freedman, V. I. Sbitnev, H. Kim, I. Santelices, A. P. Kalra, S. D. Patel, K. Shankar and L. O. Chua, Microtubules as sub-cellular memristors, *Scientific Reports* **10**, 2108 (2020).

[881] N. V. Ryzhkov, K. G. Nikolaev, A. S. Ivanov and E. V. Skorb, Infochemistry and the future of chemical information processing, *Annual Review of Chemical and Biomolecular Engineering* **12**, 63–95 (2021).

[882] A. Adamatzky and A. Gandia, Fungi anaesthesia, *Scientific Reports* **12**, 340 (2022).
[883] D. Stöckle, B. J. Reyes-Hernández, A. V. Barro, M. Nenadić, Z. Winter, S. Marc-Martin, L. Bald, R. Ursache, S. Fujita, A. Maizel *et al.*, Microtubule-based perception of mechanical conflicts controls plant organ morphogenesis, *Science Advances* **8**, eabm4974 (2022).
[884] C. Yan, F. Wang, Y. Peng, C. R. Williams, B. Jenkins, J. Wildonger, H.-J. Kim, J. B. Perr, J. C. Vaughan, M. E. Kern *et al.*, Microtubule acetylation is required for mechanosensation in Drosophila, *Cell Reports* **25**, 1051–1065 (2018).
[885] A. K. Coleman, H. C. Joca, G. Shi, W. J. Lederer and C. W. Ward, Tubulin acetylation increases cytoskeletal stiffness to regulate mechanotransduction in striated muscle, *Journal of General Physiology* **153**, e202012743 (2021).
[886] A. Falconieri, A. Coppini and V. Raffa, Microtubules as a signal hub for axon growth in response to mechanical force, *Biological Chemistry* **405**, 67–77 (2024).
[887] R. Lahoz-Beltra, S. R. Hameroff and J. E. Dayhoff, Cytoskeletal logic: a model for molecular computation via boolean operations in microtubules and microtubule-associated proteins, *BioSystems* **29**, 1–23 (1993).
[888] J. Tuszyński, S. Hameroff, M. Satarić, B. Trpisová and M. Nip, Ferroelectric behavior in microtubule dipole lattices: implications for information processing, signaling and assembly/disassembly, *Journal of Theoretical Biology* **174**, 371–380 (1995).
[889] J. Pokornỳ, Excitation of vibrations in microtubules in living cells, *Bioelectrochemistry* **63**, 321–326 (2004).
[890] B. Frieden and R. Gatenby, Signal transmission through elements of the cytoskeleton form an optimized information network in eukaryotic cells, *Scientific Reports* **9**, 6110 (2019).
[891] A. Chiolerio, T. C. Draper, R. Mayne and A. Adamatzky, On resistance switching and oscillations in tubulin microtubule droplets, *Journal of Colloid and Interface Science* **560**, 589–595 (2020).
[892] J. Lüders, Nucleating microtubules in neurons: Challenges and solutions, *Developmental Neurobiology* **81**, 273–283 (2021).
[893] J. C. Meiring, B. I. Shneyer and A. Akhmanova, Generation and regulation of microtubule network asymmetry to drive cell polarity, *Current Opinion in Cell Biology* **62**, 86–95 (2020).
[894] A. N. Rao and P. W. Baas, Polarity sorting of microtubules in the axon, *Trends in Neurosciences* **41**, 77–88 (2018).
[895] S. F. van Beuningen and C. C. Hoogenraad, Neuronal polarity: remodeling microtubule organization, *Current Opinion in Neurobiology* **39**, 1–7 (2016).

[896] L. Huang, Y. Peng, X. Tao, X. Ding, R. Li, Y. Jiang and W. Zuo, Microtubule organization is essential for maintaining cellular morphology and function, *Oxidative Medicine and Cellular Longevity* **2022**, 1623181 (2022).

[897] R. L. Margolis, Role of GTP hydrolysis in microtubule treadmilling and assembly, *Proceedings of the National Academy of Sciences USA* **78**, 1586–1590 (1981).

[898] T. C. Michaels, S. Feng, H. Liang and L. Mahadevan, Mechanics and kinetics of dynamic instability, *eLife* **9**, e54077 (2020).

[899] G. Lee, G. Leech, M. J. Rust, M. Das, R. J. McGorty, J. L. Ross and R. M. Robertson-Anderson, Myosin-driven actin-microtubule networks exhibit self-organized contractile dynamics, *Science Advances* **7**, eabe4334 (2021).

[900] M. Hemmat and D. J. Odde, Atomistic basis of microtubule dynamic instability assessed via multiscale modeling, *Annals of Biomedical Engineering* **49**, 1716–1734 (2021).

[901] J. Berezney, B. L. Goode, S. Fraden and Z. Dogic, Extensile to contractile transition in active microtubule-actin composites generates layered asters with programmable lifetimes, *Proceedings of the National Academy of Sciences USA* **119**, 5 (2022).

[902] C. M. Hough, D. N. Purschke, C. Bell, A. P. Kalra, P. J. Oliva, C. Huang, J. A. Tuszynski, B. J. Warkentin and F. A. Hegmann, Disassembly of microtubules by intense terahertz pulses, *Biomedical Optics Express* **12**, 5812–5828 (2021).

[903] R. Wu, J. Guzman-Sepulveda, A. Kalra, J. Tuszynski and A. Dogariu, Thermal hysteresis in microtubule assembly/disassembly dynamics: The aging-induced degradation of tubulin dimers, *Biochemistry and Biophysics Reports* **29**, 101199 (2022).

[904] M. d. R. Cantero, C. Villa Etchegoyen, P. L. Perez, N. Scarinci and H. F. Cantiello, Bundles of brain microtubules generate electrical oscillations, *Scientific Reports* **8**, 11899 (2018).

[905] M. del Rocío Cantero and H. F. Cantiello, Microtubule electrical oscillations and hippocampal function, *Journal of Neurology & Neuromedicine* **5**, 3 (2020).

[906] N. Scarinci, A. Priel, M. d. R. Cantero and H. F. Cantiello, Brain microtubule electrical oscillations-empirical mode decomposition analysis, *Cellular and Molecular Neurobiology* **43**, 2089–2104 (2023).

[907] B. C. Gutierrez, M. R. Pita Almenar, L. J. Martínez, M. Siñeriz Louis, V. H. Albarracín, M. d. R. Cantero and H. F. Cantiello, Honeybee brain oscillations are generated by microtubules. The concept of a brain central oscillator, *Frontiers in Molecular Neuroscience* **14**, 727025 (2021).

[908] B. C. Gutierrez, H. F. Cantiello and M. del Rocío Cantero, Electrical oscillations of isolated brain microtubules, *Biophysical Journal* **120**, 255a–256a (2021).
[909] A. Rai, T. Liu, S. Glauser, E. A. Katrukha, J. Estévez-Gallego, R. Rodríguez-García, W.-S. Fang, J. F. Díaz, M. O. Steinmetz, K.-H. Altmann *et al.*, Taxanes convert regions of perturbed microtubule growth into rescue sites, *Nature Materials* **19**, 355–365 (2020).
[910] A. Marx and E. Mandelkow, A model of microtubule oscillations, *European Biophysics Journal* **22**, 405–421 (1994).
[911] Z. Xiang, C. Tang and L. Yan, A primary model of decoherence in neuronal microtubules based on the interaction hamiltonian between microtubules and plasmon in neurons, *Chinese Physics B* **28**, 048701 (2019).
[912] M. Tegmark, Importance of quantum decoherence in brain processes, *Physical Review E* **61**, 4194 (2000).
[913] D. M. Lockwood, H. Hwang and P. J. Rossky, Electronic decoherence in condensed phases, *Chemical Physics* **268**, 285–293 (2001).
[914] J. H. Cartwright, Quantum noise may limit the mechanosensory sensitivity of cilia in the left-right organizer of the vertebrate bodyplan, *Progress in Biophysics and Molecular Biology* **180–181**, 83–86 (2022).
[915] A. R. Coates, B. W. Lovett and E. M. Gauger, Localisation determines the optimal noise rate for quantum transport, *New Journal of Physics* **23**, 123014 (2021).
[916] P. W. Jones and F. Gabbiani, Impact of neural noise on a sensory-motor pathway signaling impending collision, *Journal of Neurophysiology* **107**, 1067–1079 (2012).
[917] T. Hohmann and F. Dehghani, The cytoskeleton–a complex interacting meshwork, *Cells* **8**, 362 (2019).
[918] E. Tankou, C. B. Tabi and T. C. Kofané, Soliton-mediated ionic pulses and coupled ionic excitations in a dissipative electrical network model of microtubules, *Chaos, Solitons & Fractals* **162**, 112446 (2022).
[919] E. Tankou, C. B. Tabi, A. Mohamadou and T. C. Kofané, Transport memory effects on coupled nonlinear waves in microtubule dynamics, *Chaos, Solitons & Fractals* **181**, 114717 (2024).
[920] H. Huang, S. Gao and M. Bao, Exploring mechanical forces shaping self-organization and morphogenesis during early embryo development, *Annual Review of Cell and Developmental Biology* **40**, 75–96 (2024).

[921] R. Cao, H. Tian, Y. Tian and X. Fu, A hierarchical mechanotransduction system: from macro to micro, *Advanced Science* **11**, 2302327 (2024).

[922] S. R. Hameroff, S. A. Smith and R. C. Watt, Nonlinear electrodynamics in cytoskeletal protein lattices, in *Nonlinear Electrodynamics in Biological Systems*. Springer, 567–583 (1984).

[923] S. R. Hameroff, Quantum coherence in microtubules: A neural basis for emergent consciousness? *Journal of Consciousness Studies* **1**, 91–118 (1994).

[924] M. Jibu, S. Hagan, S. R. Hameroff, K. H. Pribram and K. Yasue, Quantum optical coherence in cytoskeletal microtubules: Implications for brain function, *BioSystems* **32**, 195–209 (1994).

[925] N. E. Mavromatos and D. V. Nanopoulos, On quantum mechanical aspects of microtubules, *International Journal of Modern Physics B* **12**, 517–542 (1998).

[926] S. R. Hameroff and R. C. Watt, Information processing in microtubules, *Journal of Theoretical Biology* **98**, 549–561 (1982).

[927] F. Baluška and M. Levin, On having no head: Cognition throughout biological systems, *Frontiers in Psychology* **7**, 902 (2016).

[928] F. Baluška, W. B. Miller Jr and A. S. Reber, Biomolecular basis of cellular consciousness via subcellular nanobrains, *International Journal of Molecular Sciences* **22**, 2545 (2021).

[929] W. B. Miller Jr, F. Baluška and A. S. Reber, A revised central dogma for the 21st century: All biology is cognitive information processing, *Progress in Biophysics and Molecular Biology* **182**, 34–48 (2023).

[930] A. Mershin, A. A. Kolomenski, H. A. Schuessler and D. V. Nanopoulos, Tubulin dipole moment, dielectric constant and quantum behavior: computer simulations, experimental results and suggestions, *BioSystems* **77**, 73–85 (2004).

[931] A. I. Nesterov, M. F. Ramírez, G. P. Berman and N. E. Mavromatos, Nonlinear dynamics of dipoles in microtubules: Pseudospin model, *Physical Review E* **93**, 062412 (2016).

[932] V. Saenko, S. E. Shirmovsky and D. Shulga, Detailed analysis of the quantum relaxation features in the cell microtubules: Markovian and non-markovian dynamics, *BioSystems* **254**, 105520 (2025).

[933] R. Ramalho, H. Soares and L. Melo, Microtubule behavior under strong electromagnetic fields, *Materials Science and Engineering: C* **27**, 1207–1210 (2007).

[934] M. Taghi, R. Gholamhosein and R.-z. Saeed, Effect of electromagnetic field on the polymerization of microtubules extracted from rat brain, *Recent Patents on Endocrine, Metabolic & Immune Drug Discovery* **6**, 251–254 (2012).

[935] M. Taghi, R. Gholamhosein and R.-z. Saeed, Effect of radio frequency waves of electromagnetic field on the tubulin, *Recent Patents on Endocrine, Metabolic & Immune Drug Discovery* **7**, 252–256 (2013).

[936] S. Sahu, S. Ghosh, D. Fujita and A. Bandyopadhyay, Live visualizations of single isolated tubulin protein self-assembly via tunneling current: effect of electromagnetic pumping during spontaneous growth of microtubule, *Scientific Reports* **4**, 7303 (2014).

[937] J. Zhou, T. Yang, B. Peng, B. Shan, M. Ding and K. Zhang, Structural water molecules confined in soft and hard nanocavities as bright color emitters, *ACS Physical Chemistry Au* **2**, 47–58 (2021).

[938] M. Hertzog, M. Wang, J. Mony and K. Börjesson, Strong light–matter interactions: A new direction within chemistry, *Chemical Society Reviews* **48**, 937–961 (2019).

[939] N. E. Mavromatos, A. Mershin and D. V. Nanopoulos, QED-cavity model of microtubules implies dissipationless energy transfer and biological quantum teleportation, *International Journal of Modern Physics B* **16**, 3623–3642 (2002).

[940] M. Steiner, H. M. Meyer, C. Deutsch, J. Reichel and M. Köhl, Single ion coupled to an optical fiber cavity, *Physical Review Letters* **110**, 043003 (2013).

[941] M. Köhl, Cavity-QED with a trapped ion in an optical fiber cavity, *in CLEO: QELS Fundamental Science*. Optica Publishing Group, pp. FW3A-1 (2015).

[942] H. Takahashi, E. Kassa, C. Christoforou and M. Keller, Strong coupling of a single ion to an optical cavity, *Physical Review Letters* **124**, 013602 (2020).

[943] O. Krivosudskỳ, P. Dráber and M. Cifra, Resolving controversy of unusually high refractive index of a tubulin, *Europhysics Letters* **117**, 38003 (2017).

[944] Z. Wang and B. Yu, Plasmonic control of refractive index without absorption in metallic photonic crystals doped with quantum dots, *Plasmonics* **13**, 567–574 (2018).

[945] N. Li, K.-X. Guo, S. Shao and G.-H. Liu, Polaron effects on the optical absorption coefficients and refractive index changes in a two-dimensional quantum pseudodot system, *Optical Materials* **34**, 1459–1463 (2012).

[946] C. Zhang, Z. Wang, Y. Liu, C. Peng and K. Guo, Polaron effects on the optical refractive index changes in asymmetrical quantum wells, *Physics Letters A* **375**, 484–487 (2011).

[947] V. G. Kravets, F. Wu, G. H. Auton, T. Yu, S. Imaizumi and A. N. Grigorenko, Measurements of electrically tunable refractive index

of MoS_2 monolayer and its usage in optical modulators, *npj 2D Materials and Applications* **3**, 36 (2019).

[948] H. A. Schuessler, A. Mershin, A. A. Kolomenskii and D. V. Nanopoulos, Surface plasmon resonance study of the actin-myosin sarcomeric complex and tubulin dimers, *Journal of Modern Optics* **50**, 2381–2391 (2003).

[949] P. Bon, S. Lécart, E. Fort and S. Lévêque-Fort, Fast label-free cytoskeletal network imaging in living mammalian cells, *Biophysical Journal* **106**, 1588–1595 (2014).

[950] R. Pizzi, G. Strini, S. Fiorentini, V. Pappalardo, M. Pregnolato *et al.*, Evidences of new biophysical properties of microtubules, *Artificial Neural Networks. Nova Science Publishers, Inc* (2010).

[951] P. Kurian, T. Obisesan and T. J. Craddock, Oxidative species-induced excitonic transport in tubulin aromatic networks: Potential implications for neurodegenerative disease, *Journal of Photochemistry and Photobiology B: Biology* **175**, 109–124 (2017).

[952] T. J. A. Craddock and J. A. Tuszynski, A critical assessment of the information processing capabilities of neuronal microtubules using coherent excitations, *Journal of Biological Physics* **36**, 53–70 (2010).

[953] L. B. Tan, O. Cotlet, A. Bergschneider, R. Schmidt, P. Back, Y. Shimazaki, M. Kroner and A. İmamoğlu, Interacting polaron-polaritons, *Physical Review X* **10**, 021011 (2020).

[954] F. J. Garcia-Vidal, C. Ciuti and T. W. Ebbesen, Manipulating matter by strong coupling to vacuum fields, *Science* **373**, eabd0336 (2021).

[955] C. Franchini, M. Reticcioli, M. Setvin and U. Diebold, Polarons in materials, *Nature Reviews Materials* **6**, 560–586 (2021).

[956] M. M. Rolls, Principles of microtubule polarity in linear cells, *Developmental Biology* **483**, 112–117 (2022).

[957] W. Nganfo, C. Kenfack-Sadem, A. Fotué, M. Ekosso, S. Wopunghwo and L. Fai, Dynamics of exciton polaron in microtubule, *Heliyon* **8**, e08897 (2022).

[958] O. Mommadi, A. El Moussaouy, M. Chnafi, M. El Hadi, A. Nougaoui and H. Magrez, Exciton–phonon properties in cylindrical quantum dot with parabolic confinement potential under electric field, *Physica E: Low-dimensional Systems and Nanostructures* **118**, 113903 (2020).

[959] W. Zhang, T. J. Craddock, Y. Li, M. Swartzlander, R. R. Alfano and L. Shi, Fano resonance line shapes in the raman spectra of tubulin and microtubules reveals active quantum effects, *Biophysical Reports* **2**, 100043 (2022).

[960] E. Del Giudice, S. Doglia and M. Milani, A collective dynamics in metabolically active cells, *Physica Scripta* **26**, 232 (1982).

[961] J. Pokorný, J. Hašek and F. Jelínek, Electromagnetic field of microtubules: Effects on transfer of mass particles and electrons, *Journal of Biological Physics* **31**, 501–514 (2005).

[962] M. Cifra, D. Havelka and M. A. Deriu, Electric field generated by longitudinal axial microtubule vibration modes with high spatial resolution microtubule model, in *Journal of Physics: Conference Series*, Vol. 329. IOP Publishing, 012013 (2011).

[963] M. d. R. Cantero, P. L. Perez, M. Smoler, C. Villa Etchegoyen and H. F. Cantiello, Electrical oscillations in two-dimensional microtubular structures, *Scientific Reports* **6**, 27143 (2016).

[964] P. Marracino, D. Havelka, J. Průša, M. Liberti, J. Tuszynski, A. T. Ayoub, F. Apollonio and M. Cifra, Tubulin response to intense nanosecond-scale electric field in molecular dynamics simulation, *Scientific Reports* **9**, 10477 (2019).

[965] A. Farhi and A. Dogariu, Coupling of electrodynamic fields to vibrational modes in helical structures, *Physical Review A* **103**, 023523 (2021).

[966] E. Prodan and C. Prodan, Topological phonon modes and their role in dynamic instability of microtubules, *Physical Review Letters* **103**, 248101 (2009).

[967] V. Subramanyan, K. L. Kirkpatrick, S. Vishveshwara and S. Vishveshwara, Are microtubules electron-based topological insulators? *Europhysics Letters* **143**, 46001 (2023).

[968] S. Shankar, A. Souslov, M. J. Bowick, M. C. Marchetti and V. Vitelli, Topological active matter, *Nature Reviews Physics* **4**, 380–398 (2022).

[969] A. Ganguly, H. Yang, R. Sharma, K. D. Patel and F. Cabral, The role of microtubules and their dynamics in cell migration, *Journal of Biological Chemistry* **287**, 43359–43369 (2012).

[970] H. V. Goodson and E. M. Jonasson, Microtubules and microtubule-associated proteins, *Cold Spring Harbor Perspectives in Biology* **10**, a022608 (2018).

[971] T. Sanchez, D. T. Chen, S. J. DeCamp, M. Heymann and Z. Dogic, Spontaneous motion in hierarchically assembled active matter, *Nature* **491**, 431–434 (2012).

[972] T. Sanchez, D. Welch, D. Nicastro and Z. Dogic, Cilia-like beating of active microtubule bundles, *Science* **333**, 456–459 (2011).

[973] P. McMillen and M. Levin, Collective intelligence: A unifying concept for integrating biology across scales and substrates, *Communications Biology* **7**, 378 (2024).

[974] M. Levin, Collective intelligence of morphogenesis as a teleonomic process, in *Evolution "On Purpose": Teleonomy in Living Systems.* Vienna Series in Theoretical Biology (2023).

[975] Y. Sumino, K. H. Nagai, Y. Shitaka, D. Tanaka, K. Yoshikawa, H. Chaté and K. Oiwa, Large-scale vortex lattice emerging from collectively moving microtubules, *Nature* **483**, 448–452 (2012).

[976] H. Ori, M. Duque, R. Frank Hayward, C. Scheibner, H. Tian, G. Ortiz, V. Vitelli and A. E. Cohen, Observation of topological action potentials in engineered tissues, *Nature Physics* **19**, 290–296 (2023).

[977] M. Levin, The biophysics of regenerative repair suggests new perspectives on biological causation, *BioEssays* **42**, 1900146 (2020).

[978] M. Levin, A. M. Pietak and J. Bischof, Planarian regeneration as a model of anatomical homeostasis: Recent progress in biophysical and computational approaches, in *Seminars in Cell & Developmental Biology*, Vol. 87. Elsevier, 125–144 (2019).

[979] M. Levin, Molecular bioelectricity: How endogenous voltage potentials control cell behavior and instruct pattern regulation in vivo, *Molecular Biology of the Cell* **25**, 3835–3850 (2014).

[980] T. Gulden and A. Kamenev, Dynamics of ion channels via non-Hermitian quantum mechanics, *Entropy* **23**, 125 (2021).

[981] K. Sone, K. Yokomizo, K. Kawaguchi and Y. Ashida, Hermitian and non-Hermitian topology in active matter, *Reports on Progress in Physics* **89**, 016501 (2026).

[982] L. Wang, E. Sadeghnezhad, P. Guan and P. Gong, Microtubules monitor calcium and reactive oxygen species signatures in signal transduction, *Plant Science* **304**, 110589 (2021).

[983] J. Sandblom and J. Galvanovskis, Electromagnetic field absorption in stochastic cellular systems: Enhanced signal detection in ion channels and calcium oscillators, *Chaos, Solitons & Fractals* **11**, 1905–1911 (2000).

[984] A. A. Saha, T. J. Craddock and J. A. Tuszynski, An investigation of the plausibility of stochastic resonance in tubulin dimers, *Biosystems* **107**, 81–87 (2012).

[985] B. McNamara and K. Wiesenfeld, Theory of stochastic resonance, *Physical Review A* **39**, 4854 (1989).

[986] P. Hänggi, Stochastic resonance in biology. How noise can enhance detection of weak signals and help improve biological information processing, *ChemPhysChem* **3**, 285–290 (2002).

[987] D. Petracchi, I. C. Gebeshuber, L. J. DeFelice and A. V. Holden, Stochastic resonance in biological systems, *Chaos Solitons and Fractals* **11**, 1819–1822 (2000).

[988] A. Dodda, A. Oberoi, A. Sebastian, T. H. Choudhury, J. M. Redwing and S. Das, Stochastic resonance in MoS_2 photodetector, *Nature Communications* **11**, 4406 (2020).

[989] M. A. Martin-Delgado, On quantum effects in a theory of biological evolution, *Scientific Reports* **2**, 302 (2012).

[990] F. Arute, K. Arya, R. Babbush, D. Bacon, J. C. Bardin, R. Barends, R. Biswas, S. Boixo, F. G. Brandao, D. A. Buell *et al.*, Quantum supremacy using a programmable superconducting processor, *Nature* **574**, 505–510 (2019).

[991] B. Fefferman, S. Ghosh, M. Gullans, K. Kuroiwa and K. Sharma, Effect of nonunital noise on random-circuit sampling, *PRX Quantum* **5**, 030317 (2024).

[992] B. Meyer, C. Ansorge and T. Nakagaki, The role of noise in self-organized decision making by the true slime mold *Physarum polycephalum, PloS One* **12**, e0172933 (2017).

[993] G. Schmid, I. Goychuk and P. Hänggi, Stochastic resonance as a collective property of ion channel assemblies, *Europhysics Letters* **56**, 22 (2001).

[994] M. D. McDonnell and D. Abbott, What is stochastic resonance? definitions, misconceptions, debates, and its relevance to biology, *PLoS Computational Biology* **5**, e1000348 (2009).

[995] B. Alicea, Stochastic resonance can drive adaptive physiological processes, *Nature Precedings*, 1–13 (2009).

[996] C. Y. Lee, W. Choi, J.-H. Han and M. S. Strano, Coherence resonance in a single-walled carbon nanotube ion channel, *Science* **329**, 1320–1324 (2010).

[997] Y. Kato and H. Nakao, Quantum coherence resonance, *New Journal of Physics* **23**, 043018 (2021).

[998] L. Colin, A. Chevallier, S. Tsugawa, F. Gacon, C. Godin, V. Viasnoff, T. E. Saunders and O. Hamant, Cortical tension overrides geometrical cues to orient microtubules in confined protoplasts, *Proceedings of the National Academy of Sciences USA* **117**, 32731–32738 (2020).

[999] H. Ma and M. Liu, The microtubule cytoskeleton acts as a sensor for stress response signaling in plants, *Molecular Biology Reports* **46**, 5603–5608 (2019).

[1000] B. Moulia, S. Douady and O. Hamant, Fluctuations shape plants through proprioception, *Science* **372**, eabc6868 (2021).

[1001] E. W. Dent, Of microtubules and memory: Implications for microtubule dynamics in dendrites and spines, *Molecular Biology of the Cell* **28**, 1–8 (2017).

[1002] M. Vala, Ł. Bujak, A. Garcia Marin, K. Holanová, V. Henrichs, M. Braun, Z. Lánský and M. Piliarik, Nanoscopic structural fluctuations of disassembling microtubules revealed by label-free super-resolution microscopy, *Small Methods* **5**, 2000985 (2021).

[1003] M. V. Satarić, S. Zdravković, T. Nemeš and B. M. Satarić, Calcium signaling modulates the dynamics of cilia and flagella, *European Biophysics Journal* **49**, 619–631 (2020).

[1004] M. V. Satarić, T. Nemeš, D. Sekulić and J. A. Tuszynski, How signals of calcium ions initiate the beats of cilia and flagella, *BioSystems* **182**, 42–51 (2019).

[1005] R. C. Weisenberg, Microtubule formation in vitro in solutions containing low calcium concentrations, *Science* **177**, 1104–1105 (1972).

[1006] M. Satarić and T. Nemeš, Calcium signaling along actin filaments in stereocilia controls hair-bundle motility, in *Nonlinear Dynamics of Nanobiophysics*. Springer, 307–344 (2022).

[1007] M. Delling, P. G. DeCaen, J. F. Doerner, S. Febvay and D. E. Clapham, Primary cilia are specialized calcium signalling organelles, *Nature* **504**, 311–314 (2013).

[1008] A. Brown and R. Zhang, Primary cilia: A closer look at the antenna of cells, *Current Biology* **30**, R1494-R1496 (2020).

[1009] V. Singla and J. F. Reiter, The primary cilium as the cell's antenna: Signaling at a sensory organelle, *Science* **313**, 629–633 (2006).

[1010] F. Zhao and Y. Long, Mechanosensing, from forces to structures, *Frontiers in Plant Science* **13**, 5047 (2022).

[1011] M. Perc and M. Marhl, Local dissipation and coupling properties of cellular oscillators: A case study on calcium oscillations, *Bioelectrochemistry* **62**, 1–10 (2004).

[1012] M. Caplow, R. Ruhlen and J. Shanks, The free energy for hydrolysis of a microtubule-bound nucleotide triphosphate is near zero: All of the free energy for hydrolysis is stored in the microtubule lattice. *The Journal of Cell Biology* **127**, 779–788 (1994).

[1013] A. R. Calixto, C. Moreira, A. Pabis, C. Kötting, K. Gerwert, T. Rudack and S. C. Kamerlin, GTP hydrolysis without an active site base: A unifying mechanism for Ras and related gtpases, *Journal of the American Chemical Society* **141**, 10684–10701 (2019).

[1014] E. Mandelkow, E.-M. Mandelkow, H. Hotani, B. Hess and S. C. Müller, Spatial patterns from oscillating microtubules, *Science* **246**, 1291–1293 (1989).

[1015] H. Obermann, E. Mandelkow, G. Lange and E. Mandelkow, Microtubule oscillations. Role of nucleation and microtubule number concentration. *Journal of Biological Chemistry* **265**, 4382–4388 (1990).

[1016] S. A. Yadav, D. Khatri, A. Soni, N. Khetan and C. A. Athale, Wave-like oscillations of clamped microtubules driven by collective dynein transport, *Biophysical Journal* **123**, 509–524 (2024).
[1017] N. Glade, J. Demongeot and J. Tabony, Microtubule self-organisation by reaction-diffusion processes causes collective transport and organisation of cellular particles, *BMC Cell Biology* **5**, 23 (2004).
[1018] N. Glade, J. Demongeot and J. Tabony, Numerical simulations of microtubule self-organisation by reaction and diffusion, *Acta Biotheoretica* **50**, 239–268 (2002).
[1019] S. Cortès, N. Glade, I. Chartier and J. Tabony, Microtubule self-organisation by reaction–diffusion processes in miniature cell-sized containers and phospholipid vesicles, *Biophysical Chemistry* **120**, 168–177 (2006).
[1020] S. Zdravkovic, Microtubules: a network for solitary waves, *Journal of the Serbian Chemical Society* **82**, 469–481 (2017).
[1021] M. Satarić, J. Tuszyński and R. Žakula, Kinklike excitations as an energy-transfer mechanism in microtubules, *Physical Review E* **48**, 589 (1993).
[1022] S. Zdravković and G. Gligorić, Kinks and bell-type solitons in microtubules, *Chaos: An Interdisciplinary Journal of Nonlinear Science* **26**, 063101 (2016).
[1023] N. Raza and Z. A. Alhussain, An explicit plethora of soliton solutions for a new microtubules transmission lines model: A fractional comparison, *Modern Physics Letters B* **35**, 2150498 (2021).
[1024] N. E. Mavromatos, Non-linear dynamics in biological microtubules: Solitons and dissipation-free energy transfer, in *Journal of Physics: Conference Series*, Vol. 880. IOP Publishing, 012010 (2017).
[1025] E. Kengne and A. Lakhssassi, Chirped modulated wave excitations in an electrical model of microtubules, *Chaos, Solitons & Fractals* **167**, 113094 (2023).
[1026] F. K. Abdullaev, S. Darmanyan and M. Djumaev, Stochastic parametric resonance of solitons, *Physics Letters A* **141**, 423–426 (1989).
[1027] A. Degasperis, Resource Letter Sol-1: Solitons, *American Journal of Physics* **66**, 486–497 (1998).
[1028] M. V. Satarić, T. Nemeš, B. Satarić, D. Sekulić and S. Zdravković, Calcium ions tune the beats of cilia and flagella, *BioSystems* **196**, 104172 (2020).
[1029] C. Hunley, D. Uribe and M. Marucho, A multi-scale approach to describe electrical impulses propagating along actin filaments in both intracellular and in vitro conditions, *RSC Advances* **8**, 12017–12028 (2018).

[1030] R. Poznanski, L. Cacha, Y. Al-Wesabi, J. Ali, M. Bahadoran, P. Yupapin and J. Yunus, Solitonic conduction of electrotonic signals in neuronal branchlets with polarized microstructure, *Scientific Reports* **7**, 2746 (2017).

[1031] J. Ali, N. Pornsuwancharoen, P. Youplao, I. Amiri, R. Poznanski, K. Chaiwong, S. Punthawanumt, S. Suwanarat and P. Yupapin, Characteristics of an on-chip polariton successively filtered circuit, *Results in Physics* **11**, 410–413 (2018).

[1032] N. Pornsuwancharoen, P. Youplao, I. Amiri, J. Ali, R. Poznanski, K. Chaiwong and P. Yupapin, On-chip polariton generation using an embedded nanograting microring circuit, *Results in Physics* **10**, 913–916 (2018).

[1033] A. Kopf, J. Renkawitz, R. Hauschild, I. Girkontaite, K. Tedford, J. Merrin, O. Thorn-Seshold, D. Trauner, H. Häcker, K.-D. Fischer *et al.*, Microtubules control cellular shape and coherence in amoeboid migrating cells, *Journal of Cell Biology* **219**, e201907154 (2020).

[1034] J. Kroll and J. Renkawitz, Principles of organelle positioning in motile and non-motile cells, *EMBO Reports* **25**, 2172–2187 (2024).

[1035] J. P. Conboy, I. Istúriz Petitjean, A. van der Net and G. H. Koenderink, How cytoskeletal crosstalk makes cells move: Bridging cell-free and cell studies, *Biophysics Reviews* **5**, 2 (2024).

[1036] V. A. Valdez, L. Neahring, S. Petry and S. Dumont, Mechanisms underlying spindle assembly and robustness, *Nature Reviews Molecular Cell Biology* **24**, 523–542 (2023).

[1037] A. Akhmanova and L. C. Kapitein, Mechanisms of microtubule organization in differentiated animal cells, *Nature Reviews Molecular Cell Biology* **23**, 541–558 (2022).

[1038] J. Zhang, Y. Ji, S. Jiang, M. Shi, W. Cai, R. J. Miron and Y. Zhang, Calcium–collagen coupling is vital for biomineralization schedule, *Advanced Science* **8**, 2100363 (2021).

[1039] I. Goychuk and P. Hänggi, Stochastic resonance in ion channels characterized by information theory, *Physical Review E* **61**, 4272 (2000).

[1040] S. M. Bezrukov and I. Vodyanoy, Noise-induced enhancement of signal transduction across voltage-dependent ion channels, *Nature* **378**, 362–364 (1995).

[1041] C. Sun, M. Shen, A. D. Chavez, A. M. Evans, X. Liu, B. Harutyunyan, N. C. Flanders, M. C. Hersam, M. J. Bedzyk, M. Olvera de la Cruz *et al.*, High aspect ratio nanotubes assembled from macrocyclic iminium salts, *Proceedings of the National Academy of Sciences* **115**, 8883–8888 (2018).

[1042] M. Seifi, A. Soltanmanesh and A. Shafiee, Quantum coherence on selectivity and transport of ion channels, *Scientific Reports* **12**, 9237 (2022).

[1043] M. Polakowski and M. Panfil, Quantum features of the transport through ion channels in the soft knock-on model, *Physical Biology* **22**, 016007 (2024).

[1044] A. Vaziri and M. B. Plenio, Quantum coherence in ion channels: resonances, transport and verification, *New Journal of Physics* **12**, 085001 (2010).

[1045] Z. Ganim, A. Tokmakoff and A. Vaziri, Vibrational excitons in ionophores: Experimental probes for quantum coherence-assisted ion transport and selectivity in ion channels, *New Journal of Physics* **13**, 113030 (2011).

[1046] S. M. Bezrukov, Sensing nature's electric fields: ion channels as active elements of linear amplification, in *AIP Conference Proceedings*, Vol. 665. American Institute of Physics, 142–149 (2003).

[1047] J. Galvanovskis and J. Sandblom, Amplification of electromagnetic signals by ion channels, *Biophysical Journal* **73**, 3056–3065 (1997).

[1048] M. Hänze, G. McMurtrie, S. Baumann, L. Malavolti, S. N. Coppersmith and S. Loth, Quantum stochastic resonance of individual fe atoms, *Science Advances* **7**, eabg2616 (2021).

[1049] I. Goychuk, P. Hänggi, J. L. Vega and S. Miret-Artés, Non-markovian stochastic resonance: Three-state model of ion channel gating, *Physical Review E* **71**, 061906 (2005).

[1050] R. Pizzi, S. Fiorentini, G. Strini and M. Pregnolato, Exploring structural and dynamical properties microtubules by means of artificial neural networks, in *Nanotechnology: Concepts, Methodologies, Tools, and Applications*. IGI Global, 191–203 (2014).

[1051] S. Li, C. Wang and P. Nithiarasu, Electromechanical vibration of microtubules and its application in biosensors, *Journal of the Royal Society Interface* **16**, 20180826 (2019).

[1052] L. A. Golbach, L. A. Portelli, H. F. Savelkoul, S. R. Terwel, N. Kuster, R. B. de Vries and B. L. Verburg-van Kemenade, Calcium homeostasis and low-frequency magnetic and electric field exposure: A systematic review and meta-analysis of *in vitro studies, Environment international* **92**, 695–706 (2016).

[1053] A. D. Rosen, Mechanism of action of moderate-intensity static magnetic fields on biological systems, *Cell Biochemistry and Biophysics* **39**, 163–173 (2003).

[1054] M. Tota, L. Jonderko, J. Witek, V. Novickij and J. Kulbacka, Cellular and molecular effects of magnetic fields, *International Journal of Molecular Sciences* **25**, 8973 (2024).

[1055] Y. Zhang, Y. Zhao, Y. Han, H. Liu and Y. Zhan, Effects of magnetic fields on the synchronization of calcium oscillations in coupled cells, *Journal of Computational and Theoretical Nanoscience* **8**, 1963–1966 (2011).

[1056] J. Galvanovskis and J. Sandblom, Periodic forcing of intracellular calcium oscillators theoretical studies of the effects of low frequency fields on the magnitude of oscillations, *Bioelectrochemistry and Bioenergetics* **46**, 161–174 (1998).

[1057] D. B. Zorov, M. Juhaszova and S. J. Sollott, Mitochondrial reactive oxygen species (ROS) and ROS-induced ROS release, *Physiological Reviews* **94**, 909–950 (2014).

[1058] S. Schuster, M. Marhl and T. Höfer, Modelling of simple and complex calcium oscillations: From single-cell responses to intercellular signalling, *European Journal of Biochemistry* **269**, 1333–1355 (2002).

[1059] J. R. Aprille, C. J. Lagace, J. Modica-Napolitano and B. A. Trimmer, Role of nitric oxide and mitochondria in control of firefly flash, *Integrative and Comparative Biology* **44**, 213–219 (2004).

[1060] R. Bychkov, M. Juhaszova, K. Tsutsui, C. Coletta, M. D. Stern, V. A. Maltsev and E. G. Lakatta, Synchronized cardiac impulses emerge from heterogeneous local calcium signals within and among cells of pacemaker tissue, *Clinical Electrophysiology* **6**, 907–931 (2020).

[1061] T. Wilson and J. Hastings, *Bioluminescence: Living Lights, Lights for Living*, Harvard University Press (2013).

[1062] A. T. Winfree, *The Geometry of Biological Time*, Vol. 2, Springer (1980).

[1063] C. Wilson and C. González-Billault, Regulation of cytoskeletal dynamics by redox signaling and oxidative stress: Implications for neuronal development and trafficking, *Frontiers in Cellular Neuroscience* **9**, 381 (2015).

[1064] X. Wu, J. Du, W. Song, M. Cao, S. Chen and R. Xia, Weak power frequency magnetic fields induce microtubule cytoskeleton reorganization depending on the epidermal growth factor receptor and the calcium related signaling, *PloS One* **13**, e0205569 (2018).

[1065] J. Ramsay and D. R. Kattnig, Radical triads, not pairs, may explain effects of hypomagnetic fields on neurogenesis, *PLoS Computational Biology* **18**, e1010519 (2022).

[1066] H. Morimoto, M. Kanastu-Shinohara, N. Ogonuki, S. Kamimura, A. Ogura, C. Yabe-Nishimura, Y. Mori, T. Morimoto, S. Watanabe, K. Otsu *et al.*, Ros amplification drives mouse spermatogonial stem cell self-renewal, *Life Science Alliance* **2**, 2 (2019).

[1067] S. S. Sabharwal and P. T. Schumacker, Mitochondrial ROS in cancer: Initiators, amplifiers or an Achilles' heel? *Nature Reviews Cancer* **14**, 709–721 (2014).
[1068] H. Pelicano, D. Carney and P. Huang, Ros stress in cancer cells and therapeutic implications, *Drug Resistance Updates* **7**, 97–110 (2004).
[1069] H. Wang and X. Zhang, Magnetic fields and reactive oxygen species, *International Journal of Molecular Sciences* **18**, 2175 (2017).
[1070] J. Zhang, X. Wang, V. Vikash, Q. Ye, D. Wu, Y. Liu and W. Dong, Ros and ros-mediated cellular signaling, *Oxidative Medicine and Cellular Longevity* **2016** (2016).
[1071] A. J. Rosenspire, A. L. Kindzelskii, B. J. Simon and H. R. Petty, Realtime control of neutrophil metabolism by very weak ultra-low frequency pulsed magnetic fields, *Biophysical Journal* **88**, 3334–3347 (2005).
[1072] S.-F. Wang, S. Chen, L.-M. Tseng and H.-C. Lee, Role of the mitochondrial stress response in human cancer progression, *Experimental Biology and Medicine* **245**, 861–878 (2020).
[1073] J. Luo, K. Mills, S. le Cessie, R. Noordam and D. van Heemst, Ageing, age-related diseases and oxidative stress: What to do next? *Ageing Research Reviews* **57**, 100982 (2020).
[1074] I. Liguori, G. Russo, F. Curcio, G. Bulli, L. Aran, D. Della-Morte, G. Gargiulo, G. Testa, F. Cacciatore, D. Bonaduce *et al.*, Oxidative stress, aging, and diseases, *Clinical Interventions in Aging* **13**, 757 (2018).
[1075] R. Chen, U. H. Lai, L. Zhu, A. Singh, M. Ahmed and N. R. Forsyth, Reactive oxygen species formation in the brain at different oxygen levels: the role of hypoxia inducible factors, *Frontiers in Cell and Developmental Biology* **6**, 132 (2018).
[1076] A. Cipriano, M. Viviano, A. Feoli, C. Milite, G. Sarno, S. Castellano and G. Sbardella, Nadph oxidases: From molecular mechanisms to current inhibitors, *Journal of Medicinal Chemistry* **66**, 11632–11655 (2023).
[1077] D. Schuermann and M. Mevissen, Manmade electromagnetic fields and oxidative stress–biological effects and consequences for health, *International Journal of Molecular Sciences* **22**, 3772 (2021).
[1078] H. Ahsan, A. Ali and R. Ali, Oxygen free radicals and systemic autoimmunity, *Clinical & Experimental Immunology* **131**, 398–404 (2003).
[1079] N. Lane, *Oxygen: The Molecule that Made the World.* Oxford University Press (2002).

[1080] T. P. Cash, Y. Pan and M. C. Simon, Reactive oxygen species and cellular oxygen sensing, *Free Radical Biology and Medicine* **43**, 1219–1225 (2007).

[1081] J. Toffaletti and C. Rackley, Monitoring oxygen status, *Advances in Clinical Chemistry* 77, 103–124 (2016).

[1082] R. Ubbink, M. A. W. Bettink, R. Janse, F. A. Harms, T. Johannes, F. M. Münker and E. G. Mik, A monitor for cellular oxygen metabolism (comet): Monitoring tissue oxygenation at the mitochondrial level, *Journal of Clinical Monitoring and Computing* **31**, 1143–1150 (2017).

[1083] K. Bedard and K.-H. Krause, The NOX family of ROS-generating NADPH oxidases: Physiology and pathophysiology, *Physiological Reviews* **87**, 245–313 (2007).

[1084] N. Demaurex and G. L. Petheö, Electron and proton transport by nadph oxidases, *Philosophical Transactions of the Royal Society B: Biological Sciences* **360**, 2315–2325 (2005).

[1085] R. Edge and T. G. Truscott, The reactive oxygen species singlet oxygen, hydroxy radicals, and the superoxide radical anion—Examples of their roles in biology and medicine, *Oxygen* **1**, 77–95 (2021).

[1086] F. Filip-Ciubotaru, C. Manciuc, G. STOLERU and F. Liliana, Nadph oxidase: Structure and activation mechanisms (review). Note i, *The Medical Surgical Journal* **120**, 29–33 (2016).

[1087] A. W. Segal, The function of the nadph oxidase of phagocytes and its relationship to other NOXs in plants, invertebrates, and mammals, *The International Journal of Biochemistry & Cell Biology* **40**, 604–618 (2008).

[1088] A. R. Cross and A. W. Segal, The nadph oxidase of professional phagocytes–prototype of the NOX electron transport chain systems, *Biochimica et Biophysica Acta (BBA)-Bioenergetics* **1657**, 1–22 (2004).

[1089] D. C. Wallace *et al.*, A mitochondrial bioenergetic etiology of disease, *The Journal of Clinical Investigation* **123**, 1405–1412 (2013).

[1090] V. Reshetnikov, J. Hahn, C. Maueröder, C. Czegley, L. E. Munoz, M. Herrmann, M. H. Hoffmann and A. Mokhir, Chemical tools for targeted amplification of reactive oxygen species in neutrophils, *Frontiers in Immunology* **9**, 1827 (2018).

[1091] A. Banstola, K. Poudel, S. Pathak, P. Shrestha, J. O. Kim, J.-H. Jeong and S. Yook, Hypoxia-mediated ROS amplification triggers mitochondriamediated apoptotic cell death via PD-L1/ROS-responsive, dual-targeted, drug-laden thioketal nanoparticles, *ACS Applied Materials & Interfaces* **13**, 22955–22969 (2021).

[1092] P. Hernansanz-Agustín, C. Choya-Foces, S. Carregal-Romero, E. Ramos, T. Oliva, T. Villa-Piña, L. Moreno, A. Izquierdo-Álvarez, J. D. Cabrera-García, A. Cortés *et al.*, Na^+ controls hypoxic signalling by the mitochondrial respiratory chain, *Nature* **586**, 287–291 (2020).
[1093] M. You, Z. Xie, N. Zhang, Y. Zhang, D. Xiao, S. Liu, W. Zhuang, L. Li and Y. Tao, Signaling pathways in cancer metabolism: Mechanisms and therapeutic targets, *Signal Transduction and Targeted Therapy* **8**, 196 (2023).
[1094] J. Wang, W. Wang, Q. Shen, L. Lan, C. Guan, X. Xu, W. Li and Y. Du, Cell-surface photochemistry mediated calcium overload for synergistic tumor therapy, *Journal of Nanobiotechnology* **21**, 335 (2023).
[1095] M. Murgia, C. Giorgi, P. Pinton and R. Rizzuto, Controlling metabolism and cell death: At the heart of mitochondrial calcium signalling, *Journal of Molecular and Cellular Cardiology* **46**, 781–788 (2009).
[1096] D. Tomar, M. Thomas, J. F. Garbincius, D. W. Kolmetzky, O. Salik, P. Jadiya, A. C. Carpenter and J. W. Elrod, Micu1 regulates mitochondrial cristae structure and function independent of the mitochondrial calcium uniporter channel, *BioRxiv*, 803213 (2019).
[1097] S. Romero-Garcia and H. Prado-Garcia, Mitochondrial calcium: Transport and modulation of cellular processes in homeostasis and cancer, *International Journal of Oncology* **54**, 1155–1167 (2019).
[1098] T. E. Gunter, D. I. Yule, K. K. Gunter, R. A. Eliseev and J. D. Salter, Calcium and mitochondria, *FEBS Letters* **567**, 96–102 (2004).
[1099] B. K. Puri, Calcium signaling and gene expression, in *Calcium Signaling*. Springer, 537–545 (2020).
[1100] C. Giorgi, S. Marchi and P. Pinton, The machineries, regulation and cellular functions of mitochondrial calcium, *Nature Reviews Molecular Cell Biology* **19**, 713–730 (2018).
[1101] F. L. BYGRAVE, Mitochondria and the control of intracellular calcium, *Biological Reviews* **53**, 43–79 (1978).
[1102] X. Liu, H. Hu, J. Chen, Y. Li, Y. Cheng, H. Wei and C.-Y. Yu, Engineering of calcium overload for state-of-the-art tumor therapy, *Chemical Engineering Journal* **501**, 157747 (2024).
[1103] R. R. Heymann, M. D. Thum, A. L. Hardee and D. E. Falvey, Visible light initiated release of calcium ions through photochemical electron transfer reactions, *Photochemical & Photobiological Sciences* **16**, 1003–1008 (2017).

[1104] L. Contreras, I. Drago, E. Zampese and T. Pozzan, Mitochondria: The calcium connection, *Biochimica et Biophysica Acta (BBA)-Bioenergetics* **1797**, 607–618 (2010).

[1105] R. R. Mould, I. Kalampouka, E. L. Thomas, G. W. Guy, A. V. Nunn and J. D. Bell, Non-chemical signalling between mitochondria, *Frontiers in Physiology* **14**, 1268075 (2023).

[1106] J. Tong, Biophoton signaling in mediation of cell-to-cell communication and radiation-induced bystander effects, *Radiation Medicine and Protection* **5**, 145–160 (2024).

[1107] B. F. Matarèse, A. Rusin, C. Seymour and C. Mothersill, Quantum biology and the potential role of entanglement and tunneling in non-targeted effects of ionizing radiation: A review and proposed model, *International Journal of Molecular Sciences* **24**, 16464 (2023).

[1108] Y. Song, Y. Zhou and X. Zhou, The role of mitophagy in innate immune responses triggered by mitochondrial stress, *Cell Communication and Signaling* **18**, 186 (2020).

[1109] E. Rosta, Qm/mm free energy calculations for enzyme catalytic reactions, *Advances in Quantum Chemistry* (2025).

[1110] F. Magnani, S. Nenci, E. Millana Fananas, M. Ceccon, E. Romero, M. W. Fraaije and A. Mattevi, Crystal structures and atomic model of nadph oxidase, *Proceedings of the National Academy of Sciences USA* **114**, 6764–6769 (2017).

[1111] T. E. DeCoursey, The voltage-gated proton channel: A riddle, wrapped in a mystery, inside an enigma, *Biochemistry* **54**, 3250–3268 (2015).

[1112] J. A. Miles, J. L. Egan, J. A. Fowler, P. Machattou, A. D. Millard, C. J. Perry, D. J. Scanlan and P. C. Taylor, The evolutionary origins of peroxynitrite signalling, *Biochemical and Biophysical Research Communications* **580**, 107–112 (2021).

[1113] M. Sedeek, R. L. Hébert, C. R. Kennedy, K. D. Burns and R. M. Touyz, Molecular mechanisms of hypertension: Role of NOX family NADPH oxidases, *Current Opinion in Nephrology and Hypertension* **18**, 122127 (2009).

[1114] M. Kellner, S. Noonepalle, Q. Lu, A. Srivastava, E. Zemskov and S. M. Black, ROS signaling in the pathogenesis of acute lung injury (ALI) and acute respiratory distress syndrome (ARDS), *Pulmonary Vyasculature Redox Signaling in Health and Disease*, 105–137 (2017).

[1115] A. Tarafdar and G. Pula, The role of NADPH oxidases and oxidative stress in neurodegenerative disorders, *International Journal of Molecular Sciences* **19**, 3824 (2018).

[1116] A. Görlach, K. Bertram, S. Hudecova and O. Krizanova, Calcium and ROS: A mutual interplay, *Redox Biology* **6**, 260–271 (2015).

[1117] D. G. Nicholls, Mitochondria and calcium signaling, *Cell Calcium* **38**, 311–317 (2005).

[1118] C. Horbinski and C. T. Chu, Kinase signaling cascades in the mitochondrion: a matter of life or death, *Free Radical Biology and Medicine* **38**, 2–11 (2005).

[1119] P. S. Brookes, Y. Yoon, J. L. Robotham, M. Anders and S.-S. Sheu, Calcium, ATP, and ROS: A mitochondrial love-hate triangle, *American Journal of Physiology-Cell Physiology* **287**, C817–C833 (2004).

[1120] F. Fontana and P. Limonta, The multifaceted roles of mitochondria at the crossroads of cell life and death in cancer, *Free Radical Biology and Medicine* **176**, 203–221 (2021).

[1121] D. F. Kashatus, The regulation of tumor cell physiology by mitochondrial dynamics, *Biochemical and Biophysical Research Communications* **500**, 9–16 (2018).

[1122] S. Nagdas and D. F. Kashatus, The interplay between oncogenic signaling networks and mitochondrial dynamics, *Antioxidants* **6**, 33 (2017).

[1123] R. Vona, A. M. Mileo and P. Matarrese, Microtubule-based mitochondrial dynamics as a valuable therapeutic target in cancer, *Cancers* **13**, 5812 (2021).

[1124] Y. Q. Tan, X. Zhang, S. Zhang, T. Zhu, M. Garg, P. E. Lobie and V. Pandey, Mitochondria: The metabolic switch of cellular oncogenic transformation, *Biochimica et Biophysica Acta (BBA)-Reviews on Cancer* **1876**, 188534 (2021).

[1125] V. Aggarwal, H. S. Tuli, A. Varol, F. Thakral, M. B. Yerer, K. Sak, M. Varol, A. Jain, M. A. Khan and G. Sethi, Role of reactive oxygen species in cancer progression: Molecular mechanisms and recent advancements, *Biomolecules* **9**, 735 (2019).

[1126] P. E. Porporato, V. L. Payen, J. Pérez-Escuredo, C. J. De Saedeleer, P. Danhier, T. Copetti, S. Dhup, M. Tardy, T. Vazeille, C. Bouzin *et al.*, A mitochondrial switch promotes tumor metastasis, *Cell Reports* **8**, 754–766 (2014).

[1127] J. Ježek, K. F. Cooper and R. Strich, Reactive oxygen species and mitochondrial dynamics: the yin and yang of mitochondrial dysfunction and cancer progression, *Antioxidants* **7**, 13 (2018).

[1128] X. Liu, Z. Liu, Z. Liu, S. Zhang, K. Bechkoum, M. Clark and L. Ren, The effects of bio-inspired electromagnetic fields on normal and cancer cells, *Journal of Bionic Engineering* **16**, 943–953 (2019).

[1129] D. K. Meijer, H. J. Geesink *et al.*, Favourable and unfavourable emf frequency patterns in cancer: Perspectives for improved therapy and prevention, *Journal of Cancer Therapy* **9**, 188 (2018).

[1130] E. G. Novoselova, V. V. Novikov, S. M. Lunin, O. V. Glushkova, T. V. Novoselova, S. B. Parfenyuk, S. V. Novoselov, M. O. Khrenov and E. E. Fesenko, Effects of low-level combined static and weak low-frequency alternating magnetic fields on cytokine production and tumor development in mice, *Electromagnetic Biology and Medicine* **38**, 74–83 (2019).

[1131] X. Guo, N. Yang, W. Ji, H. Zhang, X. Dong, Z. Zhou, L. Li, H.-M. Shen, S. Q. Yao and W. Huang, Mito-bomb: Targeting mitochondria for cancer therapy, *Advanced Materials* **33**, 2007778 (2021).

[1132] J. Pokorný, J. Pokorný and F. Borodavka, Warburg effect—damping of electromagnetic oscillations, *Electromagnetic Biology and Medicine* **36**, 270–278 (2017).

[1133] J. M. Winter, T. Yadav and J. Rutter, Stressed to death: Mitochondrial stress responses connect respiration and apoptosis in cancer, *Molecular Cell* **82**, 3321–3332 (2022).

[1134] Y. G. Lee, D. H. Park and Y. C. Chae, Role of mitochondrial stress response in cancer progression, *Cells* **11**, 771 (2022).

[1135] J. Kim and J.-H. Cheong, Role of mitochondria-cytoskeleton interactions in the regulation of mitochondrial structure and function in cancer stem cells, *Cells* **9**, 1691 (2020).

[1136] M. F. Clarke, Clinical and therapeutic implications of cancer stem cells, *New England Journal of Medicine* **380**, 2237–2245 (2019).

[1137] E. Batlle and H. Clevers, Cancer stem cells revisited, *Nature Medicine* **23**, 1124–1134 (2017).

[1138] G.-m. Wang, J.-p. Fu, W.-c. Mo, H.-t. Zhang, Y. Liu and R.-q. He, Shielded geomagnetic field accelerates glucose consumption in human neuroblastoma cells by promoting anaerobic glycolysis, *Biochemical and Biophysical Research Communications* **601**, 101–108 (2022).

[1139] S. Pavlides, D. Whitaker-Menezes, R. Castello-Cros, N. Flomenberg, A. K. Witkiewicz, P. G. Frank, M. C. Casimiro, C. Wang, P. Fortina, S. Addya *et al.*, The reverse warburg effect: Aerobic glycolysis in cancer associated fibroblasts and the tumor stroma, *Cell Cycle* **8**, 3984–4001 (2009).

[1140] J. Huang, L. Zhang, D. Wan, L. Zhou, S. Zheng, S. Lin and Y. Qiao, Extracellular matrix and its therapeutic potential for cancer treatment, *Signal Transduction and Targeted Therapy* **6**, 153 (2021).

[1141] G. Giachin, M. Jessop, R. Bouverot, S. Acajjaoui, M. Saïdi, A. Chretien, M. Bacia-Verloop, L. Signor, P. J. Mas, A. Favier *et al.*, Assembly of the mitochondrial complex I assembly complex

suggests a regulatory role for deflavination, *Angewandte Chemie International Edition* **60**, 4689–4697 (2021).
[1142] C. Gleason, S. Huang, L. F. Thatcher, R. C. Foley, C. R. Anderson, A. J. Carroll, A. H. Millar and K. B. Singh, Mitochondrial complex II has a key role in mitochondrial-derived reactive oxygen species influence on plant stress gene regulation and defense, *Proceedings of the National Academy of Sciences USA* **108**, 10768–10773 (2011).
[1143] K. Hadrava Vanova, M. Kraus, J. Neuzil and J. Rohlena, Mitochondrial complex II and reactive oxygen species in disease and therapy, *Redox Report* **25**, 26–32 (2020).
[1144] A. Bezawork-Geleta, J. Rohlena, L. Dong, K. Pacak and J. Neuzil, Mitochondrial complex II: At the crossroads, *Trends in Biochemical Sciences* **42**, 312–325 (2017).
[1145] M. Wikström and R. Springett, Thermodynamic efficiency, reversibility, and degree of coupling in energy conservation by the mitochondrial respiratory chain, *Communications Biology* **3**, 451 (2020).
[1146] A. P. West, G. S. Shadel and S. Ghosh, Mitochondria in innate immune responses, *Nature Reviews Immunology* **11**, 389–402 (2011).
[1147] B. Banoth and S. L. Cassel, Mitochondria in innate immune signaling, *Translational Research* **202**, 52–68 (2018).
[1148] C. C. Winterbourn, Reconciling the chemistry and biology of reactive oxygen species, *Nature Chemical Biology* **4**, 278–286 (2008).
[1149] W. Kühlbrandt, Structure and function of mitochondrial membrane protein complexes, *BMC Biology* **13**, 89 (2015).
[1150] H. J. Forman, Redox signaling: An evolution from free radicals to aging, *Free Radical Biology and Medicine* **97**, 398–407 (2016).
[1151] D. Pendin, R. Filadi and P. Pizzo, The concerted action of mitochondrial dynamics and positioning: New characters in cancer onset and progression, *Frontiers in Oncology* **7**, 102 (2017).
[1152] I. Shaev, V. Novikov, E. Yablokova and E. Fesenko, A brief review of the current state of research on the biological effects of weak magnetic fields, *Biophysics* **67**, 245–251 (2022).
[1153] L. Bleier and S. Dröse, Superoxide generation by complex III: from mechanistic rationales to functional consequences, *Biochimica et Biophysica Acta (BBA)-Bioenergetics* **1827**, 1320–1331 (2013).
[1154] J. A. Stuart and M. F. Brown, Mitochondrial DNA maintenance and bioenergetics, *Biochimica et Biophysica Acta (BBA)—Bioenergetics* **1757**, 79–89 (2006).
[1155] J. F. Turrens and A. Boveris, Generation of superoxide anion by the NADH dehydrogenase of bovine heart mitochondria, *Biochemical Journal* **191**, 421–427 (1980).

[1156] H. Gong, J. Li, A. Xu, Y. Tang, W. Ji, R. Gao, S. Wang, L. Yu, C. Tian, J. Li *et al.*, An electron transfer path connects subunits of a mycobacterial respiratory supercomplex, *Science* **362**, eaat8923 (2018).

[1157] M. Picard and O. S. Shirihai, Mitochondrial signal transduction, *Cell Metabolism* **34**, 1620–1653 (2022).

[1158] T. S. Fung, R. Chakrabarti and H. N. Higgs, The multiple links between actin and mitochondria, *Nature Reviews Molecular Cell Biology* **24**, 651–667 (2023).

[1159] A. S. Monzel, J. A. Enríquez and M. Picard, Multifaceted mitochondria: moving mitochondrial science beyond function and dysfunction, *Nature Metabolism* **5**, 546–562 (2023).

[1160] W. T. Schneider, R. A. Holland and O. Lindecke, Over 50 years of behavioural evidence on the magnetic sense in animals: What has been learnt and how? *The European Physical Journal Special Topics* **232**, 269–278 (2023).

[1161] H. Burda, S. Begall, J. Červený, J. Neef and P. Němec, Extremely low-frequency electromagnetic fields disrupt magnetic alignment of ruminants, *Proceedings of the National Academy of Sciences* **106**, 5708–5713 (2009).

[1162] P. Slaby, K. Tomanová and M. Vácha, Cattle on pastures do align along the north-south axis, but the alignment depends on herd density, *Journal of Comparative Physiology A* **199**, 695–701 (2013).

[1163] S. Johnsen and K. J. Lohmann, Magnetoreception in animals, *Physics Today* **61**, 29–35 (2008).

[1164] A. Pinzon-Rodriguez and R. Muheim, Zebra finches have a lightdependent magnetic compass similar to migratory birds, *The Journal of Experimental Biology* **220**, 1202–1209 (2017).

[1165] E. P. Malkemper, S. H. Eder, S. Begall, J. B. Phillips, M. Winklhofer, V. Hart and H. Burda, Magnetoreception in the wood mouse (Apodemus sylvaticus): Influence of weak frequency-modulated radio frequency fields, *Scientific Reports* **4**, 1–6 (2015).

[1166] Deutschlander, Phillips and Borland, The case for light-dependent magnetic orientation in animals, *The Journal of Experimental Biology* **202**, 891–908 (1999a).

[1167] M. E. Deutschlander, S. C. Borland and J. B. Phillip, Extraocular magnetic compass in newts, *Nature* **400**, 324–325 (1999).

[1168] K.-S. Chae, S.-C. Kim, H.-J. Kwon and Y. Kim, Human magnetic sense is mediated by a light and magnetic field resonance-dependent mechanism, *Scientific Reports* **12**, 8997 (2022).

[1169] B. Zhang, L. Wang, A. Zhan, M. Wang, L. Tian, W. Guo and Y. Pan, Long-term exposure to a hypomagnetic field attenuates

adult hippocampal neurogenesis and cognition, *Nature Communications* **12**, 1174 (2021).
[1170] G.-J. Wan, S.-L. Jiang, M. Zhang, J.-Y. Zhao, Y.-C. Zhang, W.-D. Pan, G. A. Sword and F.-J. Chen, Geomagnetic field absence reduces adult body weight of a migratory insect by disrupting feeding behavior and appetite regulation, *Insect Science* **28**, 251–260 (2021).
[1171] W.-c. Mo, Y. Liu, H. M. Cooper and R.-q. He, Altered development of xenopus embryos in a hypogeomagnetic field, *Bioelectromagnetics* **33**, 238–246 (2012).
[1172] M. Asashima, K. Shimada and C. J. Pfeiffer, Magnetic shielding induces early developmental abnormalities in the newt, cynops pyrrhogaster, *Bioelectromagnetics* **12**, 215–224 (1991).
[1173] D. L. Wang, X. S. Wang, R. Xiao, Y. Liu and R. Q. He, Tubulin assembly is disordered in a hypogeomagnetic field, *Biochemical and Biophysical Research Communications* **376**, 363–368 (2008).
[1174] W.-C. Mo, Z.-J. Zhang, D.-L. Wang, Y. Liu, P. F. Bartlett and R.-Q. He, Shielding of the geomagnetic field alters actin assembly and inhibits cell motility in human neuroblastoma cells, *Scientific Reports* **6**, 22624 (2016).
[1175] H. Zadeh-Haghighi, R. Rishabh and C. Simon, Hypomagnetic field effects as a potential avenue for testing the radical pair mechanism in biology, *Frontiers in Physics* **11**, 1026460 (2023).
[1176] R. M. Sarimov, D. A. Serov and S. V. Gudkov, Hypomagnetic conditions and their biological action, *Biology* **12**, 1513 (2023).
[1177] B. D'Autréaux and M. B. Toledano, ROS as signalling molecules: mechanisms that generate specificity in ROS homeostasis, *Nature Reviews Molecular Cell Biology* **8**, 813–824 (2007).
[1178] V. Novikov, E. Yablokova, A. Kadyrkov and E. Fesenko, The effect of pulsed magnetic fields on the production of reactive oxygen species by neutrophils, *Biophysics* **66**, 614–622 (2021).
[1179] R. Rishabh, H. Zadeh-Haghighi, D. Salahub and C. Simon, Radical pairs may explain reactive oxygen species-mediated effects of hypomagnetic field on neurogenesis, *PLoS Computational Biology* **18**, e1010198 (2022).
[1180] A. S. Parmagnani, N. Betterle, G. Mannino, S. D'Alessandro, F. F. Nocito, K. Ljumovic, G. Vigani, M. Ballottari and M. E. Maffei, The geomagnetic field (GMF) is required for lima bean photosynthesis and reactive oxygen species production, *International Journal of Molecular Sciences* **24**, 2896 (2023).
[1181] R. M. Sherrard, N. Morellini, N. Jourdan, M. El-Esawi, L.-D. Arthaut, C. Niessner, F. Rouyer, A. Klarsfeld, M. Doulazmi, J.

Witczak *et al.*, Low-intensity electromagnetic fields induce human cryptochrome to modulate intracellular reactive oxygen species, *PLoS Biology* **16**, e2006229 (2018).

[1182] F. Barnes and B. Greenebaum, Role of radical pairs and feedback in weak radio frequency field effects on biological systems, *Environmental Research* **163**, 165–170 (2018).

[1183] T. C. Player, E. D. Baxter, S. Allatt and P. Hore, Amplification of weak magnetic field effects on oscillating reactions, *Scientific Reports* **11**, 9615 (2021).

[1184] P. J. Hore, Upper bound on the biological effects of 50/60 hz magnetic fields mediated by radical pairs, *Elife* **8**, e44179 (2019).

[1185] P. Hore, Are biochemical reactions affected by weak magnetic fields? *Proceedings of the National Academy of Sciences USA* **109**, 1357–1358 (2012).

[1186] A. Zandieh, S. P. Shariatpanahi, A. A. Ravassipour, J. Azadipour, M. S. Nezamtaheri, Z. Habibi-Kelishomi, M. Ghanizadeh, A. Same-Majandeh, K. Majidzadeh-A, A. Taheri *et al.*, An amplification mechanism for weak ELF magnetic fields quantum-bio effects in cancer cells, *Scientific Reports* **15**, 2964 (2025).

[1187] J. R. Woodward, Triplet born radical pairs and the low field effect, *Applied Magnetic Resonance* **54**, 47–58 (2023).

[1188] I. R. Gould, J. A. Boiani, E. B. Gaillard, J. L. Goodman and S. Farid, Intersystem crossing in charge-transfer excited states, *The Journal of Physical Chemistry A* **107**, 3515–3524 (2003).

[1189] J. Smith, H. Zadeh Haghighi, D. Salahub and C. Simon, Radical pairs may play a role in xenon-induced general anesthesia, *Scientific Reports* **11**, 6287 (2021).

[1190] H. Zadeh-Haghighi and C. Simon, Radical pairs may play a role in microtubule reorganization, *Scientific Reports* **12**, 6109 (2022).

[1191] T. J. A. Craddock, P. Kurian, J. Preto, K. Sahu, S. R. Hameroff, M. Klobukowski and J. A. Tuszynski, Anesthetic alterations of collective terahertz oscillations in tubulin correlate with clinical potency: Implications for anesthetic action and post-operative cognitive dysfunction, *Scientific Reports* **7**, 9877 (2017).

[1192] T. J. A. Craddock, S. R. Hameroff, A. T. Ayoub, M. Klobukowski and J. A. Tuszynski, Anesthetics act in quantum channels in brain microtubules to prevent consciousness, *Current Topics in Medicinal Chemistry* **15**, 523–533 (2015).

[1193] N. Glade and J. Tabony, Brief exposure to high magnetic fields determines microtubule self-organisation by reaction–diffusion processes, *Biophysical Chemistry* **115**, 29–35 (2005).

[1194] W. Bras, G. P. Diakun, J. Diaz, G. Maret, H. Kramer, J. Bordas and F. Medrano, The susceptibility of pure tubulin to high magnetic fields: a magnetic birefringence and x-ray fiber diffraction study, *Biophysical Journal* **74**, 1509–1521 (1998).

[1195] H. Cadiou and P. A. McNaughton, Avian magnetite-based magnetoreception: A physiologist's perspective, *Journal of the Royal Society Interface* **7**, suppl_2, S193–S205 (2010).

[1196] J. Park, F. Koehler, G. Varnavides, M.-J. Antonini and P. Anikeeva, Influence of magnetic fields on electrochemical reactions of redox cofactor solutions, *Angewandte Chemie* **133**, 18443–18450 (2021).

[1197] V. N. Binhi and F. S. Prato, Biological effects of the hypomagnetic field: An analytical review of experiments and theories, *PLoS One* **12**, e0179340 (2017).

[1198] L. F. Agnez-Lima, J. T. Melo, A. E. Silva, A. H. S. Oliveira, A. R. S. Timoteo, K. M. Lima-Bessa, G. R. Martinez, M. H. Medeiros, P. Di Mascio, R. S. Galhardo *et al.*, DNA damage by singlet oxygen and cellular protective mechanisms, *Mutation Research/Reviews in Mutation Research* **751**, 15–28 (2012).

[1199] G. Bauer, The antitumor effect of singlet oxygen, *Anticancer Research* **36**, 5649–5663 (2016).

[1200] J. da Veiga Moreira, L. Schwartz and M. Jolicoeur, Targeting mitochondrial singlet oxygen dynamics offers new perspectives for effective metabolic therapies of cancer, *Frontiers in Oncology* **10**, 573399 (2020).

[1201] V. Novikov, E. Yablokova, I. Shaev and E. Fesenko, The kinetics of the production of reactive oxygen species by neutrophils after incubation in a hypomagnetic field, *Biophysics* **66**, 434–437 (2021).

[1202] V. Novikov, E. Yablokova and E. Fesenko, The effect of a "zero" magnetic field on the production of reactive oxygen species in neutrophils, *Biophysics* **63**, 365–368 (2018).

[1203] V. Novikov, E. Yablokova and E. Fesenko, The effect of weak magnetic fields on the production of reactive oxygen species in neutrophils, *Biophysics* **61**, 959–962 (2016).

[1204] M. Pooam, N. Jourdan, B. Aguida, C. Dahon, S. Baouz, C. Terry, H. Raad and M. Ahmad, Exposure to 1.8 ghz radiofrequency field modulates ROS in human HEK293 cells as a function of signal amplitude, *Communicative & Integrative Biology* **15**, 54–66 (2022).

[1205] J. C. Murray and R. W. Farndale, Modulation of collagen production in cultured fibroblasts by a low-frequency, pulsed magnetic field, *Biochimica et Biophysica Acta (BBA)-General Subjects* **838**, 98–105 (1985).

[1206] M. Jedrzejczak-Silicka, M. Kordas, M. Konopacki and R. Rakoczy, Modulation of cellular response to different parameters of the rotating magnetic field (RMF)—an in vitro wound healing study, *International Journal of Molecular Sciences* **22**, 5785 (2021).

[1207] F. Barnes and B. Greenebaum, Possible mechanism for synchronized detection of weak magnetic fields by nerve cells, *Bioelectromagnetics* **41**, 213–218 (2020).

[1208] H. Lv, J. Liu, C. Zhen, Y. Wang, Y. Wei, W. Ren and P. Shang, Magnetic fields as a potential therapy for diabetic wounds based on animal experiments and clinical trials, *Cell Proliferation* **54**, e12982 (2021).

[1209] A. Heinrich, A. Szostek, P. Meyer, F. Nees, J. Rauschenberg, J. Gröbner, M. Gilles, G. Paslakis, M. Deuschle, W. Semmler *et al.*, Cognition and sensation in very high static magnetic fields: A randomized case-crossover study with different field strengths, *Radiology* **266**, 236–245 (2013).

[1210] A. Yamagashi, T. Takeuchi, T. Hagashi and M. Date, Diamagnetic orientation of blood cells in high magnetic field, *Physica B: Condensed Matter* **177**, 523–526 (1992).

[1211] H. Okano, Effects of static magnetic fields in biology: Role of free radicals, *Frontiers in Bioscience-landmark* **13**, 6106–6125 (2008).

[1212] D. A. Smith, *Effects of Weak Electromagnetic Fields on Escherichia coli and Staphylococcus aureus*, Ph.D. thesis, University of Houston-Clear Lake (2017).

[1213] V. V. Novikov, I. M. Sheiman and E. E. Fesenko, Effect of weak static and low-frequency alternating magnetic fields on the fission and regeneration of the planarian dugesia (girardia) tigrina, *Bioelectromagnetics: Journal of the Bioelectromagnetics Society, The Society for Physical Regulation in Biology and Medicine, The European Bioelectromagnetics Association* **29**, 387–393 (2008).

[1214] H. Lei, Y. Pan, R. Wu and Y. Lv, Innate immune regulation under magnetic fields with possible mechanisms and therapeutic applications, *Frontiers in Immunology* **11**, 582772 (2020).

[1215] V. Novikov, E. Yablokova and E. Fesenko, A decrease of the respiratory burst in neutrophils after exposure to weak combined magnetic fields of a certain duration, *Biophysics* **65**, 82–87 (2020a).

[1216] V. V. Novikov, E. V. Yablokova and E. E. Fesenko, Stimulation and inhibition of respiratory burst in neutrophils as a result of action of weak combined magnetic fields adjusted to ICR of protonated water forms, *Electromagnetic Biology and Medicine* **39**, 364–373 (2020b).

[1217] V. Novikov, E. Yablokova and E. Fesenko, The role of oxygen in the priming of neutrophils on exposure to a weak magnetic field, *Biophysics* **63**, 193–196 (2018).

[1218] V. Novikov, E. Yablokova and E. Fesenko, The role of hydroxyl radicals and calcium ions in the priming of a respiratory burst in neutrophils and the increase in luminol-dependent blood chemiluminescence on exposure to combined magnetic fields with a very weak low-frequency alternating component, *Biophysics* **62**, 440–443 (2017a).

[1219] V. Novikov, E. Yablokova, G. Novikov and E. Fesenko, The role of lipid peroxidation and myeloperoxidase in priming a respiratory burst in neutrophils under the action of combined constant and alternating magnetic fields, *Biophysics* **62**, 759–763 (2017).

[1220] V. Novikov, E. Yablokova and E. Fesenko, Priming of the respiratory burst in neutrophils exposed to a combination of weak constant and alternating low-frequency magnetic fields in vitro, *Biophysics* **61**, 429–434 (2016).

[1221] V. Novikov, E. Yablokova and E. Fesenko, The action of combined magnetic fields with a very weak low-frequency alternating component on luminol-dependent chemiluminescence in mammalian blood, *Biophysics* **60**, 429–432 (2015).

[1222] V. Novikov, E. Yablokova and E. Fesenko, The effect of weak magnetic fields on the chemiluminescence of human blood, *Biophysics* **61**, 105–108 (2016).

[1223] V. Novikov, V. Ponomarev, G. Novikov, V. Kuvichkin, E. Yablokova and E. Fesenko, Effects and molecular mechanisms of the biological action of weak and extremely weak magnetic fields, *Biophysics* **55**, 565–572 (2010).

[1224] A. Patruno, A. Ferrone, E. Costantini, S. Franceschelli, M. Pesce, L. Speranza, P. Amerio, C. D'angelo, M. Felaco, A. Grilli *et al.*, Extremely low-frequency electromagnetic fields accelerates wound healing modulating MMP-9 and inflammatory cytokines, *Cell Proliferation* **51**, e12432 (2018).

[1225] G. Giorgi, C. Pirazzini, M. G. Bacalini, C. Giuliani, P. Garagnani, M. Capri, F. Bersani and B. Del Re, Assessing the combined effect of extremely low-frequency magnetic field exposure and oxidative stress on line-1 promoter methylation in human neural cells, *Radiation and Environmental Biophysics* **56**, 193–200 (2017).

[1226] F. S. Barnes and B. Greenebaum, The effects of weak magnetic fields on radical pairs, *Bioelectromagnetics* **36**, 45–54 (2015).

[1227] Q. H. Mach and M. A. Persinger, Behavioral changes with brief exposures to weak magnetic fields patterned to stimulate long-term potentiation, *Brain Research* **1261**, 45–53 (2009).

[1228] V. Ponomarev and V. Novikov, Effect of low-frequency alternating magnetic fields on the rate of biochemical reactions proceeding with formation of reactive oxygen species, *Biophysics* **54**, 163–168 (2009).

[1229] L. Makinistian, A new class of signals for magnetobiology research, *Scientific Reports* **9**, 7478 (2019).

[1230] R. Muheim, S. Sjöberg and A. Pinzon-Rodriguez, Polarized light modulates light-dependent magnetic compass orientation in birds, *Proceedings of the National Academy of Sciences* **113**, 1654–1659 (2016).

[1231] M. Grinberg, M. Mudrilov, E. Kozlova, V. Sukhov, F. Sarafanov, A. Evtushenko, N. Ilin, V. Vodeneev, C. Price and E. Mareev, Effect of extremely low-frequency magnetic fields on light-induced electric reactions in wheat, *Plant Signaling & Behavior* **17**, 2021664 (2022).

[1232] J. Martel, S.-H. Chang, G. Chevalier, D. M. Ojcius and J. D. Young, Influence of electromagnetic fields on the circadian rhythm: Implications for human health and disease, *Biomedical Journal* **46**, 48–59 (2023).

[1233] J. Martel, N. Rouleau, N. J. Murugan, W.-C. Chin, D. M. Ojcius and J. D. Young, Effects of light, electromagnetic fields and water on biological rhythms, *Biomedical Journal* **48**, 100824 (2025).

[1234] A. Elawar, A. Livache, S. Patault and D. Vila, Combined photobiomodulation and static magnetic fields to reduce side effects from laser and radiofrequency treatments for dermatological conditions, *The Journal of Clinical and Aesthetic Dermatology* **16**, 24 (2023).

[1235] A. P. Wojtovich and T. H. Foster, Optogenetic control of ros production, *Redox Biology* **2**, 368–376 (2014).

[1236] M. Abbas and P. Kurian, Quantum probes in cancer research, *Nature Reviews Cancer* **22**, 378–379 (2022).

[1237] L. Nie, A. C. Nusantara, V. G. Damle, M. V. Baranov, M. Chipaux, C. Reyes-San-Martin, T. Hamoh, C. P. Epperla, M. Guricova, P. Cigler *et al.*, Quantum sensing of free radicals in primary human dendritic cells, *Nano Letters* **22**, 1818–1825 (2021).

[1238] C. Romero-Muñiz, J. G. Vilhena, R. Pérez, J. C. Cuevas and L. A. Zotti, Recent advances in understanding the electron transport through metal-azurin-metal junctions, *Frontiers in Physics* **10**, 950929 (2022).

[1239] G. D. Scholes, A. Olaya-Castro, S. Mukamel, A. Kirrander, K.-K. Ni, G. J. Hedley and N. L. Frank, The quantum information science challenge for chemistry, *The Journal of Physical Chemistry Letters* **16**, 13761396 (2025).

[1240] P. J. Brown, Y. Qiu, E. I. Latawiec, B. T. Phelan, N. A. Tcyrulnikov, J. R. Palmer, M. D. Krzyaniak, S. M. Kopp, Y. Huang, R. M.

Young *et al.*, Enhancing photogenerated radical pair properties in donor-chromophore-acceptor systems for quantum information applications, *The Journal of Physical Chemistry A* **128**, 9371–9382 (2024).
[1241] H. Mao, G. J. Pazera, R. M. Young, M. D. Krzyaniak and M. R. Wasielewski, Quantum gate operations on a spectrally addressable photogenerated molecular electron spin-qubit pair, *Journal of the American Chemical Society* **145**, 6585–6593 (2023).
[1242] J.-S. Chen, K. J. Trerayapiwat, L. Sun, M. D. Krzyaniak, M. R. Wasielewski, T. Rajh, S. Sharifzadeh and X. Ma, Long-lived electronic spin qubits in single-walled carbon nanotubes, *Nature Communications* **14**, 848 (2023).
[1243] G. Pizzino, N. Irrera, M. Cucinotta, G. Pallio, F. Mannino, V. Arcoraci, F. Squadrito, D. Altavilla and A. Bitto, Oxidative stress: harms and benefits for human health, *Oxidative Medicine and Cellular Longevity* **2017**, 8416763 (2017).
[1244] A. Kirtonia, G. Sethi and M. Garg, The multifaceted role of reactive oxygen species in tumorigenesis, *Cellular and Molecular Life Sciences* **77**, 4459–4483 (2020).
[1245] H. Lai, Exposure to static and extremely-low frequency electromagnetic fields and cellular free radicals, *Electromagnetic Biology and Medicine* **38**, 231–248 (2019).
[1246] D. A. Chathurika, M. F. Peter and N. H. Malka, Analysis of biological effects and limits of exposure to weak magnetic fields, in *2010 Fifth International Conference on Information and Automation for Sustainability*. IEEE, 415–420 (2010).
[1247] M. Mohseni, Y. Omar, G. S. Engel and M. B. Plenio (eds.), *Quantum Effects in Biology*. Cambridge University Press (2014).
[1248] C. Di Rienzo, V. Piazza, E. Gratton, F. Beltram and F. Cardarelli, Probing short-range protein Brownian motion in the cytoplasm of living cells, *Nature Communications* **5**, 5891 (2014).
[1249] R. Tsekov and M. C. Lensen, Brownian motion and the temperament of living cells, *Chinese Physics Letters* **30**, 070501 (2013).
[1250] A. Einstein, On the movement of particles suspended in resting liquids required by the molecular kinetic theory of heat, *Annals of Physics* **17**, 549–560 (1905).
[1251] L. Cohen, The history of noise [on the 100th anniversary of its birth], *IEEE Signal Processing Magazine* **22**, 20–45 (2005).
[1252] B. Pullman and A. Reisinger, *The Atom in the History of Human Thought*. Oxford University Press (2001).
[1253] P. Ball, Water is an active matrix of life for cell and molecular biology, *Proceedings of the National Academy of Sciences* **114**, 13327–13335 (2017).

[1254] M. L. Berrens, A. Kundu, M. F. Calegari Andrade, T. A. Pham, G. Galli and D. Donadio, Nuclear quantum effects on the electronic structure of water and ice, *The Journal of Physical Chemistry Letters* **15**, 6818–6825 (2024).

[1255] N. Sieroka, T. Lossau and T. Neudecker, Emergent properties in chemistry-relating molecular properties to bulk behavior, *Chemistry–A European Journal* **30**, e202303868 (2024).

[1256] G. H. Pollack, Water, energy and life: Fresh views from the water's edge, *International Journal of Design & Nature and Ecodynamics* **5**, 27 (2010).

[1257] S. G. Hwang, J. K. Hong, A. Sharma, G. H. Pollack and G. Bahng, Exclusion zone and heterogeneous water structure at ambient temperature, *PLoS One* **13**, e0195057 (2018).

[1258] D. C. Elton, P. D. Spencer, J. D. Riches and E. D. Williams, Exclusion zone phenomena in water–A critical review of experimental findings and theories, *International Journal of Molecular Sciences* **21**, 5041 (2020).

[1259] S. Yuvan and M. Bier, Sense and nonsense about water, in *Water in Biomechanical and Related Systems*. Springer, pp. 19–36 (2021).

[1260] D. Florea, S. Musa, J. M. Huyghe and H. M. Wyss, Long-range repulsion of colloids driven by ion exchange and diffusiophoresis, *Proceedings of the National Academy of Sciences* **111**, 6554–6559 (2014).

[1261] V. A. Parsegian, *Van der Waals Forces: A Handbook for Biologists, Chemists, Engineers, and Physicists*. Cambridge University Press (2005).

[1262] G. M. Kontogeorgis, A. Holster, N. Kottaki, E. Tsochantaris, F. Topsøe, J. Poulsen, M. Bache, X. Liang, N. S. Blom and J. Kronholm, Water structure, properties and some applications–A review, *Chemical Thermodynamics and Thermal Analysis* **6**, 100053 (2022).

[1263] M. Bier and D. Pravica, Limits on quantum coherent domains in liquid water, *Acta Physica Polonica B* **49**, 9 (2018).

[1264] W.-q. An, X.-a. Yue, J.-r. Zou, L.-j. Zhang and R.-j. Yan, Direct observation of anomalous water at the solid-liquid boundary on silica surface, *Applied Surface Science* **638**, 158096 (2023).

[1265] C. Tantardini, When does a hydrogen bond become a van der Waals interaction? a topological answer, *Journal of Computational Chemistry* **40**, 937–943 (2019).

[1266] S. C. Van der Lubbe and C. Fonseca Guerra, The nature of hydrogen bonds: A delineation of the role of different energy components on hydrogen bond strengths and lengths, *Chemistry–An Asian Journal* **14**, 2760–2769 (2019).

[1267] M. Kolar, T. Kubar and P. Hobza, On the role of london dispersion forces in biomolecular structure determination, *The Journal of Physical Chemistry B* **115**, 8038–8046 (2011).

[1268] M. Stöhr and A. Tkatchenko, Quantum mechanics of proteins in explicit water: The role of plasmon-like solute-solvent interactions, *Science Advances* **5**, eaax0024 (2019).

[1269] A. Otero-de-la Roza, L. M. LeBlanc and E. R. Johnson, What is "many-body" dispersion and should I worry about it? *Physical Chemistry Chemical Physics* **22**, 8266–8276 (2020).

[1270] M. Stöhr, T. Van Voorhis and A. Tkatchenko, Theory and practice of modeling van der Waals interactions in electronic-structure calculations, *Chemical Society Reviews* **48**, 4118–4154 (2019).

[1271] K. J. Daas, E. Fabiano, F. Della Sala, P. Gori-Giorgi and S. Vuckovic, Noncovalent interactions from models for the Møller–Plesset adiabatic connection, *The Journal of Physical Chemistry Letters* **12**, 4867–4875 (2021).

[1272] H.-X. Zhou and X. Pang, Electrostatic interactions in protein structure, folding, binding, and condensation, *Chemical Reviews* **118**, 1691–1741 (2018).

[1273] D. V. Fedorov, M. Sadhukhan, M. Stöhr and A. Tkatchenko, Quantum-mechanical relation between atomic dipole polarizability and the van der Waals radius, *Physical Review Letters* **121**, 183401 (2018).

[1274] A. Jones, F. Cipcigan, V. P. Sokhan, J. Crain and G. Martyna, Electronically coarse-grained model for water, *Physical Review Letters* **110**, 227801 (2013).

[1275] T. Morawietz, A. Singraber, C. Dellago and J. Behler, How van der Waals interactions determine the unique properties of water, *Proceedings of the National Academy of Sciences USA* **113**, 8368–8373 (2016).

[1276] J. Hermann, R. A. DiStasio Jr and A. Tkatchenko, First-principles models for van der Waals interactions in molecules and materials: Concepts, theory, and applications, *Chemical Reviews* **117**, 4714–4758 (2017).

[1277] D. J. Cole and N. D. Hine, Applications of large-scale density functional theory in biology, *Journal of Physics: Condensed Matter* **28**, 393001 (2016).

[1278] S. Kristyán and P. Pulay, Can (semi) local density functional theory account for the london dispersion forces? *Chemical Physics Letters* **229**, 175–180 (1994).

[1279] O. Gritsenko, Ł. Mentel and E. Baerends, On the errors of local density (LDA) and generalized gradient (GGA) approximations to the kohn-sham potential and orbital energies, *The Journal of Chemical Physics* **144**, 20 (2016).

[1280] M. Elstner, D. Porezag, G. Jungnickel, J. Elsner, M. Haugk, T. Frauenheim, S. Suhai and G. Seifert, Self-consistent-charge density-functional tight-binding method for simulations of complex materials properties, *Physical Review B* **58**, 7260 (1998).

[1281] M. Elstner, T. Frauenheim and S. Suhai, An approximate DFT method for QM/MM simulations of biological structures and processes, *Journal of Molecular Structure: THEOCHEM* **632**, 29–41 (2003).

[1282] M. Elstner, The SCC-DFTB method and its application to biological systems, *Theoretical Chemistry Accounts: Theory, Computation, & Modeling* **116**, 316–325 (2006).

[1283] M. Gori, P. Kurian and A. Tkatchenko, Second quantization of many-body dispersion interactions for chemical and biological systems, *Nature Communications* **14**, 8218 (2023).

[1284] R. A. Friesner and M. D. Beachy, Quantum mechanical calculations on biological systems, *Current Opinion in Structural Biology* **8**, 257–262 (1998).

[1285] M. Elstner and G. Seifert, Density functional tight binding, *Philosophical Transactions of the Royal Society A: Mathematical, Physical and Engineering Sciences* **372**, 20120483 (2014).

[1286] F. Spiegelman, N. Tarrat, J. Cuny, L. Dontot, E. Posenitskiy, C. Martí, A. Simon and M. Rapacioli, Density-functional tight-binding: Basic concepts and applications to molecules and clusters, *Advances in Physics: X* **5** 1710252 (2020).

[1287] M. Gaus, Q. Cui and M. Elstner, DFTB3: Extension of the self-consistent-charge density-functional tight-binding method (scc-dftb), *Journal of Chemical Theory and Computation* **7**, 931–948 (2011).

[1288] J. Cao and B. Berne, Many-body dispersion forces of polarizable clusters and liquids, *The Journal of Chemical Physics* **97**, 8628–8636 (1992).

[1289] V. A. Adhav and K. Saikrishnan, The realm of unconventional noncovalent interactions in proteins: their significance in structure and function, *ACS Omega* **8**, 22268–22284 (2023).

[1290] A. G. Donchev, A. G. Taube, E. Decolvenaere, C. Hargus, R. T. McGibbon, K.-H. Law, B. A. Gregersen, J.-L. Li, K. Palmo, K. Siva *et al.*, Quantum chemical benchmark databases of gold-standard dimer interaction energies, *Scientific Data* **8**, 55 (2021).

[1291] J. Chen, Q. Peng, X. Peng, H. Zhang and H. Zeng, Probing and manipulating noncovalent interactions in functional polymeric systems, *Chemical Reviews* **122**, 14594–14678 (2022).

[1292] E. Palos, H. Zhao, K. J. Daas, E. Fabiano, F. Paesani and S. Vuckovic, Møller–plesset adiabatic connection theory for diverse noncovalent interactions, *The Journal of Physical Chemistry Letters* **16**, 7898–7908 (2025).

[1293] P. Xu, M. Alkan and M. S. Gordon, Many-body dispersion, *Chemical Reviews* **120**, 12343–12356 (2020).

[1294] S. Piana, A. G. Donchev, P. Robustelli and D. E. Shaw, Water dispersion interactions strongly influence simulated structural properties of disordered protein states, *The Journal of Physical Chemistry B* **119**, 5113–5123 (2015).

[1295] L. Kronik and A. Tkatchenko, Understanding molecular crystals with dispersion-inclusive density functional theory: Pairwise corrections and beyond, *Accounts of Chemical Research* **47**, 3208–3216 (2014).

[1296] A. Ambrosetti, N. Ferri, R. A. DiStasio Jr and A. Tkatchenko, Wavelike charge density fluctuations and van der Waals interactions at the nanoscale, *Science* **351**, 1171–1176 (2016).

[1297] J. F. Dobson, Beyond pairwise additivity in London dispersion interactions, *International Journal of Quantum Chemistry* **114**, 1157–1161 (2014).

[1298] J. Klimeš and A. Michaelides, Perspective: Advances and challenges in treating van der Waals dispersion forces in density functional theory, *The Journal of Chemical Physics* **137**, 12 (2012).

[1299] K. Carter-Fenk, M. Liu, L. Pujal, M. Loipersberger, M. Tsanai, R. M. Vernon, J. D. Forman-Kay, M. Head-Gordon, F. Heidar-Zadeh and T. Head-Gordon, The energetic origins of pi–pi contacts in proteins, *Journal of the American Chemical Society* **145**, 24836–24851 (2023).

[1300] F. Häse and M. Zacharias, Free energy analysis and mechanism of base pair stacking in nicked DNA, *Nucleic Acids Research* **44**, 7100–7108 (2016).

[1301] A. K. Singh, C. Wen, S. Cheng and N. Q. Vinh, Long-range DNA-water interactions, *Biophysical Journal* **120**, 4966–4979 (2021).

[1302] G. Zundel, W. Lubos and K. Kölkenbeck, Proton dispersion forces: Secondary-structure stabilizing forces between the hydrogen bonds of the polynucleotides, *Biophysical Journal* **12**, 1509–1514 (1972).

[1303] V. Guallar and K. W. Borrelli, A binding mechanism in protein–nucleotide interactions: implication for U1A RNA binding, *Proceedings of the National Academy of Sciences USA* **102**, 3954–3959 (2005).

[1304] M. Chawla, K. Kalra, Z. Cao, L. Cavallo and R. Oliva, Occurrence and stability of anion–π interactions between phosphate and nucleobases in functional RNA molecules, *Nucleic Acids Research* **50**, 11455–11469 (2022).

[1305] P. Yakovchuk, E. Protozanova and M. D. Frank-Kamenetskii, Base-stacking and base-pairing contributions into thermal stability of the DNA double helix, *Nucleic Acids Research* **34**, 564–574 (2006).

[1306] K. Carter-Fenk and J. M. Herbert, Electrostatics does not dictate the slip-stacked arrangement of aromatic $\pi - \pi$ interactions, *Chemical Science* **11**, 6758–6765 (2020).

[1307] K. Carter-Fenk and J. M. Herbert, Reinterpreting π-stacking, *Physical Chemistry Chemical Physics* **22**, 24870–24886 (2020b).

[1308] K. Carter-Fenk, K. U. Lao and J. M. Herbert, Predicting and understanding non-covalent interactions using novel forms of symmetry-adapted perturbation theory, *Accounts of Chemical Research* **54**, 3679–3690 (2021).

[1309] J. M. Herbert, Neat, simple, and wrong: Debunking electrostatic fallacies regarding noncovalent interactions, *The Journal of Physical Chemistry A* **125**, 7125–7137 (2021).

[1310] M. Gray and J. M. Herbert, Origins of offset-stacking in porous frameworks, *The Journal of Physical Chemistry C* **127**, 2675–2686 (2023).

[1311] B. Schramm, M. Gray and J. M. Herbert, Substituent and heteroatom effects on π-π interactions: Evidence that parallel-displaced π-stacking is not driven by quadrupolar electrostatics, *Journal of the American Chemical Society* **147**, 3243–3260 (2025).

[1312] C.-M. Chang, A. C. Neto and A. Bishop, Long-range charge transfer in periodic DNA through polaron diffusion, *Chemical Physics* **303**, 189–196 (2004).

[1313] H. Yamada and K. Iguchi, Some effective tight-binding models for electrons in DNA conduction: A review, *Advances in Condensed Matter Physics* **2010**, 380710 (2010).

[1314] J. C. Genereux and J. K. Barton, Mechanisms for DNA charge transport, *Chemical Reviews* **110**, 1642–1662 (2010).

[1315] J. D. Slinker, N. B. Muren, S. E. Renfrew and J. K. Barton, Dna charge transport over 34 nm, *Nature Chemistry* **3**, 228–233 (2011).

[1316] D. B. Bucher, B. M. Pilles, T. Carell and W. Zinth, Charge separation and charge delocalization identified in long-living states of photoexcited DNA, *Proceedings of the National Academy of Sciences* **111**, 4369–4374 (2014).

[1317] A. R. Arnold, M. A. Grodick and J. K. Barton, Dna charge transport: From chemical principles to the cell, *Cell Chemical Biology* **23**, 183–197 (2016).

[1318] B. Demir, C. Akin Gultakti, Z. Koker, M. Anantram and E. E. Oren, Electronic properties of DNA origami nanostructures revealed by in silico calculations, *The Journal of Physical Chemistry B* **128**, 4646–4654 (2024).

[1319] D. N. Beratan, Why are DNA and protein electron transfer so different? *Annual Review of Physical Chemistry* **70**, 71–97 (2019).

[1320] P. J. Hore, A DNA-based magnetic sensor, **4**, 318–320 (2018).

[1321] T. J. Zwang, E. C. Tse and J. K. Barton, Sensing DNA through DNA charge transport, *ACS Chemical Biology* **13**, 1799–1809 (2018).

[1322] P. A. Sontz, N. B. Muren and J. K. Barton, DNA charge transport for sensing and signaling, *Accounts of Chemical Research* **45**, 1792–1800 (2012).

[1323] E. C. Tse, T. J. Zwang, S. Bedoya and J. K. Barton, Effective distance for DNA-mediated charge transport between repair proteins, *ACS Central Science* **5**, 65–72 (2019).

[1324] S. Puig, L. Ramos-Alonso, A. M. Romero and M. T. Martínez-Pastor, The elemental role of iron in DNA synthesis and repair, *Metallomics* **9**, 1483–1500 (2017).

[1325] L. Slocombe, M. Winokan, J. Al-Khalili and M. Sacchi, Quantum tunnelling effects in the guanine-thymine wobble misincorporation via tautomerism, *The Journal of Physical Chemistry Letters* **14**, 9–15 (2023).

[1326] L. Slocombe, M. Sacchi and J. Al-Khalili, An open quantum systems approach to proton tunnelling in DNA, *Communications Physics* **5**, 109 (2022).

[1327] L. Slocombe, J. S. Al-Khalili and M. Sacchi, Quantum and classical effects in DNA point mutations: Watson–crick tautomerism in AT and GC base pairs, *Physical Chemistry Chemical Physics* **23**, 4141–4150 (2021).

[1328] F. Ghasemi and A. Tirandaz, Environment assisted quantum model for studying RNA-DNA-error correlation created due to the base tautomery, *Scientific Reports* **13**, 10788 (2023).

[1329] J. McFadden and J. Al-Khalili, A quantum mechanical model of adaptive mutation, *BioSystems* **50**, 203–211 (1999).

[1330] J. Al-Khalili and J. McFadden, Quantum coherence and the search for the first replicator, in *Quantum Aspects of Life*. Imperial College Press London, 33–47 (2008).

[1331] V. V. Ogryzko, A quantum-theoretical approach to the phenomenon of directed mutations in bacteria (hypothesis), *BioSystems* **43**, 83–95 (1997).

[1332] R. Siebert, O. Ammerpohl, M. Rossini, D. Herb, S. Rau, M. B. Plenio, F. Jelezko and J. Ankerhold, A quantum physics layer of epigenetics: A hypothesis deduced from charge transfer and chirality-induced spin selectivity of DNA, *Clinical Epigenetics* **15**, 145 (2023).

[1333] R. W. Carthew, Gene regulation and cellular metabolism: An essential partnership, *Trends in Genetics* **37**, 389–400 (2021).

[1334] C. M. Metallo and M. G. Vander Heiden, Understanding metabolic regulation and its influence on cell physiology, *Molecular cell* **49**, 388–398 (2013).

[1335] Y. Li, E. Grotewold and N. Dudareva, Enough is enough: Feedback control of specialized metabolism, *Trends in Plant Science* **29**, 514–523 (2024).

[1336] J. Hermann, D. Alfe and A. Tkatchenko, Nanoscale π–π stacked molecules are bound by collective charge fluctuations, *Nature Communications* **8**, 14052 (2017).

[1337] P. Kurian, A. Capolupo, T. Craddock and G. Vitiello, Water-mediated correlations in DNA-enzyme interactions, *Physics Letters A* **382**, 33–43 (2018).

[1338] E. Faraji, V. Calandrini, P. Kurian, R. Franzosi, S. Mancini, E. Floriani, G. Pettini and M. Pettini, Electrodynamic forces driving DNA-protein interactions at large distances, *Frontiers of Physics* **20**, 061200 (2025).

[1339] A. M. Reilly and A. Tkatchenko, Role of dispersion interactions in the polymorphism and entropic stabilization of the aspirin crystal, *Physical Review Letters* **113**, 055701 (2014).

[1340] S. Kumari, The role of water molecules in CT-DNA interactions: Implications for drug design, *Journal of Molecular Liquids*, 125997 (2024).

[1341] V. N. Uversky, Intrinsically disordered proteins and their "mysterious" (meta)physics, *Frontiers in Physics* **7**, 10 (2019).

[1342] I. L. Morgan, R. Avinery, G. Rahamim, R. Beck and O. A. Saleh, Glassy dynamics and memory effects in an intrinsically disordered protein construct, *Physical Review Letters* **125**, 058001 (2020).

[1343] P. L. Silvestrelli, E. Martin, M. Boero, A. Bouzid, G. Ori and C. Massobrio, Atomic structure of glassy $GeTe_4$ as a playground to assess the performances of density functional schemes accounting for dispersion forces, *The Journal of Physical Chemistry B* **124**, 11273–11279 (2020).

[1344] E. Martin, I. B. Amiehe Essomba, K. Ishisone, M. Boero, G. Ori and C. Massobrio, Impact of dispersion force schemes on liquid systems: Comparing efficiency and drawbacks for well-targeted test cases, *Molecules* **27**, 9034 (2022).
[1345] F. Eisenhaber, B. Persson and P. Argos, Protein structure prediction: Recognition of primary, secondary, and tertiary structural features from amino acid sequence, *Critical Reviews in Biochemistry and Molecular Biology* **30**, 1–94 (1995).
[1346] S. Dagar, A. Panwar, V. Sharma, I. Sheikh, V. Kushwaha, D. Kaur, A. K. Sharma and S. Kant, Basic structure of proteins: Current paradigms, trends, and perspective, in *Bioinformatics: Drug Discovery.* Walter de Gruyter GmbH & Co KG, 151 (2024).
[1347] C. N. Pace, J. M. Scholtz and G. R. Grimsley, Forces stabilizing proteins, *FEBS Letters* **588**, 2177–2184 (2014).
[1348] S. Pal, M. Bhattacharya, S.-S. Lee and C. Chakraborty, Quantum computing in the next-generation computational biology landscape: From protein folding to molecular dynamics, *Molecular Biotechnology* **66**, 1–16 (2024).
[1349] A. V. Melkikh and D. K. Meijer, On a generalized levinthal's paradox: The role of long-and short range interactions in complex biomolecular reactions, including protein and DNA folding, *Progress in Biophysics and Molecular Biology* **132**, 57–79 (2018).
[1350] R. W. Newberry and R. T. Raines, Secondary forces in protein folding, *ACS Chemical Biology* **14**, 1677–1686 (2019).
[1351] P. Davies, A quantum recipe for life, *Nature* **437**, 819–819 (2005).
[1352] P. C. Davies, A quantum origin of life? in *Quantum Aspects of Life.* Imperial College Press, 3–18 (2008).
[1353] R. Zwanzig, A. Szabo and B. Bagchi, Levinthal's paradox, *Proceedings of the National Academy of Sciences* **89**, 20–22 (1992).
[1354] Z. Yang, X. Zeng, Y. Zhao and R. Chen, AlphaFold2 and its applications in the fields of biology and medicine, *Signal Transduction and Targeted Therapy* **8**, 115 (2023).
[1355] J. Abramson, J. Adler, J. Dunger, R. Evans, T. Green, A. Pritzel, O. Ronneberger, L. Willmore, A. J. Ballard, J. Bambrick *et al.*, Accurate structure prediction of biomolecular interactions with AlphaFold 3, *Nature* **630**, 493–500 (2024).
[1356] E. Callaway, Developers of AlphaFold win chemistry Nobel, *Nature* **634**, 525–526 (2024).
[1357] C. B. Anfinsen, E. Haber, M. Sela and F. White Jr, The kinetics of formation of native ribonuclease during oxidation of the reduced polypeptide chain, *Proceedings of the National Academy of Sciences USA* **47**, 1309–1314 (1961).

[1358] P. Chawhan and I. Singh, A hybrid quantum machine learning for the prediction of protein folding, in *International Conference on Computing and Communication Systems for Industrial Applications*. Springer, 21–32 (2024).

[1359] C. Chakraborty, M. Bhattacharya, K. Dhama and S.-S. Lee, Quantum computing on nucleic acid research: Approaching towards next-generation computing, *Molecular Therapy-Nucleic Acids* **33**, 53–56 (2023).

[1360] P. S. Emani, J. Warrell, A. Anticevic, S. Bekiranov, M. Gandal, M. J. McConnell, G. Sapiro, A. Aspuru-Guzik, J. T. Baker, M. Bastiani *et al.*, Quantum computing at the frontiers of biological sciences, *Nature Methods* **18**, 701–709 (2021).

[1361] D. M. Fox, C. M. MacDermaid, A. M. Schreij, M. Zwierzyna and R. C. Walker, RNA folding using quantum computers, *PLoS Computational Biology* **18**, e1010032 (2022).

[1362] J. Fallmann, S. Will, J. Engelhardt, B. Grüning, R. Backofen and P. F. Stadler, Recent advances in RNA folding, *Journal of Biotechnology* **261**, 97–104 (2017).

[1363] C. Hyeon and D. Thirumalai, Capturing the essence of folding and functions of biomolecules using coarse-grained models, *Nature Communications* **2**, 487 (2011).

[1364] R. L. Barnett, P. Maragakis, A. Turner, M. Fyta and E. Kaxiras, Multiscale model of electronic behavior and localization in stretched dry DNA, *Journal of Materials Science* **42**, 8894–8903 (2007).

[1365] D. Herb, M. Rossini and J. Ankerhold, Ultrafast excitonic dynamics in DNA: Bridging correlated quantum dynamics and sequence dependence, *Physical Review E* **109**, 064413 (2024).

[1366] J. H. Olshansky, M. D. Krzyaniak, R. M. Young and M. R. Wasielewski, Photogenerated spin-entangled qubit (radical) pairs in DNA hairpins: Observation of spin delocalization and coherence, *Journal of the American Chemical Society* **141**, 2152–2160 (2019).

[1367] M. D'Acunto, Quantum biology. π–π entanglement signatures in protein-DNA interactions, *Physical Biology* **19**, 036003 (2022).

[1368] P. Kurian, G. Dunston and J. Lindesay, How quantum entanglement in DNA synchronizes double-strand breakage by Type II restriction endonucleases, *Journal of Theoretical Biology* **391**, 102–112 (2016).

[1369] T. Bollinger, The DNA chromatin superradiance hypothesis, (2024), http://medium.com/@terrybollinger/the-dna-chromatin-superradiance-hypothesis-360187a8bc8f.

[1370] R. A. DiStasio, O. A. von Lilienfeld and A. Tkatchenko, Collective manybody van der Waals interactions in molecular systems,

Proceedings of the National Academy of Sciences **109**, 14791–14795 (2012).
[1371] A. Ambrosetti, A. M. Reilly, R. A. DiStasio and A. Tkatchenko, Long-range correlation energy calculated from coupled atomic response functions, *The Journal of Chemical Physics* **140**, 18A508 (2014).
[1372] M. Cobb, 60 years ago, Francis Crick changed the logic of biology, *PLoS Biology* **15**, e2003243 (2017).
[1373] P. Ball, Life does not run like clockwork, (2022), https://www.chemistryworld.com/opinion/causal-emergence-might-explain-how-living-systems-can-operate/4015345.article.
[1374] V. Piras, M. Tomita and K. Selvarajoo, Is central dogma a global property of cellular information flow? *Frontiers in Physiology* **3**, 439 (2012).
[1375] C. D. Kocher and K. A. Dill, Origins of life: The protein folding problem all over again? *Proceedings of the National Academy of Sciences* **121**, e2315000121 (2024).
[1376] Y. Ezerzer, M. Frenkel-Pinter, R. Kolodny and N. Ben-Tal, A building blocks perspective on protein emergence and evolution, *Current Opinion in Structural Biology* **91**, 102996 (2025).
[1377] J. D. Watson and F. H. Crick, Molecular structure of nucleic acids: a structure for deoxyribose nucleic acid, *Nature* **171**, 737–738 (1953).
[1378] R. Dahm, Friedrich miescher and the discovery of DNA, *Developmental Biology* **278**, 274–288 (2005).
[1379] Y. A. Çengel, On entropy, information, and conservation of information, *Entropy* **23**, 779 (2021).
[1380] L. Boltzmann, On certain questions of the theory of gases. *Nature* **51**, 1322 (1895).
[1381] S. Chapman, Boltzmann's H-theorem, *Nature* **139**, 931–931 (1937).
[1382] G. B. Lesovik, A. V. Lebedev, I. A. Sadovskyy, M. Suslov and V. M. Vinokur, H-theorem in quantum physics, *Scientific Reports* **6**, 32815 (2016).
[1383] J. von Neumann, Proof of the ergodic theorem and the H-theorem in quantum mechanics: Translation of: Beweis des ergodensatzes und des H-theorems in der neuen mechanik, *The European Physical Journal H* **35**, 201–237 (2010).
[1384] J. Gemmer, M. Michel and G. Mahler, *Quantum Thermodynamics: Emergence of Thermodynamic Behavior within Composite Quantum Systems*, Vol. 784. Springer (2009).
[1385] G. Aquino, L. Tweedy, D. Heinrich and R. G. Endres, Memory improves precision of cell sensing in fluctuating environments, *Scientific Reports* **4**, 5688 (2014).

[1386] F. Ormeño and I. J. General, Convergence and equilibrium in molecular dynamics simulations, *Communications Chemistry* **7**, 26 (2024).

[1387] R. Kaschner and D. Hohl, Density functional theory and biomolecules: A study of glycine, alanine, and their oligopeptides, *The Journal of Physical Chemistry A* **102**, 5111–5116 (1998).

[1388] P. Wendering and Z. Nikoloski, Model-driven insights into the effects of temperature on metabolism, *Biotechnology Advances* **67**, 108203 (2023).

[1389] M. Ceriotti, G. Bussi and M. Parrinello, Langevin equation with colored noise for constant-temperature molecular dynamics simulations, *Physical Review Letters* **102**, 020601 (2009).

[1390] H. C. Andersen, Molecular dynamics simulations at constant pressure and/or temperature, *The Journal of Chemical Physics* **72**, 2384–2393 (1980).

[1391] H. J. Berendsen, J. v. Postma, W. F. Van Gunsteren, A. DiNola and J. R. Haak, Molecular dynamics with coupling to an external bath, *The Journal of Chemical Physics* **81**, 3684–3690 (1984).

[1392] D. J. Evans and B. L. Holian, The Nose-Hoover thermostat, *The Journal of Chemical Physics* **83**, 4069–4074 (1985).

[1393] D. Suh, A. R. Arattu Thodika, S. Kim, K. Nam and W. Im, CHARMM-GUI QM/MM interfacer for a quantum mechanical and molecular mechanical (QM/MM) simulation setup: 1. Semiempirical methods, *Journal of Chemical Theory and Computation* **20**, 5337–5351 (2024).

[1394] A. Hospital, J. R. Goñi, M. Orozco and J. L. Gelpí, Molecular dynamics simulations: advances and applications, *Advances and Applications in Bioinformatics and Chemistry*, 37–47 (2015).

[1395] S. A. Hollingsworth and R. O. Dror, Molecular dynamics simulation for all, *Neuron* **99**, 1129–1143 (2018).

[1396] Y. Cong, M. Li, G. Feng, Y. Li, X. Wang and L. Duan, Trypsin-ligand binding affinities calculated using an effective interaction entropy method under polarized force field, *Scientific Reports* **7**, 17708 (2017).

[1397] V. Govind Kumar, A. Polasa, S. Agrawal, T. K. S. Kumar and M. Moradi, Binding affinity estimation from restrained umbrella sampling simulations, *Nature Computational Science* **3**, 59–70 (2023).

[1398] J. Fan, Y. Pan, Z. Gao and H. Qu, Molecular dynamic simulations of the physical properties of four ionic liquids, *International Journal of Molecular Sciences* **25**, 11217 (2024).

[1399] J. A. Harrison, J. D. Schall, S. Maskey, P. T. Mikulski, M. T. Knippenberg and B. H. Morrow, Review of force fields and intermolecular potentials used in atomistic computational materials research, *Applied Physics Reviews* **5**, 031104 (2018).

[1400] J. A. Lemkul, J. Huang, B. Roux and A. D. MacKerell Jr, An empirical polarizable force field based on the classical drude oscillator model: development history and recent applications, *Chemical Reviews* **116**, 4983–5013 (2016).

[1401] Q. Mao, M. Feng, X. Z. Jiang, Y. Ren, K. H. Luo and A. C. van Duin, Classical and reactive molecular dynamics: Principles and applications in combustion and energy systems, *Progress in Energy and Combustion Science* **97**, 101084 (2023).

[1402] G. Náray-Szabó, J. Oláh and B. Krámos, Quantum mechanical modeling: A tool for the understanding of enzyme reactions, *Biomolecules* **3**, 662–702 (2013).

[1403] T. Tomaru, H. Takahashi, T. Hirano, S. Tahara and F. Sato, Chemical reaction simulator on quantum computers by first quantization–basic treatment: Theoretical, *AIP Advances* **14**, 125306 (2024).

[1404] S. Faraji, D. Picconi and E. Palacino-González, Advanced quantum and semiclassical methods for simulating photoinduced molecular dynamics and spectroscopy, *Wiley Interdisciplinary Reviews: Computational Molecular Science* **14**, e1731 (2024).

[1405] D. Picconi, M. F. S. Menger, E. Palacino-González, E. X. Salazar and S. Faraji, Implementation of quasiclassical mapping approaches for nonadiabatic molecular dynamics in the PySurf package, *Physical Chemistry Chemical Physics* **27**, 19105–19122 (2025).

[1406] O. T. Unke, M. Stöhr, S. Ganscha, T. Unterthiner, H. Maennel, S. Kashubin, D. Ahlin, M. Gastegger, L. Medrano Sandonas, J. T. Berryman *et al.*, Biomolecular dynamics with machine-learned quantum-mechanical force fields trained on diverse chemical fragments, *Science Advances* **10**, eadn4397 (2024).

[1407] J. Nochebuena, S. Naseem-Khan and G. A. Cisneros, Development and application of quantum mechanics/molecular mechanics methods with advanced polarizable potentials, *Wiley Interdisciplinary Reviews: Computational Molecular Science* **11**, e1515 (2021).

[1408] T. Yu and H. Guo, Understanding enzyme catalysis mechanism using QM/MM simulation methods, in *Mechanistic Enzymology: Bridging Structure and Function.* ACS Publications, 121–137 (2020).

[1409] X. Wang, R. Li, W. Cui, Q. Li and J. Yao, QM/MM free energy simulations of an efficient gluten hydrolase (Kuma030) implicate for a reactant-state based protein-design strategy for general acid/base catalysis, *Scientific Reports* **8**, 7042 (2018).

[1410] T. H. Rod and U. Ryde, Accurate QM/MM free energy calculations of enzyme reactions: Methylation by catechol o-methyltransferase, *Journal of Chemical Theory and Computation* **1**, 1240–1251 (2005).

[1411] Y. Zhang, H. Liu and W. Yang, Free energy calculation on enzyme reactions with an efficient iterative procedure to determine minimum energy paths on a combined *ab initio* QM/MM potential energy surface, *The Journal of Chemical Physics* **112**, 3483–3492 (2000).

[1412] M. R. Blomberg and P. E. Siegbahn, Quantum chemistry as a tool in bioenergetics, *Biochimica et Biophysica Acta (BBA)-Bioenergetics* **1797**, 129–142 (2010).

[1413] J. M. Herbert, Fantasy versus reality in fragment-based quantum chemistry, *The Journal of Chemical Physics* **151**, 170901 (2019).

[1414] Y. Mochizuki, K. Yamashita, T. Nakano, Y. Okiyama, K. Fukuzawa, N. Taguchi and S. Tanaka, Higher-order correlated calculations based on fragment molecular orbital scheme, *Theoretical Chemistry Accounts* **130**, 515–530 (2011).

[1415] C. Greenwell, J. Řezáč and G. J. Beran, Spin-component-scaled and dispersion-corrected second-order Møller-Plesset perturbation theory: a path toward chemical accuracy, *Physical Chemistry Chemical Physics* **24**, 3695–3712 (2022).

[1416] G. J. Beran, C. Greenwell, C. Cook and J. Rezac, Improved description of intra-and intermolecular interactions through dispersion-corrected second-order Møller-Plesset perturbation theory, *Accounts of Chemical Research* **56**, 3525–3534 (2023).

[1417] K. J. Daas, H. Zhao, E. Polak and S. Vuckovic, Exact Møller-Plesset adiabatic connection correlation energy densities, *Journal of Chemical Theory and Computation* **21**, 5501–5513 (2025).

[1418] Y. Nishimoto and D. G. Fedorov, The fragment molecular orbital method combined with density-functional tight-binding and the polarizable continuum model, *Physical Chemistry Chemical Physics* **18**, 22047–22061 (2016).

[1419] Y. Mochizuki, S. Tanaka and K. Fukuzawa, *Recent Advances of the Fragment Molecular Orbital Method*, Springer (2021).

[1420] K. Fukuzawa and S. Tanaka, Fragment molecular orbital calculations for biomolecules, *Current Opinion in Structural Biology* **72**, 127–134 (2022).

[1421] K. Akisawa, R. Hatada, K. Okuwaki, Y. Mochizuki, K. Fukuzawa, Y. Komeiji and S. Tanaka, Interaction analyses of SARS-CoV-2 spike protein based on fragment molecular orbital calculations, *RSC Advances* **11**, 3272–3279 (2021).

[1422] V. Vennelakanti, A. Nazemi, R. Mehmood, A. H. Steeves and H. J. Kulik, Harder, better, faster, stronger: Large-scale QM and QM/MM for predictive modeling in enzymes and proteins, *Current Opinion in Structural Biology* **72**, 9–17 (2022).

[1423] C. M. Clemente, L. Capece and M. A. Martí, Best practices on QM/MM simulations of biological systems, *Journal of Chemical Information and Modeling* **63**, 2609–2627 (2023).

[1424] B. Raghavan, M. Paulikat, K. Ahmad, L. Callea, A. Rizzi, E. Ippoliti, D. Mandelli, L. Bonati, M. De Vivo and P. Carloni, Drug design in the exascale era: A perspective from massively parallel QM/MM simulations, *Journal of Chemical Information and Modeling* **63**, 3647–3658 (2023).

[1425] R. Van Noorden, Computer modellers secure chemistry Nobels, *Nature* (2013).

[1426] A. Warshel and M. Karplus, Calculation of ground and excited state potential surfaces of conjugated molecules. I. Formulation and parametrization, *Journal of the American Chemical Society* **94**, 5612–5625 (1972a).

[1427] A. Warshel and M. Karplus, Vibrational structure of electronic transitions in conjugated molecules, *Chemical Physics Letters* **17**, 7–14 (1972).

[1428] M. Levitt and A. Warshel, Computer simulation of protein folding, *Nature* **253**, 694–698 (1975).

[1429] A. Warshel, Molecular dynamics simulations of biological reactions, *Accounts of Chemical Research* **35**, 385–395 (2002).

[1430] P. Priya, M. Kesheri, R. P. Sinha and S. Kanchan, Molecular dynamics simulations for biological systems, in *Biomedical Image Analysis and Mining Techniques for Improved Health Outcomes.* IGI Global Scientific Publishing, 286–313 (2016).

[1431] B. Mennucci and S. Corni, Multiscale modelling of photoinduced processes in composite systems, *Nature Reviews Chemistry* **3**, 315–330 (2019).

[1432] H. M. Senn and W. Thiel, QM/MM studies of enzymes, *Current Opinion in Chemical Biology* **11**, 182–187 (2007).

[1433] S. F. Sousa, A. J. Ribeiro, R. P. Neves, N. F. Brás, N. M. Cerqueira, P. A. Fernandes and M. J. Ramos, Application of quantum mechanics/molecular mechanics methods in the study of enzymatic reaction mechanisms, *Wiley Interdisciplinary Reviews: Computational Molecular Science* **7**, e1281 (2017).

[1434] O. Acevedo and W. L. Jorgensen, Advances in quantum and molecular mechanical (QM/MM) simulations for organic and enzymatic reactions, *Accounts of Chemical Research* **43**, 142–151 (2010).

[1435] R. A. Mata, H.-J. Werner, S. Thiel and W. Thiel, Toward accurate barriers for enzymatic reactions: QM/MM case study on p-hydroxybenzoate hydroxylase, *The Journal of Chemical Physics* **128**, 025104 (2008).

[1436] C. Liu, J. Gao and M. Liu, Tutorial on umbrella sampling simulation with a combined QM/MM potential: The potential of mean force for an S_N2 reaction in water, *The Journal of Physical Chemistry B* **128**, 10506–10514 (2024).

[1437] J. Chen, J. Kato, J. B. Harper, Y. Shao and J. Ho, On the accuracy of QM/MM models: A systematic study of intramolecular proton transfer reactions of amino acids in water, *The Journal of Physical Chemistry B* **125**, 9304–9316 (2021).

[1438] M. W. Van Der Kamp and A. J. Mulholland, Combined quantum mechanics/molecular mechanics (QM/MM) methods in computational enzymology, *Biochemistry* **52**, 2708–2728 (2013).

[1439] M. Kowalewski, B. P. Fingerhut, K. E. Dorfman, K. Bennett and S. Mukamel, Simulating coherent multidimensional spectroscopy of nonadiabatic molecular processes: From the infrared to the x-ray regime, *Chemical Reviews* **117**, 12165–12226 (2017).

[1440] M. Dittrich, P. L. Freddolino and K. Schulten, When light falls in LOV: a quantum mechanical/molecular mechanical study of photoexcitation in Phot-LOV1 of chlamydomonas reinhardtii, *The Journal of Physical Chemistry B* **109**, 13006–13013 (2005).

[1441] P. J. Ollitrault, G. Mazzola and I. Tavernelli, Nonadiabatic molecular quantum dynamics with quantum computers, *Physical Review Letters* **125**, 260511 (2020).

[1442] Z.-L. Li, S. Pei, Z. Chen, T.-Y. Huang, X.-D. Wang, L. Shen, X. Chen, Q.Q. Wang, D.-X. Wang and Y.-F. Ao, Machine learning-assisted amidasecatalytic enantioselectivity prediction and rational design of variants for improving enantioselectivity, *Nature Communications* **15**, 8778 (2024).

[1443] P. Kouba, P. Kohout, F. Haddadi, A. Bushuiev, R. Samusevich, J. Sedlar, J. Damborsky, T. Pluskal, J. Sivic and S. Mazurenko, Machine learning-guided protein engineering, *ACS Catalysis* **13**, 13863–13895 (2023).

[1444] J. Yang, F.-Z. Li and F. H. Arnold, Opportunities and challenges for machine learning-assisted enzyme engineering, *ACS Central Science* **10**, 226–241 (2024).

[1445] E. Orsi, L. Schada von Borzyskowski, S. Noack, P. I. Nikel and S. N. Lindner, Automated in vivo enzyme engineering accelerates biocatalyst optimization, *Nature Communications* **15**, 3447 (2024).

[1446] L. Böselt, M. Thürlemann and S. Riniker, Machine learning in QM/MM molecular dynamics simulations of condensed-phase systems, *Journal of Chemical Theory and Computation* **17**, 2641–2658 (2021).

[1447] B. A. Cordier, N. P. Sawaya, G. G. Guerreschi and S. K. McWeeney, Biology and medicine in the landscape of quantum advantages, *Journal of the Royal Society Interface* **19**, 20220541 (2022).

[1448] P. Ettenhuber, M. B. Hansen, P. P. Poier, I. Shaik, S. E. Rasmussen, N. K. Madsen, M. Majland, F. Jensen, L. Olsen and N. T. Zinner, Calculating the energy profile of an enzymatic reaction on a quantum computer, *Journal of Chemical Theory and Computation* **21**, 3493–3503 (2025).

[1449] J. Damborsky, P. Kouba, J. Sivic, D. Bednar and S. Mazurenko, Accelerating enzyme engineering with quantum power: Quest for quantum advantage in biocatalysis, *ChemRxiv pwkq5* (2024).

[1450] P. Reinholdt, E. Kjellgren, K. M. Ziems, S. Coriani, S. P. A. Sauer and J. Kongsted, Self-consistent quantum linear response with a polarizable embedding environment, *Journal of Physical Chemistry A* **129**, 1504–1515 (2025).

[1451] M. Capone, M. Romanelli, D. Castaldo, G. Parolin, A. Bello, G. Gil and M. Vanzan, A vision for the future of multiscale modeling, *ACS Physical Chemistry Au* **4**, 202–225 (2024).

[1452] W. Li, Z. Huang, C. Cao, Y. Huang, Z. Shuai, X. Sun, J. Sun, X. Yuan and D. Lv, Toward practical quantum embedding simulation of realistic chemical systems on near-term quantum computers, *Chemical Science* **13**, 8953–8962 (2022).

[1453] E. Gull, P. Werner, S. Fuchs, B. Surer, T. Pruschke and M. Troyer, Continuous-time quantum Monte Carlo impurity solvers, *Computer Physics Communications* **182**, 1078–1082 (2011).

[1454] A. Menczer and O. Legeza, Massively parallel tensor network state algorithms on hybrid CPU-GPU based architectures, *Journal of Chemical Theory and Computation* **21**, 1572–1587 (2025).

[1455] S. Sharma, K. Sivalingam, F. Neese and G. K.-L. Chan, Low-energy spectrum of iron-sulfur clusters directly from many-particle quantum mechanics, *Nature Chemistry* **6**, 927–933 (2014).

[1456] T. N. Georges, M. Bothe, C. Sünderhauf, B. K. Berntson, R. Izsák and A. V. Ivanov, Quantum simulations of chemistry in first quantization with any basis set, *npj Quantum Information* **11**, 55 (2025).

[1457] D. W. Berry, C. Gidney, M. Motta, J. R. McClean and R. Babbush, Qubitization of arbitrary basis quantum chemistry leveraging sparsity and low rank factorization, *Quantum* **3**, 208 (2019).

[1458] S. Lee, J. Lee, H. Zhai, Y. Tong, A. M. Dalzell, A. Kumar, P. Helms, J. Gray, Z.-H. Cui, W. Liu *et al.*, Evaluating the evidence for exponential quantum advantage in ground-state quantum chemistry, *Nature Communications* **14**, 1952 (2023).

[1459] E. Morosan, D. Natelson, A. H. Nevidomskyy and Q. Si, Strongly correlated materials, *Advanced Materials* **24**, 4896–4923 (2012).

[1460] B. Keimer and J. E. Moore, The physics of quantum materials, *Nature Physics* **13**, 1045–1055 (2017).

[1461] K. Head-Marsden, J. Flick, C. J. Ciccarino and P. Narang, Quantum information and algorithms for correlated quantum matter, *Chemical Reviews* **121**, 3061–3120 (2020).

[1462] L. Sá, P. Ribeiro and T. Prosen, Lindbladian dissipation of strongly-correlated quantum matter, *Physical Review Research* **4**, L022068 (2022).

[1463] A. Menczer, M. van Damme, A. Rask, L. Huntington, J. Hammond, S. S. Xantheas, M. Ganahl and O. Legeza, Parallel implementation of the density matrix renormalization group method achieving a quarter petaFLOPS performance on a single DGX-H100 GPU node, *Journal of Chemical Theory and Computation* **20**, 8397–8404 (2024).

[1464] R. Shinde, C. Filippi, A. Scemama and W. Jalby, Shifting sands of hardware and software in exascale quantum mechanical simulations, *Nature Reviews Physics* 7, 378–387 (2025).

[1465] C. Outeiral, M. Strahm, J. Shi, G. M. Morris, S. C. Benjamin and C. M. Deane, The prospects of quantum computing in computational molecular biology, *Wiley Interdisciplinary Reviews: Computational Molecular Science* **11**, e1481 (2021).

[1466] E. Linscott, *Accounting for Strong Electronic Correlation in Metalloproteins*, Ph.D. thesis, University of Cambridge (2019).

[1467] J. Liu, S. Chakraborty, P. Hosseinzadeh, Y. Yu, S. Tian, I. Petrik, A. Bhagi and Y. Lu, Metalloproteins containing cytochrome, iron-sulfur, or copper redox centers, *Chemical Reviews* **114**, 4366–4469 (2014).

[1468] S. Mayda, Z. Kandemir, N. Bulut and S. Maekawa, Magnetic mechanism for the biological functioning of hemoglobin, *Scientific Reports* **10**, 8569 (2020).

[1469] M. G. Quesne, R. Latifi, L. E. Gonzalez-Ovalle, D. Kumar and S. P. De Visser, Quantum mechanics/molecular mechanics study on the oxygen binding and substrate hydroxylation step in AlkB repair enzymes, *Chemistry-A European Journal* **20**, 435–446 (2013).

[1470] C. Weber, D. D. O'Regan, N. D. Hine, P. B. Littlewood, G. Kotliar and M. C. Payne, Importance of many body effects in the kernel of hemoglobin for ligand binding, *Physical Review Letters* **110**, 106402 (2013).

[1471] T. L. Poulos, Heme enzyme structure and function, *Chemical Reviews* **114**, 3919–3962 (2014).

[1472] C. J. Reedy and B. R. Gibney, Heme protein assemblies, *Chemical Reviews* **104**, 617–650 (2004).

[1473] T. Li, H. L. Bonkovsky and J.-t. Guo, Structural analysis of heme proteins: implications for design and prediction, *BMC Structural Biology* **11**, 13 (2011).

[1474] C. Weber, D. J. Cole, D. D. O'Regan and M. C. Payne, Renormalization of myoglobin-ligand binding energetics by quantum many-body effects, *Proceedings of the National Academy of Sciences* **111**, 5790–5795 (2014).

[1475] V. Kardam and K. D. Dubey, Tyr 118-mediated electron transfer is key to the chlorite decomposition in heme-dependent chlorite dismutase, *Inorganic Chemistry* **62**, 18322–18330 (2023).

[1476] B. I. Fedeles, V. Singh, J. C. Delaney, D. Li and J. M. Essigmann, The AlkB family of Fe(II)/α-ketoglutarate-dependent dioxygenases: Repairing nucleic acid alkylation damage and beyond, *Journal of Biological Chemistry* **290**, 20734–20742 (2015).

[1477] Y. Fu, B. Wang and Z. Cao, Coordination dynamics of iron center enables the C–H bond activation: QM/MM insight into the catalysis of hydroxyglutarate synthase (HglS), *Journal of Catalysis* **428**, 115149 (2023).

[1478] C.-G. Yang, C. Yi, E. M. Duguid, C. T. Sullivan, X. Jian, P. A. Rice and C. He, Crystal structures of DNA/RNA repair enzymes AlkB and ABH2 bound to dsDNA, *Nature* **452**, 961–965 (2008).

[1479] X. Guo, J. Zhang, L. Han, J. Lee, S. C. Williams, A. Forsberg, Y. Xu, R. N. Austin and L. Feng, Structure and mechanism of the alkane-oxidizing enzyme AlkB, *Nature Communications* **14**, 2180 (2023).

[1480] V. Genna, E. Donati and M. De Vivo, The catalytic mechanism of DNA and RNA polymerases, *ACS Catalysis* **8**, 11103–11118 (2018).

[1481] S. Fomichev, K. Hejazi, M. S. Zini, M. Kiser, J. Fraxanet, P. A. M. Casares, A. Delgado, J. Huh, A.-C. Voigt, J. E. Mueller *et al.*, Initial state preparation for quantum chemistry on quantum computers, *PRX Quantum* **5**, 040339 (2024).

[1482] J. Shee, M. Loipersberger, D. Hait, J. Lee and M. Head-Gordon, Revealing the nature of electron correlation in transition metal complexes with symmetry breaking and chemical intuition, *The Journal of Chemical Physics* **154**, 19 (2021).

[1483] M. Otten, M. R. Hermes, R. Pandharkar, Y. Alexeev, S. K. Gray and L. Gagliardi, Localized quantum chemistry on quantum computers, *Journal of Chemical Theory and Computation* **18**, 7205–7217 (2022).

[1484] C.-F. Kuo, D. E. McRee, C. L. Fisher, S. F. O'Handley, R. P. Cunningham and J. A. Tainer, Atomic structure of the DNA repair [4Fe-4S] enzyme endonuclease III, *Science* **258**, 434–440 (1992).

[1485] J. Mu, J. Wang, W. Ji, Z. Wei, D. Yang and H. Liu, Insights into application research of the density functional theory in iron sulfur compounds: A bibliometric mapping analysis, *ACS Omega* **9**, 46679–46697 (2024).

[1486] Ş. K. Özdemir, S. Rotter, F. Nori and L. Yang, Parity-time symmetry and exceptional points in photonics, *Nature Materials* **18**, 783–798 (2019).

[1487] D. Tzeli, P. Golub, J. Brabec, M. Matoušek, K. Pernal, L. Veis, S. Raugei and S. S. Xantheas, Importance of electron correlation on the geometry and electronic structure of [2Fe-2S] systems, *Journal of Chemical Theory and Computation* **20**, 10406–10423 (2024).

[1488] Y. Pang and R. Bjornsson, Understanding the electronic structure basis for N_2 binding to FeMoco: A systematic quantum mechanics/molecular mechanics investigation, *Inorganic Chemistry* **62**, 5357–5375 (2023).

[1489] B. Benediktsson and R. Bjornsson, Analysis of the geometric and electronic structure of spin coupled iron-sulfur dimers with broken-symmetry DFT: implications for FeMoco, *Journal of Chemical Theory and Computation* **18**, 1437–1457 (2022).

[1490] A. K. Almansori, K. Al-Ameed, M. Obies and F. S. Abd-Suhail, Magnetic exchange coupling in macrocyclic cobalt (II) complexes: The influence of bridging ligands and choice of the computational methodologies, *ChemistrySelect* **8**, e202302629 (2023).

[1491] L. Amico, R. Fazio, A. Osterloh and V. Vedral, Entanglement in many-body systems, *Reviews of Modern Physics* **80**, 517 (2008).

[1492] S. J. Bennie, M. W. van der Kamp, R. C. Pennifold, M. Stella, F. R. Manby and A. J. Mulholland, A projector-embedding approach for multiscale coupled-cluster calculations applied to citrate synthase, *Journal of Chemical Theory and Computation* **12**, 2689–2697 (2016).

[1493] M. Brütting, H. Bahmann and S. Kümmel, Predicting fundamental gaps accurately from density functional theory with non-empirical local range separation, *The Journal of Chemical Physics* **160**, 18 (2024).

[1494] P. Lykos and G. Pratt, Discussion on the Hartree-Fock approximation, *Reviews of Modern Physics* **35**, 496 (1963).
[1495] S. J. Lee, M. Welborn, F. R. Manby and T. F. Miller III, Projection-based wavefunction-in-DFT embedding, *Accounts of Chemical Research* **52**, 1359–1368 (2019).
[1496] M. W. van der Kamp, J. Zurek, F. R. Manby, J. N. Harvey and A. J. Mulholland, Testing high-level QM/MM methods for modeling enzyme reactions: acetyl-CoA deprotonation in citrate synthase, *The Journal of Physical Chemistry B* **114**, 11303–11314 (2010).
[1497] P. R. Nagy, State-of-the-art local correlation methods enable affordable gold standard quantum chemistry for up to hundreds of atoms, *Chemical Science* **15**, 14556–14584 (2024).
[1498] M. Gerenkamp and S. Grimme, Spin-component scaled second-order Møller-Plesset perturbation theory for the calculation of molecular geometries and harmonic vibrational frequencies, *Chemical Physics Letters* **392**, 229–235 (2004).
[1499] R. Buenker and S. Krebs, The configuration-driven approach for multireference configuration interaction calculations, in *Recent Advances in Multireference Methods.* World Scientific, 1–29 (1999).
[1500] A. Dreuw and M. Head-Gordon, Single-reference ab initio methods for the calculation of excited states of large molecules, *Chemical Reviews* **105**, 4009–4037 (2005).
[1501] Y. Shikano, H. C. Watanabe, K. M. Nakanishi and Y.-y. Ohnishi, Post-Hartree-Fock method in quantum chemistry for quantum computer, *The European Physical Journal Special Topics* **230**, 1037–1051 (2021).
[1502] M. Kaupp, A. Wodyński, A. V. Arbuznikov, S. Fürst and C. J. Schattenberg, Toward the next generation of density functionals: Escaping the zero-sum game by using the exact-exchange energy density, *Accounts of Chemical Research* **57**, 1815–1826 (2024).
[1503] D. A. Wappett and L. Goerigk, Benchmarking density functional theory methods for metalloenzyme reactions: The introduction of the MME55 set, *Journal of Chemical Theory and Computation* **19**, 8365–8383 (2023).
[1504] J. P. Perdew, Climbing the ladder of density functional approximations, *MRS Bulletin* **38**, 743–750 (2013).
[1505] T. Yanai, M. Saitow, X.-G. Xiong, C. Jakub, Y. Kurashige, S. Guo and S. Sharma, Multistate complete-active-space second-order perturbation theory based on density matrix renormalization group reference states, *Journal of Chemical Theory and Computation* **13**, 4829–4840 (2017).

[1506] A. Bachhar and N. J. Mayhall, Restricted open-shell cluster mean-field theory for strongly correlated systems, *The Journal of Physical Chemistry A* **128**, 9015–9027 (2024).

[1507] M. Matoušek, N. Vu, N. Govind, J. J. Foley IV and L. Veis, Polaritonic chemistry using the density matrix renormalization group method, *Journal of Chemical Theory and Computation* **20**, 9424–9434 (2024).

[1508] M. Ruggenthaler, D. Sidler and A. Rubio, Understanding polaritonic chemistry from ab initio quantum electrodynamics, *Chemical Reviews* **123**, 11191–11229 (2023).

[1509] M. Ruggenthaler, J. Flick, C. Pellegrini, H. Appel, I. V. Tokatly and A. Rubio, Quantum-electrodynamical density-functional theory: Bridging quantum optics and electronic-structure theory, *Physical Review A* **90**, 012508 (2014).

[1510] I. Y. Zhang and X. Xu, On the top rung of Jacob's ladder of density functional theory: Toward resolving the dilemma of SIE and NCE, *Wiley Interdisciplinary Reviews: Computational Molecular Science* **11**, e1490 (2021).

[1511] H. Zhai, S. Lee, Z.-H. Cui, L. Cao, U. Ryde and G. K.-L. Chan, Multireference protonation energetics of a dimeric model of nitrogenase ironsulfur clusters, *The Journal of Physical Chemistry A* **127**, 99749984 (2023).

[1512] H. Nakano and K. Hirao, Recent advances in multireference-based perturbation theory, *Bulletin of the Korean Chemical Society* **24**, 812–816 (2003).

[1513] M. Radoń, G. Drabik, M. Hodorowicz and J. Szklarzewicz, Performance of quantum chemistry methods for a benchmark set of spin-state energetics derived from experimental data of 17 transition metal complexes (SSE17), *Chemical Science* **15**, 20189–20204 (2024).

[1514] U. Ermler, W. Grabarse, S. Shima, M. Goubeaud and R. K. Thauer, Active sites of transition-metal enzymes with a focus on nickel, *Current Opinion in Structural Biology* **8**, 749–758 (1998).

[1515] M. Cococcioni, The LDA+U approach: A simple Hubbard correction for correlated ground states, in *Correlated Electrons: From Models to Materials Modeling and Simulation*, Vol. 2. Forschungszentrum Julich Julich, Germany (2012).

[1516] S. J. Bennie, D. Collison and J. J. McDouall, Electronic and magnetic properties of Kremer's tris-hydroxo bridged chromium dimer: A challenge for DFT, *Journal of Chemical Theory and Computation* **8**, 4915–4921 (2012).

[1517] D. Vollhardt, Dynamical mean-field theory of strongly correlated electron systems, in *Proceedings of the International Conference on Strongly Correlated Electron Systems (SCES2019)*, 011001 (2020).

[1518] S. Biermann, Dynamical mean field theory-based electronic structure calculations for correlated materials, in *First Principles Approaches to Spectroscopic Properties of Complex Materials.* Springer, 303–345 (2014).

[1519] V. Turkowski, A. Kabir, N. Nayyar and T. S. Rahman, Dynamical meanfield theory for molecules and nanostructures, *The Journal of Chemical Physics* **136**, 11 (2012).

[1520] M. Schüler, S. Barthel, T. Wehling, M. Karolak, A. Valli and G. Sangiovanni, Realistic theory of electronic correlations in nanoscopic systems, *The European Physical Journal Special Topics* **226**, 2615–2640 (2017).

[1521] H. Lee, C. Weber and E. B. Linscott, Many-body study of iron (III)-bound human serum transferrin, *The Journal of Physical Chemistry Letters* **13**, 4419–4425 (2022).

[1522] D. Novoselov, D. M. Korotin and V. I. Anisimov, Spin state transition in the active center of the hemoglobin molecule: DFT + DMFT study, *JETP Letters* **103**, 658–662 (2016).

[1523] W. Metzner and D. Vollhardt, Correlated lattice fermions in $d = \infty$ dimensions, *Physical Review Letters* **62**, 324 (1989).

[1524] W. Metzner and D. Vollhardt, Correlated lattice fermions in high dimensions, *Helvetica Physica Acta* **63**, 364–376 (1990).

[1525] A. Georges, G. Kotliar, W. Krauth and M. J. Rozenberg, Dynamical mean-field theory of strongly correlated fermion systems and the limit of infinite dimensions, *Reviews of Modern Physics* **68**, 13 (1996).

[1526] M. Potthoff, Cluster extensions of dynamical mean-field theory, *DMFT: From Infinite Dimensions to Real Materials* **8**, 5–1 (2018).

[1527] G. Kotliar, S. Y. Savrasov, K. Haule, V. S. Oudovenko, O. Parcollet and C. Marianetti, Electronic structure calculations with dynamical meanfield theory, *Reviews of Modern Physics* **78**, 865–951 (2006).

[1528] Q. Sun and G. K.-L. Chan, Quantum embedding theories, *Accounts of Chemical Research* **49**, 2705–2712 (2016).

[1529] A. Georges and G. Kotliar, Hubbard model in infinite dimensions, *Physical Review B* **45**, 6479 (1992).

[1530] C. Melnick, P. Sémon, K. Yu, N. D'Imperio, A.-M. Tremblay and G. Kotliar, Accelerated impurity solver for DMFT and its diagrammatic extensions, *Computer Physics Communications* **267**, 108075 (2021).

[1531] J. Chen, A. J. Millis and C. A. Marianetti, Density functional plus dynamical mean-field theory of the spin-crossover molecule $Fe(phen)_2(NCS)_2$, *Physical Review B* **91**, 241111 (2015).

[1532] C. Weber, D. D. O'Regan, N. D. Hine, M. C. Payne, G. Kotliar and P. B. Littlewood, Vanadium dioxide: A Peierls-Mott insulator stable against disorder, *Physical Review Letters* **108**, 256402 (2012).

[1533] D. Case, B. H. Huynh and M. Karplus, Binding of oxygen and carbon monoxide to hemoglobin. an analysis of the ground and excited states, *Journal of the American Chemical Society* **101**, 4433–4453 (1979).

[1534] L. Pauling and C. D. Coryell, The magnetic properties and structure of hemoglobin, oxyhemoglobin and carbonmonoxyhemoglobin, *Proceedings of the National Academy of Sciences USA* **22**, 210–216 (1936).

[1535] D. A. Pantazis, Missing pieces in the puzzle of biological water oxidation, *ACS Catalysis* **8**, 9477–9507 (2018).

[1536] S. Chu, D. Bovi, F. Cappelluti, A. G. Orellana, H. Martin and L. Guidoni, Effects of static correlation between spin centers in multicenter transition metal complexes, *Journal of Chemical Theory and Computation* **13**, 4675–4683 (2017).

[1537] D. A. Pantazis, Meeting the challenge of magnetic coupling in a triplybridged chromium dimer: Complementary broken-symmetry density functional theory and multireference density matrix renormalization group perspectives, *Journal of Chemical Theory and Computation* **15**, 938–948 (2019).

[1538] D.-T. Chen, J. Chen, X.-G. Li, G. Christou, S. Hill, X.-G. Zhang and H.P. Cheng, Long-range magnetic exchange pathways in complex clusters from first principles, *The Journal of Physical Chemistry C* **125**, 11124–11131 (2021).

[1539] R. Dutta, *Lessons From Electron Pairing Models for Postmodern Quantum Chemistry*, Ph.D. thesis, Rice University (2023).

[1540] Y. Cheng and H. Ma, Renormalized-residue-based multireference configuration interaction method for strongly correlated systems, *Journal of Chemical Theory and Computation* **20**, 1988–2009 (2024).

[1541] K. Liao, L. Ding and C. Schilling, Quantum information orbitals (QIO): Unveiling intrinsic many-body complexity by compressing single-body triviality, *The Journal of Physical Chemistry Letters* **15**, 6782–6790 (2024).

[1542] J. Li and T. Zhu, Interacting-bath dynamical embedding for capturing nonlocal electron correlation in solids, *Physical Review Letters* **133**, 216402 (2024).

[1543] R. Pandharkar, M. R. Hermes, C. J. Cramer and L. Gagliardi, Localized active space-state interaction: A multireference method for chemical insight, *Journal of Chemical Theory and Computation* **18**, 6557–6566 (2022).

[1544] R. U. Pandharkar, *Localized Multireference Methods for Strongly Correlated Systems*, Ph.D. thesis, The University of Chicago (2022).

[1545] T. N. Lan, A. A. Kananenka and D. Zgid, Communication: Towards ab initio self-energy embedding theory in quantum chemistry, *The Journal of Chemical Physics* **143**, 24 (2015).

[1546] Y. Lu, K. Sen, C. Yong, D. S. Gunn, J. A. Purton, J. Guan, A. Desmoutier, J. A. Nasir, X. Zhang, L. Zhu *et al.*, Multiscale QM/MM modelling of catalytic systems with chemshell, *Physical Chemistry Chemical Physics* **25**, 21816–21835 (2023).

[1547] B. Kang, P. Semon, C. Melnick, M. Han, S. Mo, H. Lee, G. Kotliar and S. Choi, Comdmft v. 2.0: Fully self-consistent ab initio GW+EDMFT for the electronic structure of correlated quantum materials, *Computer Physics Communications* **308**, 109447 (2025).

[1548] E. Deumens, V. F. Lotrich, A. S. Perera, R. J. Bartlett, N. Jindal and B. A. Sanders, The super instruction architecture: A framework for high-productivity parallel implementation of coupled-cluster methods on petascale computers, in *Annual Reports in Computational Chemistry*, Vol. 7. Elsevier, 179–191 (2011).

[1549] F. A. Evangelista and V. S. Batista, Quantum computing for chemistry, Journal of Chemical Theory and Computation **19**, 7435–7436 (2023).

[1550] J. D. Weidman, M. Sajjan, C. Mikolas, Z. J. Stewart, J. Pollanen, S. Kais and A. K. Wilson, Quantum computing and chemistry, *Cell Reports Physical Science* **5**, 102105 (2024).

[1551] T. J. Penfold, J. O. Johansson and J. Eng, Towards understanding and controlling ultrafast dynamics in molecular photomagnets, *Coordination Chemistry Reviews* **494**, 215346 (2023).

[1552] A. Miessen, P. J. Ollitrault, F. Tacchino and I. Tavernelli, Quantum algorithms for quantum dynamics, *Nature Computational Science* **3**, 25–37 (2023).

[1553] M. Motta, G. O. Jones, J. E. Rice, T. P. Gujarati, R. Sakuma, I. Liepuoniute, J. M. Garcia and Y.-y. Ohnishi, Quantum chemistry simulation of ground-and excited-state properties of the sulfonium cation on a superconducting quantum processor, *Chemical Science* **14**, 2915–2927 (2023).

[1554] B. Smith and A. V. Akimov, Modeling nonadiabatic dynamics in condensed matter materials: Some recent advances and applications, *Journal of Physics: Condensed Matter* **32**, 073001 (2019).

[1555] P. J. Ollitrault, A. Miessen and I. Tavernelli, Molecular quantum dynamics: A quantum computing perspective, *Accounts of Chemical Research* **54**, 4229–4238 (2021).

[1556] J. Whitlow, Z. Jia, Y. Wang, C. Fang, J. Kim and K. R. Brown, Simulating conical intersections with trapped ions, *Nature Chemistry* (2023).

[1557] D. Varsano, S. Caprasecca and E. Coccia, Theoretical description of protein field effects on electronic excitations of biological chromophores, *Journal of Physics: Condensed Matter* **29**, 013002 (2016).

[1558] P. You, D. Chen, C. Lian, C. Zhang and S. Meng, First-principles dynamics of photoexcited molecules and materials towards a quantum description, *Wiley Interdisciplinary Reviews: Computational Molecular Science* **11**, e1492 (2021).

[1559] H. Lischka, D. Nachtigallova, A. J. Aquino, P. G. Szalay, F. Plasser, F. B. Machado and M. Barbatti, Multireference approaches for excited states of molecules, *Chemical Reviews* **118**, 7293–7361 (2018).

[1560] N. Sheng, *Multiscale Methods for Quantum Many-Body Systems*, Ph.D. thesis, The University of Chicago (2023).

[1561] S. Bravyi, A. Chowdhury, D. Gosset and P. Wocjan, Quantum Hamiltonian complexity in thermal equilibrium, *Nature Physics* **18**, 1367–1370 (2022).

[1562] P. J. Ollitrault, C. L. Cortes, J. F. Gonthier, R. M. Parrish, D. Rocca, G.L. Anselmetti, M. Degroote, N. Moll, R. Santagati and M. Streif, Enhancing initial state overlap through orbital optimization for faster molecular electronic ground-state energy estimation, *Physical Review Letters* **133**, 250601 (2024).

[1563] B. Fauseweh, Quantum many-body simulations on digital quantum computers: State-of-the-art and future challenges, *Nature Communications* **15**, 2123 (2024).

[1564] L. H. Delgado-Granados, T. J. Krogmeier, L. M. Sager-Smith, I. Avdic, Z. Hu, M. Sajjan, M. Abbasi, S. E. Smart, P. Narang, S. Kais *et al.*, Quantum algorithms and applications for open quantum systems, *Chemical Reviews* **125**, 1823–1839 (2025).

[1565] V. Kashyap, G. Styliaris, S. Mouradian, J. I. Cirac and R. Trivedi, Accuracy guarantees and quantum advantage in analog open quantum simulation with and without noise, *Physical Review X* **15**, 021017 (2025).

[1566] M. P. Andersson, M. N. Jones, K. V. Mikkelsen, F. You and S. S. Mansouri, Quantum computing for chemical and biomolecular product design, *Current Opinion in Chemical Engineering* **36**, 100754 (2022).

[1567] B. Bonde, P. Patil and B. Choubey, The future of drug development with quantum computing, *High Performance Computing for Drug Discovery and Biomedicine*, 153–179 (2023).
[1568] M. Zinner, F. Dahlhausen, P. Boehme, J. Ehlers, L. Bieske and L. Fehring, Toward the institutionalization of quantum computing in pharmaceutical research, *Drug Discovery Today* **27**, 378–383 (2022).
[1569] V. Marx, Biology begins to tangle with quantum computing, *Nature Methods* **18**, 715–719 (2021).
[1570] B. Bauer, S. Bravyi, M. Motta and G. K.-L. Chan, Quantum algorithms for quantum chemistry and quantum materials science, *Chemical Reviews* **120**, 12685–12717 (2020).
[1571] D. Basov, R. Averitt and D. Hsieh, Towards properties on demand in quantum materials, *Nature Materials* **16**, 1077–1088 (2017).
[1572] A. Einstein, Physics and reality, *The Journal of the Franklin Institute* **221**, 349–382 (1936).
[1573] M. H. Abdel-kader, The journey of PDT throughout history: PDT from pharos to present, in *Photodynamic Medicine: From Bench to Clinic*, 1–21 (2016).
[1574] J. J. Anders, R. J. Lanzafame and P. R. Arany, Low-level light/laser therapy versus photobiomodulation therapy, *Photomedicine and Laser Surgery* **33**, 183 (2015).
[1575] W. Glinkowski, Low-level energy laser therapy in the musculoskeletal system, *Lasers in the Musculoskeletal System*, 188–198 (2001).
[1576] P. Gøtzsche, Niels finsen's treatment for lupus vulgaris, *Journal of the Royal Society of Medicine* **104**, 41–42 (2011).
[1577] A. Grzybowski and K. Pietrzak, From patient to discoverer—Niels Ryberg Finsen (1860–1904)—the founder of phototherapy in dermatology, *Clinics in Dermatology* **30**, 451–455 (2012).
[1578] The treatment of disease by light, *Nature* **64**, 525–526 (1901).
[1579] P. Avci, G. K. Gupta, J. Clark, N. Wikonkal and M. R. Hamblin, Lowlevel laser (light) therapy (LLLT) for treatment of hair loss, *Lasers in Surgery and Medicine* **46**, 144–151 (2014).
[1580] M. R. Hamblin, Photobiomodulation or low-level laser therapy, *Journal of Biophotonics* **9**, 1122 (2016).
[1581] M. Staelens, E. Di Gregorio, A. P. Kalra, H. T. Le, N. Hosseinkhah, M. Karimpoor, L. Lim and J. A. Tuszyński, Near-infrared photobiomodulation of living cells, tubulin, and microtubules *in vitro*, *Frontiers in Medical Technology* **4**, 871196 (2022).
[1582] C. Dompe, L. Moncrieff, J. Matys, K. Grzech-Leśniak, I. Kocherova, A. Bryja, M. Bruska, M. Dominiak, P. Mozdziak, T. H. I. Skiba *et al.*, Photobiomodulation-underlying mechanism and clinical applications, *Journal of Clinical Medicine* **9**, 1724 (2020).

[1583] C. H. B. Reis, D. V. Buchaim, A. d. C. Ortiz, S. O. M. Fideles, J. A. Dias, M. A. Miglino, D. d. B. Teixeira, E. d. S. B. M. Pereira, M. R. da Cunha and R. L. Buchaim, Application of fibrin associated with photobiomodulation as a promising strategy to improve regeneration in tissue engineering: A systematic review, *Polymers* **14**, 3150 (2022).

[1584] H. B. Cotler, R. T. Chow, M. R. Hamblin and J. Carroll, The use of low level laser therapy (LLLT) for musculoskeletal pain, *MOJ Orthopedics & Rheumatology* **2**, 5 (2015).

[1585] L. Santana-Blank and E. Rodríguez-Santana, Photobiomodulation in light of our biological clock's inner workings, *Photomedicine and Laser Surgery* **36**, 119–121 (2018).

[1586] R. Fekrazad, S. Asefi, M. B. Eslaminejad, L. Taghiar, S. Bordbar and M. R. Hamblin, Photobiomodulation with single and combination laser wavelengths on bone marrow mesenchymal stem cells: proliferation and differentiation to bone or cartilage, *Lasers in Medical Science* **34**, 115–126 (2019).

[1587] Y. Ruan, H. Kato, Y. Taguchi, N. Yamauchi and M. Umeda, Irradiation by high-intensity red light-emitting diode enhances human bone marrow mesenchymal stem cells osteogenic differentiation and mineralization through Wnt/β-catenin signaling pathway, *Lasers in Medical Science* **36**, 55–65 (2021).

[1588] T. I. Karu, Mitochondrial signaling in mammalian cells activated by red and near-IR radiation, *Photochemistry and Photobiology* **84**, 10911099 (2008).

[1589] M. T. Wong-Riley, H. L. Liang, J. T. Eells, B. Chance, M. M. Henry, E. Buchmann, M. Kane and H. T. Whelan, Photobiomodulation directly benefits primary neurons functionally inactivated by toxins: Role of cytochrome c oxidase, *Journal of Biological Chemistry* **280**, 47614771 (2005).

[1590] M. Yang, Z. Yang, P. Wang and Z. Sun, Current application and future directions of photobiomodulation in central nervous diseases, *Neural Regeneration Research* **16**, 1177 (2021).

[1591] G. M. Cho, S.-Y. Lee, J. H. Park, M. J. Kim, K.-J. Park, B. T. Choi, Y.I. Shin, N. G. Kim and H. K. Shin, Photobiomodulation using a low-level light-emitting diode improves cognitive dysfunction in the 5XFAD mouse model of Alzheimer's disease, *The Journals of Gerontology: Series A* **75**, 631–639 (2020).

[1592] L. M. Ailioaie, C. Ailioaie and G. Litscher, Photobiomodulation in Alzheimer's disease-A complementary method to state-of-the-art pharmaceutical formulations and nanomedicine? *Pharmaceutics* **15**, 916 (2023).

[1593] L. Joy, R. Jolien, C. Marithé, E. Stijn, S. Laura, L. Hilde, B. Sandra, N. Wendy, H. Ruth, R. Liesbeth *et al.*, The use of photobiomodulation therapy for the prevention of chemotherapy-induced peripheral neuropathy: A randomized, placebo-controlled pilot trial (neurolaser trial), *Supportive Care in Cancer* **30**, 5509–5517 (2022).

[1594] C. Teng, S. Egger, P. L. Blinman and J. L. Vardy, Evaluating laser photobiomodulation for chemotherapy-induced peripheral neuropathy: A randomised phase II trial, *Supportive Care in Cancer* **31**, 52 (2023).

[1595] T. P. Prado, F. C. Zanchetta, B. Barbieri, C. Aparecido, M. H. Melo Lima and E. P. Araujo, Photobiomodulation with blue light on wound healing: A scoping review, *Life* **13**, 575 (2023).

[1596] T. Karu, Photobiology of low-power laser effects, *Health Physics* **56**, 691–704 (1989).

[1597] F. Rossi, G. Magni, F. Tatini, M. Banchelli, F. Cherchi, M. Rossi, E. Coppi, A. M. Pugliese, D. Rossi degl'Innocenti, D. Alfieri *et al.*, Photobiomodulation of human fibroblasts and keratinocytes with blue light: implications in wound healing, *Biomedicines* **9**, 41 (2021).

[1598] G. Magni, F. Tatini, M. Rossi, G. de Siena, R. Cicchi, F. S. Pavone, S. Bacci and F. Rossi, In vivo evidence of blue LED light photobiomodulation, in *Medical Sciences Forum*, Vol. 21. MDPI, 6 (2023).

[1599] T. Karu *et al.*, Molecular mechanism of the therapeutic effect of low-intensity laser radiation, *Lasers Life Sci* **2**, 53–74 (1988).

[1600] U. Cagin, O. F. Duncan, A. P. Gatt, M. S. Dionne, S. T. Sweeney and J. M. Bateman, Mitochondrial retrograde signaling regulates neuronal function, *Proceedings of the National Academy of Sciences USA* **112**, E6000–E6009 (2015).

[1601] A. Liebert, W. Capon, V. Pang, D. Vila, B. Bicknell, C. McLachlan and H. Kiat, Photophysical mechanisms of photobiomodulation therapy as precision medicine, *Biomedicines* **11**, 237 (2023).

[1602] P. L. Lima, C. V. Pereira, N. Nissanka, T. Arguello, G. Gavini, C. M. da Costa Maranduba, F. Diaz and C. T. Moraes, Photobiomodulation enhancement of cell proliferation at 660 nm does not require cytochrome *c* oxidase, *Journal of Photochemistry and Photobiology B: Biology* **194**, 71–75 (2019).

[1603] C. Moro, A. Valverde, M. Dole, J. Hoh Kam, C. Hamilton, A. Liebert, B. Bicknell, A.-L. Benabid, P. Magistretti and J. Mitrofanis, The effect of photobiomodulation on the brain during wakefulness and sleep, *Frontiers in Neuroscience* **16**, 942536 (2022).

[1604] L. F. De Freitas and M. R. Hamblin, Proposed mechanisms of photobiomodulation or low-level light therapy, *IEEE Journal of Selected Topics in Quantum Electronics* **22**, 348–364 (2016).

[1605] I. S. Raja, C. Kim, N. Oh, J.-H. Park, S. W. Hong, M. S. Kang, C. Mao and D.-W. Han, Tailoring photobiomodulation to enhance tissue regeneration, *Biomaterials* **309**, 122623 (2024).

[1606] K. E. Kennedy, C. C. Wills, C. Holt and M. A. Grandner, A randomized, sham-controlled trial of a novel near-infrared phototherapy device on sleep and daytime function, *Journal of Clinical Sleep Medicine* **19**, 1669–1675 (2023).

[1607] D. Si, B. Su, J. Zhang, K. Zhao, J. Li, D. Chen, S. Hu and X. Wang, Low-level laser therapy with different irradiation methods modulated the response of bone marrow mesenchymal stem cells in vitro, *Lasers in Medical Science* **37**, 3509–3516 (2022).

[1608] S. A. Mohamad, M. R. Milward, M. A. Hadis, S. A. Kuehne and P. R. Cooper, Photobiomodulation of mineralisation in mesenchymal stem cells, *Photochemical & Photobiological Sciences* **20**, 699–714 (2021).

[1609] S. Chae, B. Y. Ahn, K. Byun, Y. M. Cho, M.-H. Yu, B. Lee, D. Hwang and K. S. Park, A systems approach for decoding mitochondrial retrograde signaling pathways, *Science Signaling* **6**, rs4 (2013).

[1610] S. Yu, C.-C. E. Lan and H.-S. Yu, Mechanisms of repigmentation induced by photobiomodulation therapy in vitiligo, *Experimental Dermatology* **28**, 10–14 (2019).

[1611] G. Bagkos, K. Koufopoulos and C. Piperi, Mitochondrial emitted electromagnetic signals mediate retrograde signaling, *Medical Hypotheses* **85**, 810–818 (2015).

[1612] M. P. Horan and D. N. Cooper, The emergence of the mitochondrial genome as a partial regulator of nuclear function is providing new insights into the genetic mechanisms underlying age-related complex disease, *Human Genetics* **133**, 435–458 (2014).

[1613] R. S. Ribeiro, J. A. Rodrigues, A. L. Mencalha and A. d. S. da Fonseca, Low-power blue LED and red laser effects on plasmid DNA, survival, and photolyase mRNA levels in Escherichia coli cultures, *Laser Physics Letters* **20**, 085601 (2023).

[1614] J. D. Lee and M. J. M. Oh, Light-emitting diode (LED) and Arndt-Schultz curve, in *Lasers in Dermatology: Parameters and Choice: With Special Reference to the Asian Population*. Springer, 149–154 (2023).

[1615] S. Ravera, N. Bertola, C. Pasquale, S. Bruno, S. Benedicenti, S. Ferrando, A. Zekiy, P. Arany and A. Amaroli, 808-nm photobiomodulation affects the viability of a head and neck squamous carcinoma cellular model, acting on energy metabolism and oxidative stress production, *Biomedicines* **9**, 1717 (2021).

[1616] C. Mendes, D. P. dos Santos Haupenthal, R. P. Zaccaron, G. de Bem Silveira, M. E. A. B. Corrêa, L. de Roch Casagrande, S. de Sousa Mariano, J. I. de Souza Silva, T. A. M. de Andrade, P. E. Feuser *et al.*, Effects of the association between photobiomodulation and hyaluronic acid linked gold nanoparticles in wound healing, *ACS Biomaterials Science & Engineering* **6**, 5132–5144 (2020).

[1617] Y. Mayer, J. A. Shibli, H. A. Saada, M. Melo, E. Gabay, S. Barak and O. Ginesin, Pulsed electromagnetic therapy: Literature review and current update, *Brazilian Dental Journal* **35**, e24–6109 (2024).

[1618] S. Zhong, S. Yao, Q. Zhao, Z. Wang, Z. Liu, L. Li and Z. L. Wang, Electricity-assisted cancer therapy: From traditional clinic applications to emerging methods integrated with nanotechnologies, *Advanced NanoBiomed Research* **3**, 2200143 (2023).

[1619] R. H. Funk, T. Monsees and N. Özkucur, Electromagnetic effects-from cell biology to medicine, *Progress in Histochemistry and Cytochemistry* **43**, 177–264 (2009).

[1620] V. N. Binhi and A. B. Rubin, Theoretical concepts in magnetobiology after 40 years of research, *Cells* **11**, 274 (2022).

[1621] V. Martynyuk, Y. V. Tseyslyer and N. Temuryants, Interference of the mechanisms of influence that weak extremely low-frequency electromagnetic fields have on the human body and animals, *Izvestiya, Atmospheric and Oceanic Physics* **48**, 832–846 (2012).

[1622] T. Saliev, D. Begimbetova, A.-R. Masoud and B. Matkarimov, Biological effects of non-ionizing electromagnetic fields: Two sides of a coin, *Progress in Biophysics and Molecular Biology* **141**, 25–36 (2019).

[1623] P. Piszczek, K. Wójcik-Piotrowicz, K. Gil and J. Kaszuba-Zwoińska, Immunity and electromagnetic fields, *Environmental Research* **200**, 111505 (2021).

[1624] C. L. Ross, D. C. Ang and G. Almeida-Porada, Targeting mesenchymal stromal cells/pericytes (MSCs) with pulsed electromagnetic field (PEMF) has the potential to treat rheumatoid arthritis, *Frontiers in Immunology* **10**, 266 (2019).

[1625] C. L. Ross, M. Siriwardane, G. Almeida-Porada, C. D. Porada, P. Brink, G. J. Christ and B. S. Harrison, The effect of low-frequency electromagnetic field on human bone marrow stem/progenitor cell differentiation, *Stem Cell Research* **15**, 96–108 (2015).

[1626] A. Sendera, B. Pikuła and A. Banaś-Ząbczyk, Preconditioning of mesenchymal stem cells with electromagnetic fields and its impact on biological responses and "fate"-potential use in therapeutic applications, *Frontiers in Bioscience-Landmark* **28**, 285 (2023).

[1627] C. L. Ross, M. J. Pettenati, J. Procita, L. Cathey, S. K. George and G. Almeida-Porada, Evaluation of cytotoxic and genotoxic effects of extremely low-frequency electromagnetic field on mesenchymal stromal cells, *Global Advances in Health and Medicine* **7**, 2164956118777472 (2018).

[1628] Y. Shen, R. Xia, H. Jiang, Y. Chen, L. Hong, Y. Yu, Z. Xu and Q. Zeng, Exposure to 50 Hz-sinusoidal electromagnetic field induces DNA damage-independent autophagy, *The International Journal of Biochemistry & Cell Biology* **77**, 72–79 (2016).

[1629] N. V. Bobkova, V. V. Novikov, N. I. Medvinskaya, I. Y. Aleksandrova, I. V. Nesterova and E. E. Fesenko, Effect of weak combined static and extremely low-frequency alternating magnetic fields on spatial memory and brain amyloid-β in two animal models of Alzheimer's disease, *Electromagnetic Biology and Medicine* **37**, 127–137 (2018).

[1630] Y. Yang, P. Wang, A. Liu, X. Wu, Z. Yan, S. Dai, J. Wei, Z. Zhang, X. Li, P. Luo *et al.*, Pulsed electromagnetic field protects against brain injury after intracerebral hemorrhage: Involvement of anti-inflammatory processes and hematoma clearance via CD36, *Journal of Molecular Neuroscience* **72**, 2150–2161 (2022).

[1631] G. Romero, J. Park, F. Koehler, A. Pralle and P. Anikeeva, Modulating cell signalling in vivo with magnetic nanotransducers, *Nature Reviews Methods Primers* **2**, 92 (2022).

[1632] M. A. İnevi, *Magnetic Levitation of Cells from Bone Marrow Origin*, Ph.D. thesis, Izmir Institute of Technology (Turkey) (2021).

[1633] C. Khatua, D. Bhattacharya and V. K. Balla, In situ electrical stimulation for enhanced bone growth: A mini-review, *Medical Devices & Sensors* **3**, e10090 (2020).

[1634] M. Iwasaka, Deformation of cellular components of bone forming cells when exposed to a magnetic field, *AIP Advances* **9**, 035327 (2019).

[1635] J. Zhang, C. Ding, L. Ren, Y. Zhou and P. Shang, The effects of static magnetic fields on bone, *Progress in Biophysics and Molecular Biology* **114**, 146–152 (2014).

[1636] V. Thöni, D. Mauracher, A. Ramalingam, B. Fiechtner, A. M. Sandbichler and M. Egg, Quantum based effects of therapeutic nuclear magnetic resonance persistently reduce glycolysis, *iScience* **25**, 105536 (2022).

[1637] H. Jimenez, D. Gibo, S. Sharma, M. Pennison, L. D. Miller, M. Wang, K. Sheffield *et al.*, Treatment of glioblastoma with tumor-specific amplitude-modulated radiofrequency electromagnetic fields, *Oncotarget* **16**, 741–757 (2025).

[1638] J. W. Zimmerman, M. J. Pennison, I. Brezovich, N. Yi, C. T. Yang, R. Ramaker, D. Absher, R. M. Myers, N. Kuster, F. P. Costa *et al.*, Cancer cell proliferation is inhibited by specific modulation frequencies, *British Journal of Cancer* **106**, 307–313 (2012).

[1639] A. Barbault, F. P. Costa, B. Bottger, R. F. Munden, F. Bomholt, N. Kuster and B. Pasche, Amplitude-modulated electromagnetic fields for the treatment of cancer: Discovery of tumor-specific frequencies and assessment of a novel therapeutic approach, *Journal of Experimental & Clinical Cancer Research* **28**, 1–10 (2009).

[1640] M. Vadalà, J. C. Morales-Medina, A. Vallelunga, B. Palmieri, C. Laurino and T. Iannitti, Mechanisms and therapeutic effectiveness of pulsed electromagnetic field therapy in oncology, *Cancer Medicine* **5**, 3128–3139 (2016).

[1641] S. Crocetti, C. Beyer, G. Schade, M. Egli, J. Fröhlich and A. Franco-Obregón, Low intensity and frequency pulsed electromagnetic fields selectively impair breast cancer cell viability, *PloS One* **8**, e72944 (2013).

[1642] A.-C. Francisco, S.-A. M. del Mar, C. Irene, R.-A. Sandra, L. Josefa, R.M. Elisa, O. Nicolás and N. M. Isabel, Could radiotherapy effectiveness be enhanced by electromagnetic field treatment? *International Journal of Molecular Sciences* **14**, 14974–14995 (2013).

[1643] M. Tanori, A. Casciati, A. Zambotti, R. Pinto, I. Gianlorenzi, A. Pannicelli, P. Giardullo, B. Benassi, C. Marino, M. Mancuso *et al.*, Microsecond pulsed electric fields: An effective way to selectively target and radiosensitize medulloblastoma cancer stem cells, *International Journal of Radiation Oncology* **109**, 1495–1507 (2021).

[1644] N. Kong, H. Zhang, C. Feng, C. Liu, Y. Xiao, X. Zhang, L. Mei, J. S. Kim, W. Tao and X. Ji, Arsenene-mediated multiple independently targeted reactive oxygen species burst for cancer therapy, *Nature Communications* **12**, 4777 (2021).

[1645] B. Perillo, M. Di Donato, A. Pezone, E. Di Zazzo, P. Giovannelli, G. Galasso, G. Castoria and A. Migliaccio, ROS in cancer therapy: The bright side of the moon, *Experimental & Molecular Medicine* **52**, 192–203 (2020).

[1646] H. Yang, R. M. Villani, H. Wang, M. J. Simpson, M. S. Roberts, M. Tang and X. Liang, The role of cellular reactive oxygen species in cancer chemotherapy, *Journal of Experimental & Clinical Cancer Research* **37**, 266 (2018).

[1647] V. E. Orel, O. Dasyukevich, O. Rykhalskyi, V. B. Orel, A. Burlaka and S. Virko, Magneto-mechanical effects of magnetite nanoparticles on Walker-256 carcinosarcoma heterogeneity, redox state and growth modulated by an inhomogeneous stationary magnetic field, *Journal of Magnetism and Magnetic Materials* **538**, 168314 (2021).

[1648] Z. P. Germon, J. R. Sillar, A. Mannan, R. J. Duchatel, D. Staudt, H. C. Murray, I. J. Findlay, E. R. Jackson, H. P. McEwen, A. M. Douglas *et al.*, Blockade of ROS production inhibits oncogenic signaling in acute myeloid leukemia and amplifies response to precision therapies, *Science Signaling* **16**, eabp9586 (2023).

[1649] S. Falone, S. Santini Jr, V. Cordone, G. Di Emidio, C. Tatone, M. Cacchio and F. Amicarelli, Extremely low-frequency magnetic fields and redox-responsive pathways linked to cancer drug resistance: Insights from co-exposure-based in vitro studies, *Frontiers in Public Health* **6**, 33 (2018).

[1650] S. Gupta, R. S. Sharma and R. Singh, Non-ionizing radiation as possible carcinogen, *International Journal of Environmental Health Research* **32**, 916–940 (2022).

[1651] B. López-Díaz, S. Mercado-Sáenz, A. M. Burgos-Molina, A. GonzálezVidal, F. Sendra-Portero and M. J. Ruiz-Gómez, Genomic DNA damage induced by co-exposure to DNA damaging agents and pulsed magnetic field, *International Journal of Radiation Biology* **99**, 853-865 (2023).

[1652] L.-Q. Yuan, C. Wang, D.-F. Lu, X.-D. Zhao, L.-H. Tan and X. Chen, Induction of apoptosis and ferroptosis by a tumor suppressing magnetic field through ROS-mediated DNA damage, *Aging (Albany NY)* **12**, 3662 (2020).

[1653] J. Pokorný, C. Vedruccio, M. Cifra and O. Kučera, Cancer physics: diagnostics based on damped cellular elastoelectrical vibrations in microtubules, *European Biophysics Journal* **40**, 747–759 (2011).

[1654] K. Kose, Physical and technical aspects of human magnetic resonance imaging: present status and 50 years historical review, *Advances in Physics: X* **6**, 1885310 (2021).

[1655] M. Capiglioni, F. Turco, R. Wiest and C. Kiefer, Analysis of the robustness and dynamics of spin-locking preparations for the detection of oscillatory magnetic fields, *Scientific Reports* **12**, 16965 (2022).

[1656] V. Kotoula, J. W. Evans, C. E. Punturieri and C. A. Zarate Jr, The use of functional magnetic resonance imaging (FMRI) in clinical trials and experimental research studies for depression, *Frontiers in Neuroimaging* **2**, 1110258 (2023).

[1657] C. Vedruccio and A. Meessen, EM cancer detection by means of nonlinear resonance interaction, *Progress in Electromagnetic Research Symposium* (2004).

[1658] L. Cormio, C. Vedruccio, G. Leucci, P. Massenio, G. Di Fino, V. Cavaliere and G. Carrieri, Noninvasive electromagnetic detection of bladder cancer, *International Scholarly Research Notices* **2014** (2014).
[1659] M. P. Dore, M. O. Tufano, G. M. Pes, M. Cuccu, V. Farina, A. Manca and D. Y. Graham, Tissue resonance interaction accurately detects colon lesions: A double-blind pilot study, *World Journal of Gastroenterology: WJG* **21**, 7851 (2015).
[1660] G. Boryczka, M. Hartleb and M. Janik, Detection of hepatocellular carcinoma by tissue resonance interaction method (TRIM), *Gastroenterology Review/Przeglad Gastroenterologiczny* **13**, 40–46 (2018).
[1661] S.-C. Lee, H. Poptani, E. J. Delikatny, S. Pickup, D. S. Nelson, S. J. Schuster, S. D. Nasta, J. Svoboda, S. C. Goldstein, S. G. Wallace *et al.*, Nmr metabolic and physiological markers of therapeutic response, in *Oxygen Transport to Tissue XXXII.* Springer, 129–135 (2011).
[1662] A. Mann, B. Steinecker-Frohnwieser, A. Naghilou, F. Millesi, P. Supper, L. Semmler, S. Wolf, L. Marinova, L. Weigl, T. Weiss *et al.*, Nuclear magnetic resonance treatment accelerates the regeneration of dorsal root ganglion neurons in vitro, *Frontiers in Cellular Neuroscience* **16** (2022).
[1663] V. Thöni, R. Oliva, D. Mauracher and M. Egg, Therapeutic nuclear magnetic resonance affects the core clock mechanism and associated hypoxia-inducible factor-1, *Chronobiology International* **38**, 11201134 (2021).
[1664] S. Ghodbane, A. Lahbib, M. Sakly and H. Abdelmelek, Bioeffects of static magnetic fields: Oxidative stress, genotoxic effects, and cancer studies, *BioMed Research International* **2013** (2013).
[1665] L. A. Portelli, T. E. Schomay and F. S. Barnes, Inhomogeneous background magnetic field in biological incubators is a potential confounder for experimental variability and reproducibility, *Bioelectromagnetics* **34**, 337–348 (2013).
[1666] R. J. Usselman, I. Hill, D. J. Singel and C. F. Martino, Spin biochemistry modulates reactive oxygen species (ROS) production by radio frequency magnetic fields, *PloS One* **9**, e93065 (2014).
[1667] F. Barnes and B. Greenebaum, Setting guidelines for electromagnetic exposures and research needs, *Bioelectromagnetics* **41**, 392–397 (2020).
[1668] E. Di Gregorio, S. Israel, M. Staelens, G. Tankel, K. Shankar and J. A. Tuszyński, The distinguishing electrical properties of cancer cells, *Physics of Life Reviews* **43**, 139–188 (2022).

[1669] V. Salari, S. Barzanjeh, M. Cifra, C. Simon, F. Scholkmann, Z. Alirezaei and J. A. Tuszynski, Electromagnetic fields and optomechanics in cancer diagnostics and treatment, *Frontiers in Bioscience-Landmark* **23**, 1391–1406 (2018).
[1670] M. Levin, Bioelectrical approaches to cancer as a problem of the scaling of the cellular self, *Progress in Biophysics and Molecular Biology* **165**, 102–113 (2021).
[1671] G. Thrivikraman, S. K. Boda and B. Basu, Unraveling the mechanistic effects of electric field stimulation towards directing stem cell fate and function: A tissue engineering perspective, *Biomaterials* **150**, 60–86 (2018).
[1672] G. Kandasamy, Y. Karuppasamy and U. M. Krishnan, Emerging trends in nano-driven immunotherapy for treatment of cancer, *Vaccines* **11**, 458 (2023).
[1673] V. B. Orel, L. A. Syvak and V. E. Orel, Remote control of magnetic nanocomplexes for delivery and destruction of cancer cells, *Journal of Biomaterials Applications* **36**, 872–881 (2021).
[1674] G. Sun, J. Li, W. Zhou, R. G. Hoyle and Y. Zhao, Electromagnetic interactions in regulations of cell behaviors and morphogenesis, *Frontiers in Cell and Developmental Biology* **10**, 1014030 (2022).
[1675] U. Chilakamarthi and L. Giribabu, Photodynamic therapy: Past, present and future, *The Chemical Record* **17**, 775–802 (2017).
[1676] E. H. White, J. Wiecko and D. F. Roswell, Photochemistry without light, *Journal of the American Chemical Society* **91**, 5194–5196 (1969).
[1677] O. C. Correspondent, Chemical change: Photochemistry without light, *Nature* **224**, 110 (1969).
[1678] K. Kalyanasundaram and M. Neumann-Spallart, Photophysical and redox properties of water-soluble porphyrins in aqueous media, *The Journal of Physical Chemistry* **86**, 5163–5169 (1982).
[1679] M. Roncel, M. Hervás, J. A. Navarro, M. A. De la Rosa and G. Tollin, Flavin-photosensitized oxidation of reduced c-type cytochromes: Reaction mechanism and comparison with photoreduction of oxidized cytochromes by flavin semiquinones, *European Journal of Biochemistry* **191**, 531–536 (1990).
[1680] J. Kou, D. Dou and L. Yang, Porphyrin photosensitizers in photodynamic therapy and its applications, *Oncotarget* **8**, 81591 (2017).
[1681] D. Trauner, Richard Willstätter and the 1915 Nobel Prize in chemistry, *Angewandte Chemie (International Ed. in English)* **54**, 1191011916 (2015).
[1682] G. Hendry and O. Jones, Haems and chlorophylls: Comparison of function and formation. *Journal of Medical Genetics* **17**, 1 (1980).

[1683] S. Granick and S. I. Beale, Hemes, chlorophylls, and related compounds: Biosynthesis and metabolic regulation, in *Advances in Enzymology and Related Areas of Molecular Biology*, Vol. 46, 33–203 (1978).
[1684] P. Rothemund, Hemin and chlorophyll: The two most important pigments for life on earth, *The Ohio Journal of Science* **56**, 193–202 (1956).
[1685] G. Gunaydin, M. E. Gedik and S. Ayan, Photodynamic therapy for the treatment and diagnosis of cancer–A review of the current clinical status, *Frontiers in Chemistry* **9**, 686303 (2021).
[1686] H. von Tappeiner, The photodynamic phenomenon (sensitization by fluorescent substances), *Results of Physiology* **8**, 698–741 (1909).
[1687] H. von Tappeiner and H. Jesionek, Therapeutische versuche mit fluoreszierenden stoffen, *Münchener Medizinische Wochenschrift* **50**, 2042–2044 (1903).
[1688] H. von Tappeiner and H. Jesionek, Zur behandlung der hautcarcinome mit fluoreszierenden stoffen, *Deutsches Archiv für Klinische Medizin* **82**, 223–226 (1905).
[1689] J. Moan and Q. Peng, An outline of the hundred-year history of PDT. *Anticancer Research* **23**, 3591–3600 (2003).
[1690] H. Kato, History of photodynamic therapy–past, present and future, *Gan To Kagaku Ryoho* **23**, 8–15 (1996).
[1691] D. Kessel, Photodynamic therapy: A brief history, *Journal of Clinical Medicine* **8**, 1581 (2019).
[1692] S. Kwiatkowski, B. Knap, D. Przystupski, J. Saczko, E. Kędzierska, K. Knap-Czop, J. Kotlińska, O. Michel, K. Kotowski and J. Kulbacka, Photodynamic therapy-mechanisms, photosensitizers and combinations, *Biomedicine & Pharmacotherapy* **106**, 1098–1107 (2018).
[1693] M. Kang, Z. Zhang, N. Song, M. Li, P. Sun, X. Chen, D. Wang and B. Z. Tang, Aggregation-enhanced theranostics: AIE sparkles in biomedical field, *Aggregate* **1**, 80–106 (2020).
[1694] D. E. Dolmans, D. Fukumura and R. K. Jain, Photodynamic therapy for cancer, *Nature Reviews Cancer* **3**, 380–387 (2003).
[1695] D. Bartusik-Aebisher, A. Żołyniak, E. Barnaś, A. Machorowska-Pieniążek, P. Oleś, A. Kawczyk-Krupka and D. Aebisher, The use of photodynamic therapy in the treatment of brain tumors–A review of the literature, *Molecules* **27**, 6847 (2022).
[1696] X. Wang, Y. Song, G. Pan, W. Han, B. Wang, L. Cui, H. Ma, Z. An, Z. Xie, B. Xu *et al.*, Exploiting radical-pair intersystem crossing for maximizing singlet oxygen quantum yields in pure organic fluorescent photosensitizers, *Chemical Science* **11**, 10921–10927 (2020).

[1697] P. R. Ogilby, Singlet oxygen: There is indeed something new under the sun, *Chemical Society Reviews* **39**, 3181–3209 (2010).

[1698] D. Wang, M. L. Kuzma, X. Tan, T.-C. He, C. Dong, Z. Liu and J. Yang, Phototherapy and optical waveguides for the treatment of infection, *Advanced Drug Delivery Reviews* **179**, 114036 (2021).

[1699] Z. Zhou, J. Song, L. Nie and X. Chen, Reactive oxygen species generating systems meeting challenges of photodynamic cancer therapy, *Chemical Society Reviews* **45**, 6597–6626 (2016).

[1700] J. Wang, B. Zhang, J. Sun, Y. Wang and H. Wang, Nanomedicine-enabled modulation of tumor hypoxic microenvironment for enhanced cancer therapy, *Advanced Therapeutics* **3**, 1900083 (2020).

[1701] J. D. Spikes, New trends in photobiology: Chlorins as photosensitizers in biology and medicine, *Journal of Photochemistry and Photobiology B: Biology* **6**, 259–274 (1990).

[1702] T. C. Pham, V.-N. Nguyen, Y. Choi, S. Lee and J. Yoon, Recent strategies to develop innovative photosensitizers for enhanced photodynamic therapy, *Chemical Reviews* **121**, 13454–13619 (2021).

[1703] L. Gourdon, K. Cariou and G. Gasser, Phototherapeutic anticancer strategies with first-row transition metal complexes: A critical review, *Chemical Society Reviews* **51**, 1167–1195 (2022).

[1704] M. Hussain, S. S. Razi, T. Tao and F. Hartl, Triplet-triplet annihilation photon up-conversion: Accessing triplet excited states with minimum energy loss, *Journal of Photochemistry and Photobiology C: Photochemistry Reviews*, 100618 (2023).

[1705] T. Schloemer, P. Narayanan, Q. Zhou, E. Belliveau, M. Seitz and D. N. Congreve, Nanoengineering triplet triplet annihilation upconversion: from materials to real-world applications, *ACS Nano* **17**, 3259–3288 (2023).

[1706] K. Riehemann, S. W. Schneider, T. A. Luger, B. Godin, M. Ferrari and H. Fuchs, Nanomedicine-challenge and perspectives, *Angewandte Chemie International Edition* **48**, 872–897 (2009).

[1707] X. Duan, P. Wang, L. He, Z. He, S. Wang, F. Yang, C. Gao, W. Ren, J. Lin, T. Chen *et al.*, Peptide-functionalized inorganic oxide nanomaterials for solid cancer imaging and therapy, *Advanced Materials* **36**, 2311548 (2024).

[1708] B. Y. Kim, J. T. Rutka and W. C. Chan, Nanomedicine, *New England Journal of Medicine* **363**, 2434–2443 (2010).

[1709] S. Logothetidis, Nanomedicine: The medicine of tomorrow, in *Nanomedicine and Nanobiotechnology*. Springer, 1–26 (2011).

[1710] J. Shi, P. W. Kantoff, R. Wooster and O. C. Farokhzad, Cancer nanomedicine: Progress, challenges and opportunities, *Nature Reviews Cancer* **17**, 20–37 (2017).

[1711] C. Zhang, L. Yan, X. Wang, S. Zhu, C. Chen, Z. Gu and Y. Zhao, Progress, challenges, and future of nanomedicine, *Nano Today* **35**, 101008 (2020).

[1712] Z. Lu, L. Zhou, Y. Yao, F. Li, Z. Hong, Y. Wu and X. Li, Reaction mechanisms in platinum-mediated electrodynamic therapy, *Advanced Functional Materials* **33**, 2214749 (2023).

[1713] W. Liu, N. Wu, J. Song, L. Xie, J. Zhang, J. Liu and A. Dong, Electrodynamic therapy: Novel nanodynamic strategy for anticancer and anti-infection therapy, *Chemical Engineering Journal* **494**, 152882 (2024).

[1714] K. Tachibana, L. B. Feril Jr and Y. Ikeda-Dantsuji, Sonodynamic therapy, *Ultrasonics* **48**, 253–259 (2008).

[1715] M. Trendowski, The promise of sonodynamic therapy, *Cancer and Metastasis Reviews* **33**, 143–160 (2014).

[1716] D. Costley, C. Mc Ewan, C. Fowley, A. P. McHale, J. Atchison, N. Nomikou and J. F. Callan, Treating cancer with sonodynamic therapy: A review, *International Journal of Hyperthermia* **31**, 107–117 (2015).

[1717] A. P. McHale, J. F. Callan, N. Nomikou, C. Fowley and B. Callan, Sonodynamic therapy: Concept, mechanism and application to cancer treatment, *Therapeutic ultrasound*, 429–450 (2016).

[1718] X. Cao, M. Li, Q. Liu, J. Zhao, X. Lu and J. Wang, Inorganic sonosensitizers for sonodynamic therapy in cancer treatment, *Small* **19**, 2303195 (2023).

[1719] D. Tzerkovsky, A. Mazurenko, F. Borychevsky, D. Shashkouski *et al.*, Radiodynamic therapy with photosensitizers: Mini-review of experimental and clinical studies, *Journal of Analytical Oncology* **11**, 79–85 (2022).

[1720] S. Zhu, F. Yan, L. Yang, B. Li, R. Xue, W. Yu, Y. Wang, L. Huang, L. Wang, R. Han *et al.*, Low-dose x-ray radiodynamic therapy solely based on gold nanoclusters for efficient treatment of deep hypoxic solid tumors combined with enhanced antitumor immune response, *Theranostics* **13**, 1042 (2023).

[1721] B. Zhang, Y. Huang and Y. Huang, Advances in nanodynamic therapy for cancer treatment, *Nanomaterials* **14**, 648 (2024).

[1722] Y. Chen, Y. Deng, Y. Li, Y. Qin, Z. Zhou, H. Yang and Y. Sun, Oxygenindependent radiodynamic therapy: Radiation-boosted chemodynamics for reprogramming the tumor immune environment and enhancing antitumor immune response, *ACS Applied Materials & Interfaces* **16**, 21546–21556 (2024).

[1723] A. Haleem, M. Javaid, R. P. Singh, S. Rab and R. Suman, Applications of nanotechnology in medical field: A brief review, *Global Health Journal* **7**, 70–77 (2023).

[1724] R. A. Freitas Jr, What is nanomedicine? *Nanomedicine: Nanotechnology, Biology and Medicine* **1**, 2–9 (2005).

[1725] L. M. Liz-Marzán, A. E. Nel, C. J. Brinker, W. C. Chan, C. Chen, X. Chen, D. Ho, T. Hu, K. Kataoka, N. A. Kotov *et al.*, What do we mean when we say nanomedicine? *ACS Nano* **16**, 13257–13259 (2022).

[1726] B. Pelaz, C. Alexiou, R. A. Alvarez-Puebla, F. Alves, A. M. Andrews, S. Ashraf, L. P. Balogh, L. Ballerini, A. Bestetti, C. Brendel *et al.*, Diverse applications of nanomedicine, *ACS Nano* **11**, 2313–2381 (2017).

[1727] L. Wang, K. Song, C. Jiang, S. Liu, S. Huang, H. Yang, X. Li and F. Zhao, Metal-coordinated polydopamine structures for tumor imaging and therapy, *Advanced Healthcare Materials* **13**, 2401451 (2024a).

[1728] J. Wang, W. Zhao, Z. Zhang, X. Liu, T. Xie, L. Wang, Y. Xue and Y. Zhang, A journey of challenges and victories: A bibliometric worldview of nanomedicine since the 21st century, *Advanced Materials* **36**, 2308915 (2024b).

[1729] R. Feynman, There's plenty of room at the bottom, *Science* **254**, 1300–1301 (1991).

[1730] A. J. Thorley and T. D. Tetley, New perspectives in nanomedicine, *Pharmacology & Therapeutics* **140**, 176–185 (2013).

[1731] Q. Liu, J. Zou, Z. Chen, W. He and W. Wu, Current research trends of nanomedicines, *Acta Pharmaceutica Sinica B* **13**, 4391–4416 (2023).

[1732] C. Jia, Y. Guo and F.-G. Wu, Chemodynamic therapy via Fenton and Fenton-like nanomaterials: Strategies and recent advances, *Small* **18**, 2103868 (2022).

[1733] H. J. H. Fenton, Lxxiii.—Oxidation of tartaric acid in presence of iron, *Journal of the Chemical Society, Transactions* **65**, 899–910 (1894).

[1734] Y. Lin, J. Qiao, Y. Sun and H. Dong, The profound review of Fenton process: What's the next step? *Journal of Environmental Sciences* **147**, 114–130 (2025).

[1735] C. Zhang, W. Bu, D. Ni, S. Zhang, Q. Li, Z. Yao, J. Zhang, H. Yao, Z. Wang and J. Shi, Synthesis of iron nanometallic glasses and their application in cancer therapy by a localized Fenton reaction, *Angewandte Chemie International Edition* **55**, 2101–2106 (2016).

[1736] P. Zhao, H. Li and W. Bu, A forward vision for chemodynamic therapy: Issues and opportunities, *Angewandte Chemie International Edition* **62**, e202210415 (2023).

[1737] X. Wang, X. Zhong, Z. Liu and L. Cheng, Recent progress of chemodynamic therapy-induced combination cancer therapy, *Nano Today* **35**, 100946 (2020).
[1738] X. Li, Y. Lan, X. Fu, X. Luo, J. Chen, W. Zhang, B. Zuo, T. Yang, B. Liu, C. Zhang *et al.*, DNA nanomachine-driven chemodynamic therapy against glioblastoma, *Aggregate* **5**, e603 (2024).
[1739] Y. Zhu, W. R. Archer, K. F. Morales, M. D. Schulz, Y. Wang and J. B. Matson, Enzyme-triggered chemodynamic therapy via a peptide-H_2S donor conjugate with complexed Fe^{2+}, *Angewandte Chemie* **135**, e202302303 (2023).
[1740] X. Cai, D. Cai, X. Wang, D. Zhang, L. Qiu, Z. Diao, Y. Liu, J. Sun, D. Cui, Y. Liu *et al.*, Manganese self-boosting hollow nanoenzymes with glutathione depletion for synergistic cancer chemo-chemodynamic therapy, *Biomaterials Science* **12**, 3622–3632 (2024).
[1741] W. Zhang, J. Liu, X. Li, Y. Zheng, L. Chen, D. Wang, M. F. Foda, Z. Ma, Y. Zhao and H. Han, Precise chemodynamic therapy of cancer by trifunctional bacterium-based nanozymes, *ACS Nano* **15**, 1932119333 (2021).
[1742] M. Sun, P. Zhou, S. Meng, P. Zhang, Y. Sun, C. Zhou, S. Su, C.-S. He, Y. Liu, H. Zhang *et al.*, New insights into photo-fenton chemistry: The overlooked role of excited $iron^{III}$ species, *Environmental Science & Technology* **58**, 10817–10827 (2024).
[1743] Y. Zhang, P.-H. Chen, B. Li, H. Guo, J. Zhu, Z. Dang, S. Lei, P. Huang and J. Lin, Comprehensively optimizing Fenton reaction factors for antitumor chemodynamic therapy by charge-reversal theranostics, *ACS Nano* **17**, 16743–16756 (2023).
[1744] J. Qin, J. Su, H. Li, L. Zhang and K. Jiang, Fenton chemistry activation in metal-organic frameworks for synergistic bacteria eradication, *Chemical Engineering Journal* **497**, 154413 (2024).
[1745] L. Zhao, X. Zhang, X. Wang, X. Guan, W. Zhang and J. Ma, Recent advances in selective photothermal therapy of tumor, *Journal of Nanobiotechnology* **19**, 335 (2021).
[1746] L. Zou, H. Wang, B. He, L. Zeng, T. Tan, H. Cao, X. He, Z. Zhang, S. Guo and Y. Li, Current approaches of photothermal therapy in treating cancer metastasis with nanotherapeutics, *Theranostics* **6**, 762 (2016).
[1747] G. Gao, X. Sun and G. Liang, Nanoagent-promoted mild-temperature photothermal therapy for cancer treatment, *Advanced Functional Materials* **31**, 2100738 (2021).
[1748] J. Weng, G. Zheng, J. Wen, J. Yang, Q. Yang, X. Zheng and Q. Yan, Construction and application of microneedle-mediated

photothermal therapy and immunotherapy combined anti-tumor drug delivery system, *Drug Delivery* **30**, 2232950 (2023).

[1749] J. Chen, C. Ning, Z. Zhou, P. Yu, Y. Zhu, G. Tan and C. Mao, Nanomaterials as photothermal therapeutic agents, *Progress in Materials Science* **99**, 1–26 (2019).

[1750] J. M. Margatho, C. C. Costa, A. de Camargo Tardio, G. S. Toldo and J. Cancino-Bernardi, Light responsive plasmonic nanomaterials applied to cancer phototherapy, in *Nanophototherapy*. Elsevier, pp. 99–129 (2025).

[1751] H. Sun, Q. Zhang, J. Li, S. Peng, X. Wang and R. Cai, Near-infrared photoactivated nanomedicines for photothermal synergistic cancer therapy, *Nano Today* **37**, 101073 (2021).

[1752] D. Han, X. Liu and S. Wu, Metal organic framework-based antibacterial agents and their underlying mechanisms, *Chemical Society Reviews* **51**, 7138–7169 (2022).

[1753] W.-K. Wong, Y. Ren and F. K.-C. Leung, Photothermal-chemotherapy: The emerging supramolecular photothermal molecules and the recent advances, in *Nanophototherapy*. Elsevier, 463–499 (2025).

[1754] Y. Xiao, T. Zhu, Q. Zeng, Q. Tan, G. Jiang and X. Huang, Functionalized biomimetic nanoparticles combining programmed death-1/programmed death-ligand 1 blockade with photothermal ablation for enhanced colorectal cancer immunotherapy, *Acta Biomaterialia* **157**, 451–466 (2023).

[1755] S. Devi, M. Kumar, A. Tiwari, V. Tiwari, D. Kaushik, R. Verma, S. Bhatt, B. M. Sahoo, T. Bhattacharya, S. Alshehri *et al.*, Quantum dots: An emerging approach for cancer therapy, *Frontiers in Materials* **8**, 798440 (2022).

[1756] M. M. Zorab, N. Mohammadjani, M. Ashengroph and M. Alavi, Biosynthesis of quantum dots and their therapeutic applications in the diagnosis and treatment of cancer and SARS-CoV-2, *Advanced Pharmaceutical Bulletin* **13**, 411 (2022).

[1757] L. K. Boopathy, T. Gopal, A. Roy, R. R. K. Kandy and M. K. Arumugam, Recent trends in macromolecule-conjugated hybrid quantum dots for cancer theranostic applications, *RSC Advances* **13**, 18760–18774 (2023).

[1758] J. Ahmad, A. Garg, G. Mustafa, M. Z. Ahmad, M. Aslam and A. Mishra, Hybrid quantum dot as promising tools for theranostic application in cancer, *Electronics* **12**, 972 (2023).

[1759] J. Kwon, S. Jun, S. Choi, X. Mao, J. Kim, E. Koh, Y.-H. Kim, S.-K. Kim, D. Hwang, C.-S. Kim *et al.*, FeSe quantum dots for in vivo multiphoton biomedical imaging, *Science Advances* **5**, eaay0044 (2019).

[1760] N. Iqbal and N. Iqbal, Human epidermal growth factor receptor 2 (HER2) in cancers: Overexpression and therapeutic implications, *Molecular Biology International* **2014**, 852748 (2014).
[1761] Z. Fu, S. Li, S. Han, C. Shi and Y. Zhang, Antibody drug conjugate: The "biological missile" for targeted cancer therapy, *Signal Transduction and Targeted Therapy* **7**, 93 (2022).
[1762] F. Remacle and R. Levine, Architecture with designer atoms: Simple theoretical considerations, *Proceedings of the National Academy of Sciences USA* **97**, 553–558 (2000).
[1763] T. L. Rapp and C. A. DeForest, Targeting drug delivery with light: A highly focused approach, *Advanced Drug Delivery Reviews* **171**, 94107 (2021).
[1764] M. Ruzycka-Ayoush, P. Kowalik, A. Kowalczyk, P. Bujak, A. M. Nowicka, M. Wojewodzka, M. Kruszewski and I. P. Grudzinski, Quantum dots as targeted doxorubicin drug delivery nanosystems in human lung cancer cells, *Cancer Nanotechnology* **12**, 8 (2021).
[1765] P. Kobauri, F. J. Dekker, W. Szymanski and B. L. Feringa, Rational design in photopharmacology with molecular photoswitches, *Angewandte Chemie International Edition* **62**, e202300681 (2023).
[1766] P. Ma and R. J. Mumper, Paclitaxel nano-delivery systems: A comprehensive review, *Journal of Nanomedicine & Nanotechnology* **4**, 1000164 (2013).
[1767] L. Rao, Y. Yuan, X. Shen, G. Yu and X. Chen, Designing nanotheranostics with machine learning, *Nature Nanotechnology* **19**, 1769–1781 (2024).
[1768] J. Zhou, L. Rao, G. Yu, T. R. Cook, X. Chen and F. Huang, Supramolecular cancer nanotheranostics, *Chemical Society Reviews* **50**, 2839–2891 (2021).
[1769] S. Mura and P. Couvreur, Nanotheranostics for personalized medicine, *Advanced Drug Delivery Reviews* **64**, 1394–1416 (2012).
[1770] R. B. Ladju, Z. S. Ulhaq and G. V. Soraya, Nanotheranostics: A powerful next-generation solution to tackle hepatocellular carcinoma, *World Journal of Gastroenterology* **28**, 176 (2022).
[1771] J. L. Corchero and A. Villaverde, Biomedical applications of distally controlled magnetic nanoparticles, *Trends in Biotechnology* **27**, 468–476 (2009).
[1772] A. A. Bukhvostov, A. S. Dvornikov, K. V. Ermakov, P. B. Kurapov and D. A. Kuznetsov, Retinoblastoma: Magnetic isotope effects might make a difference in the current anti-cancer research strategy, *Acta Medica* **60**, 93–96 (2017).
[1773] A. P. Demchenko, Phototheranostics: Combining targeting, imaging, therapy, in *Introduction to Fluorescence Sensing: Volume 2: Target Recognition and Imaging.* Springer, 649–691 (2023).

[1774] X. Yin, Y. Cheng, Y. Feng, W. R. Stiles, S. H. Park, H. Kang and H. S. Choi, Phototheranostics for multifunctional treatment of cancer with fluorescence imaging, *Advanced Drug Delivery Reviews* **189**, 114483 (2022).

[1775] Z. Zhang, M. Kang, H. Tan, N. Song, M. Li, P. Xiao, D. Yan, L. Zhang, D. Wang and B. Z. Tang, The fast-growing field of photo-driven theranostics based on aggregation-induced emission, *Chemical Society Reviews* (2022).

[1776] K. W. Lam, J. H. Chau, E. Y. Yu, F. Sun, J. W. Lam, D. Ding, R. T. Kwok, J. Sun, X. He and B. Z. Tang, An alkaline phosphatase-responsive aggregation-induced emission photosensitizer for selective imaging and photodynamic therapy of cancer cells, *ACS Nano* **17**, 7145–7156 (2023).

[1777] X. Ruan, M. Wei, X. He, L. Wang, D. Yang, Y. Cai, J. Shao and X. Dong, Asymmetric aza-bodipy photosensitizer for photoacoustic/photothermal imaging-guided synergistic photodynamic/photothermal therapy, *Colloids and Surfaces B: Biointerfaces* **231**, 113547 (2023).

[1778] B. Li, Y. Gan, K. Yang, E. Pang, X. Ren, S. Zhao, D. He, F. Zhao, B. Wang, P. Yin *et al.*, Acceptor-donor-acceptor structured phototheranostics for near-infrared II fluorescent and photoacoustic imaging-guided photodynamic and photothermal synergistic therapy, *Science China Materials* **66**, 385–394 (2023).

[1779] S. Yan, P. Sun, N. Niu, Z. Zhang, W. Xu, S. Zhao, L. Wang, D. Wang and B. Z. Tang, "One stone, four birds" ion engineering to fabricate versatile core-shell organosilica nanoparticles for intelligent nanotheranostics, *ACS Nano* **16**, 9785–9798 (2022).

[1780] O. Tour, R. M. Meijer, D. A. Zacharias, S. R. Adams and R. Y. Tsien, Genetically targeted chromophore-assisted light inactivation, *Nature Biotechnology* **21**, 1505–1508 (2003).

[1781] C. Yu, W. Jiang, B. Li, Y. Hu and D. Liu, The role of integrins for mediating nanodrugs to improve performance in tumor diagnosis and treatment, *Nanomaterials* **13**, 1721 (2023).

[1782] T. Ishimoto and H. Mori, Control of actin polymerization via reactive oxygen species generation using light or radiation, *Frontiers in Cell and Developmental Biology* **10**, 1014008 (2022).

[1783] P. Gödtel, J. Starrett and Z. Pianowski, Heterocyclic hemipiperazines: Water-compatible peptide-derived photoswitches, *Chemistry-A European Journal* **29**, e202204009 (2023).

[1784] F. Borys, P. Tobiasz, J. Sobel and H. Krawczyk, Synthesis and study of Dibenzo[b, f]oxepine combined with fluoroazobenzenes—new photoswitches for application in biological systems, *Molecules* **27**, 5836 (2022).

[1785] S. Kirchner, A.-L. Leistner, P. Gödtel, A. Seliwjorstow, S. Weber, J. Karcher, M. Nieger and Z. Pianowski, Hemipiperazines as peptidederived molecular photoswitches with low-nanomolar cytotoxicity, *Nature Communications* **13**, 6066 (2022).

[1786] L. Gao, J. C. Meiring, A. Varady, I. E. Ruider, C. Heise, M. Wranik, C. D. Velasco, J. A. Taylor, B. Terni, T. Weinert *et al.*, In vivo photocontrol of microtubule dynamics and integrity, migration and mitosis, by the potent GFP-imaging-compatible photoswitchable reagents SBTubA4P and SBTub2M, *Journal of the American Chemical Society* **144**, 5614–5628 (2022).

[1787] S. Kirchner and Z. Pianowski, Photopharmacology of antimitotic agents, *International Journal of Molecular Sciences* **23**, 5657 (2022).

[1788] D. Bates and A. Eastman, Microtubule destabilising agents: Far more than just antimitotic anticancer drugs, *British Journal of Clinical Pharmacology* **83**, 255–268 (2017).

[1789] M. Reynders, M. Garścia, A. Müller-Deku, M. Wranik, K. Krauskopf, K. Schaffer, A. Jötten, A. Rode, V. Stierle, Y. Kraus *et al.*, A photo-SAR study of photoswitchable azobenzene tubulin-inhibiting antimitotics identifying a general method for near-quantitative photocontrol, *Chemical Science* **15**, 12301–12309 (2024).

[1790] J. Kim, J. Kim and J.-S. Bae, ROS homeostasis and metabolism: A critical liaison for cancer therapy, *Experimental & Molecular Medicine* **48**, e269 (2016).

[1791] S. J. Hack, W. S. Beane and K. A.-S. Tseng, Biophysics at the edge of life and death: Radical control of apoptotic mechanisms, *Frontiers in Cell Death* **2**, 1147605 (2023).

[1792] M. Wranik, T. Weinert, C. Slavov, T. Masini, A. Furrer, N. Gaillard, D. Gioia, M. Ferrarotti, D. James, H. Glover *et al.*, Watching the release of a photopharmacological drug from tubulin using time-resolved serial crystallography, *Nature Communications* **14**, 903 (2023).

[1793] P. L. Apopa, Y. Qian, R. Shao, N. L. Guo, D. Schwegler-Berry, M. Pacurari, D. Porter, X. Shi, V. Vallyathan, V. Castranova *et al.*, Iron oxide nanoparticles induce human microvascular endothelial cell permeability through reactive oxygen species production and microtubule remodeling, *Particle and Fibre Toxicology* **6**, 1 (2009).

[1794] S. J. Dixon and B. R. Stockwell, The role of iron and reactive oxygen species in cell death, *Nature Chemical Biology* **10**, 9–17 (2014).

[1795] P. Whissell and M. Persinger, Emerging synergisms between drugs and physiologically-patterned weak magnetic fields: Implications for neuropharmacology and the human population in the twenty-first century, *Current Neuropharmacology* **5**, 278–288 (2007).

[1796] M. Platt, G. Muthukrishnan, W. O. Hancock and M. E. Williams, Millimeter scale alignment of magnetic nanoparticle functionalized microtubules in magnetic fields, *Journal of the American Chemical Society* **127**, 15686–15687 (2005).

[1797] V. Emiliani, E. Entcheva, R. Hedrich, P. Hegemann, K. R. Konrad, C. Lüscher, M. Mahn, Z.-H. Pan, R. R. Sims, J. Vierock *et al.*, Optogenetics for light control of biological systems, *Nature Reviews Methods Primers* **2**, 55 (2022).

[1798] J. Joshi, M. Rubart and W. Zhu, Optogenetics: Background, methodological advances and potential applications for cardiovascular research and medicine, *Frontiers in Bioengineering and Biotechnology* **7**, 466 (2020).

[1799] T. M. Courtney and A. Deiters, Optical control of protein phosphatase function, *Nature Communications* **10**, 4384 (2019).

[1800] S. M. Castillo-Hair, E. A. Baerman, M. Fujita, O. A. Igoshin and J. J. Tabor, Optogenetic control of Bacillus subtilis gene expression, *Nature Communications* **10**, 3099 (2019).

[1801] L. Zhu, H. M. McNamara and J. E. Toettcher, Light-switchable transcription factors obtained by direct screening in mammalian cells, *Nature Communications* **14**, 3185 (2023).

[1802] G. Ma, L. He, S. Liu, J. Xie, Z. Huang, J. Jing, Y.-T. Lee, R. Wang, H. Luo, W. Han *et al.*, Optogenetic engineering to probe the molecular choreography of STIM1-mediated cell signaling, *Nature Communications* **11**, 1039 (2020).

[1803] D. Krueger, E. Izquierdo, R. Viswanathan, J. Hartmann, C. Pallares Cartes and S. De Renzis, Principles and applications of optogenetics in developmental biology, *Development* **146**, dev175067 (2019).

[1804] W. Chen, C. Li, W. Liang, Y. Li, Z. Zou, Y. Xie, Y. Liao, L. Yu, Q. Lin, M. Huang *et al.*, The roles of optogenetics and technology in neurobiology: A review, *Frontiers in Aging Neuroscience* **14**, 867–863 (2022).

[1805] H. Ye and M. Fussenegger, Optogenetic medicine: Synthetic therapeutic solutions precision-guided by light, *Cold Spring Harbor Perspectives in Medicine* **9**, a034371 (2019).

[1806] R. Prasad, A. Ghosh, V. Patel, B. Peng, B. B. Mendes, E. H. A. Win, L. G. Delogu, J. Y. Wong, K. J. Pischel, J. R. Bellare *et al.*, Voices of nanomedicine: Blueprint guidelines for collaboration in addressing global unmet medical needs, *ACS Nano* **19**, 2979–2991 (2025).

[1807] F. Facchin, E. Bianconi, S. Canaider, V. Basoli, P. M. Biava and C. Ventura, Tissue regeneration without stem cell transplantation:

Self-healing potential from ancestral chemistry and physical energies, *Stem Cells International* **2018** (2018).
[1808] K. L. McKinley, M. T. Longaker and S. Naik, Emerging frontiers in regenerative medicine, *Science* **380**, 796–798 (2023).
[1809] P.-S. Welz, Clock regulation of skin regeneration in stem cell aging, *Journal of Investigative Dermatology* **141**, 1024–1030 (2021).
[1810] I. V. Kholodenko and K. N. Yarygin, Cellular mechanisms of liver regeneration and cell-based therapies of liver diseases, *BioMed Research International* **2017**, 8910821 (2017).
[1811] G. J. Christ, J. M. Saul, M. E. Furth and K.-E. Andersson, The pharmacology of regenerative medicine, *Pharmacological Reviews* **65**, 1091–1133 (2013).
[1812] S. Anthony and C. V. Borlongan, Recent progress in regenerative medicine for brain disorders, *Brain Circulation* **3**, 121–123 (2017).
[1813] T. C. Burns and A. Quinones-Hinojosa, Regenerative medicine for neurological diseases-will regenerative neurosurgery deliver? *British Medical Journal* **373**, n955 (2021).
[1814] A. Esmaeili, A. Eteghadi, F. S. Landi, S. F. Yavari and N. Taghipour, Recent approaches in regenerative medicine in the fight against neurodegenerative disease, *Brain Research* **1825**, 148688 (2024).
[1815] M. Z. Ratajczak, K. Bujko, A. Mack, M. Kucia and J. Ratajczak, Cancer from the perspective of stem cells and misappropriated tissue regeneration mechanisms, *Leukemia* **32**, 2519–2526 (2018).
[1816] N. J. Oviedo and W. S. Beane, Regeneration: The origin of cancer or a possible cure? in *Seminars in Cell & Developmental Biology*, Vol. 20. Elsevier, 557–564 (2009).
[1817] S. Petrus-Reurer, M. Romano, S. Howlett, J. L. Jones, G. Lombardi and K. Saeb-Parsy, Immunological considerations and challenges for regenerative cellular therapies, *Communications Biology* **4**, 798 (2021).
[1818] L. Pezzanite, L. Barrachina and L. Schnabel, Links between regenerative medicine and immunotherapy: How cellular therapies modulate immune responses for improved outcomes, *Frontiers in Veterinary Science* **11**, 1462218 (2024).
[1819] M. W. Sajjad, F. Muzamil, M. Sabir and U. A. Ashfaq, Regenerative medicine and nanotechnology approaches against cardiovascular diseases: Recent advances and future prospective, *Current Stem Cell Research & Therapy* **20**, 50–71 (2025).
[1820] S. Dash, A. Shakeel and S. Mohanty, Impact of nanotechnology on the realm of stem cells and regenerative medicine, *ChemNanoMat* **8**, e202200177 (2022).

[1821] Y. Yang, A. Chawla, J. Zhang, A. Esa, H. L. Jang and A. Khademhosseini, Applications of nanotechnology for regenerative medicine; healing tissues at the nanoscale, in *Principles of Regenerative Medicine*. Elsevier, 485–504 (2019).

[1822] K. Chaudhury, V. Kumar, J. Kandasamy and S. RoyChoudhury, Regenerative nanomedicine: Current perspectives and future directions, *International Journal of Nanomedicine*, 4153–4167 (2014).

[1823] P. Arora, A. Sindhu, N. Dilbaghi, A. Chaudhury, G. Rajakumar and A. A. Rahuman, Nano-regenerative medicine towards clinical outcome of stem cell and tissue engineering in humans, *Journal of Cellular and Molecular Medicine* **16**, 1991–2000 (2012).

[1824] Š. Kubinová and E. Syková, Nanotechnologies in regenerative medicine, *Minimally Invasive Therapy & Allied Technologies* **19**, 144–156 (2010).

[1825] D. Khang, J. Carpenter, Y. W. Chun, R. Pareta and T. J. Webster, Nanotechnology for regenerative medicine, *Biomedical Microdevices* **12**, 575–587 (2010).

[1826] E. Engel, A. Michiardi, M. Navarro, D. Lacroix and J. A. Planell, Nanotechnology in regenerative medicine: The materials side, *Trends in Biotechnology* **26**, 39–47 (2008).

[1827] A. Garmany and A. Terzic, Artificial intelligence powers regenerative medicine into predictive realm, *Regenerative Medicine* **19**, 611–616 (2024).

[1828] A. D. Henn, M. Wolff, K. Mitra, K. Alm, I. Finger-Baker, S. R. Bauer, A. Terzic and R. Yerden, Smart biomanufacturing for health equity in regenerative medicine therapies, *Regenerative Medicine Reports* **2**, 31–35 (2025).

[1829] P. Zhu, E. M. Pfrender, A. W. Steffeck, C. R. Reczek, Y. Zhou, A. V. Thakkar, N. R. Gupta, A. Kupai, A. Willbanks, R. L. Lieber *et al.*, Immunomodulatory role of the stem cell circadian clock in muscle repair, *Science Advances* **11**, eadq8538 (2025).

[1830] C. L. Ruby, R. J. Major and R. D. Hinrichsen, Regulation of tissue regeneration by the circadian clock, *European Journal of Neuroscience* **53**, 3576–3597 (2021).

[1831] E. Paatela, D. Munson and N. Kikyo, Circadian regulation in tissue regeneration, *International Journal of Molecular Sciences* **20**, 2263 (2019).

[1832] A. P. Patton and M. H. Hastings, The mammalian circadian timekeeping system, *Journal of Huntington's Disease* **12**, 91–104 (2023).

[1833] J. F. Duffy and C. A. Czeisler, Effect of light on human circadian physiology, *Sleep Medicine Clinics* **4**, 165–177 (2009).

[1834] M. H. Vitaterna, J. S. Takahashi and F. W. Turek, Overview of circadian rhythms, *Alcohol Research & Health* **25**, 85 (2001).

[1835] S. Wahl, M. Engelhardt, P. Schaupp, C. Lappe and I. V. Ivanov, The inner clock-blue light sets the human rhythm, *Journal of Biophotonics* **12**, e201900102 (2019).
[1836] H. E. Watts, A. R. Robart, T. P. Hahn and R. Gomulkiewicz, Environmental cue integration and phenology in a changing world, *Integrative and Comparative Biology* **62**, 972–979 (2022).
[1837] J. M. McNamara, Z. Barta, M. Klaassen and S. Bauer, Cues and the optimal timing of activities under environmental changes, *Ecology Letters* **14**, 1183–1190 (2011).
[1838] C. Vetter, P. M. Pattison, K. Houser, M. Herf, A. J. Phillips, K. P. Wright, D. J. Skene, G. C. Brainard, D. B. Boivin and G. Glickman, A review of human physiological responses to light: Implications for the development of integrative lighting solutions, *Leukos* **18**, 387–414 (2022).
[1839] G. Yadav, S. Malik, S. Rani and V. Kumar, Role of light wavelengths in synchronization of circadian physiology in songbirds, *Physiology & Behavior* **140**, 164–171 (2015).
[1840] R. W. Logan and C. A. McClung, Rhythms of life: circadian disruption and brain disorders across the lifespan, *Nature Reviews Neuroscience* **20**, 49–65 (2019).
[1841] Y. Leng, E. S. Musiek, K. Hu, F. P. Cappuccio and K. Yaffe, Association between circadian rhythms and neurodegenerative diseases, *The Lancet Neurology* **18**, 307–318 (2019).
[1842] C. Scheiermann, Y. Kunisaki and P. S. Frenette, Circadian control of the immune system, *Nature Reviews Immunology* **13**, 190–198 (2013).
[1843] R. Allada and J. Bass, Circadian mechanisms in medicine, *New England Journal of Medicine* **384**, 550–561 (2021).
[1844] F. Fagiani, D. Di Marino, A. Romagnoli, C. Travelli, D. Voltan, L. Di Cesare Mannelli, M. Racchi, S. Govoni and C. Lanni, Molecular regulations of circadian rhythm and implications for physiology and diseases, *Signal Transduction and Targeted Therapy* **7**, 41 (2022).
[1845] N. Meyer, A. G. Harvey, S. W. Lockley and D.-J. Dijk, Circadian rhythms and disorders of the timing of sleep, *The Lancet* **400**, 1061–1078 (2022).
[1846] B. Shen, C. Ma, G. Wu, H. Liu, L. Chen and G. Yang, Effects of exercise on circadian rhythms in humans, *Frontiers in Pharmacology* **14**, 1282357 (2023).
[1847] A. Naik, K. M. Forrest, O. Paul, Y. Issah, U. K. Valekunja, S. Y. Tang, A. B. Reddy, E. J. Hennessy, T. G. Brooks, F. Chaudhry *et al.*, Circadian regulation of lung repair and regeneration, *JCI Insight* **8**, e164720 (2023).

[1848] T. Dean, A. V. Koffi, E. Goldstein, J. Ghaemmaghami and V. Gallo, Endogenous circadian clock machinery in cortical NG2-Glia regulates cellular proliferation, *eNeuro* **9**, 5 (2022).

[1849] N. P. Hoyle, E. Seinkmane, M. Putker, K. A. Feeney, T. P. Krogager, J. E. Chesham, L. K. Bray, J. M. Thomas, K. Dunn, J. Blaikley *et al.*, Circadian actin dynamics drive rhythmic fibroblast mobilization during wound healing, *Science Translational Medicine* **9**, eaal2774 (2017).

[1850] E. H. Rogers, J. A. Hunt and V. Pekovic-Vaughan, Adult stem cell maintenance and tissue regeneration around the clock: Do impaired stem cell clocks drive age-associated tissue degeneration? *Biogerontology* **19**, 497–517 (2018).

[1851] D. A. Golombek and R. E. Rosenstein, Physiology of circadian entrainment, *Physiological Reviews* **90**, 1063–1102 (2010).

[1852] Z. Kaffash, R. Saha, S. A. Mirbozorgi, M. Faezipour and S. A. Mirbozorgi, Photobiomodulation therapy; survey and principal study leading to design rules for implants, *IEEE Transactions on Biomedical Engineering* **72**, 2237–2248 (2025).

[1853] B. Ladoux and R.-M. Mège, Mechanobiology of collective cell behaviours, *Nature Reviews Molecular Cell Biology* **18**, 743–757 (2017).

[1854] C. W. Rennolds and A. E. Bely, Integrative biology of injury in animals, *Biological Reviews* **98**, 34–62 (2023).

[1855] F. Ferreira, V. Raghunathan, G. Luxardi, K. Zhu and M. Zhao, Early redox activities modulate xenopus tail regeneration, *Nature Communications* **9**, 4296 (2018).

[1856] M. Levin, Bioelectric signaling: Reprogrammable circuits underlying embryogenesis, regeneration, and cancer, *Cell* **184**, 1971–1989 (2021).

[1857] S. A. Eming, P. J. Murray and E. J. Pearce, Metabolic orchestration of the wound healing response, *Cell Metabolism* **33**, 1726–1743 (2021).

[1858] J. A. Goldman and K. D. Poss, Gene regulatory programmes of tissue regeneration, *Nature Reviews Genetics* **21**, 511–525 (2020).

[1859] A. S. Mehta and A. Singh, Insights into regeneration tool box: An animal model approach, *Developmental Biology* **453**, 111–129 (2019).

[1860] X.-Y. Ge, X. Han, Y.-L. Zhao, G.-S. Cui and Y.-G. Yang, An insight into planarian regeneration, *Cell Proliferation* **55**, e13276 (2022).

[1861] J. C. Rink, Stem cell systems and regeneration in planaria, *Development Genes and Evolution* **223**, 67–84 (2013).

[1862] H.-P. P. Wu and M. A. Persinger, Increased mobility and stem-cell proliferation rate in dugesia tigrina induced by 880 nm light emitting diode, *Journal of Photochemistry and Photobiology B: Biology* **102**, 156160 (2011).

[1863] T. R. Paskin, J. Jellies, J. Bacher and W. S. Beane, Planarian phototactic assay reveals differential behavioral responses based on wavelength, *PloS One* **9**, e114708 (2014).

[1864] S. C. de Souza, E. Munin, L. P. Alves, M. A. C. Salgado and M. T. T. Pacheco, Low power laser radiation at 685 nm stimulates stem-cell proliferation rate in dugesia tigrina during regeneration, *Journal of Photochemistry and Photobiology B: Biology* **80**, 203–207 (2005).

[1865] L. J. Kinsey, A. V. Van Huizen and W. S. Beane, Weak magnetic fields modulate superoxide to control planarian regeneration, *Frontiers in Physics* **10**, 1356 (2023).

[1866] C. Gauron, C. Rampon, M. Bouzaffour, E. Ipendey, J. Teillon, M. Volovitch and S. Vriz, Sustained production of ROS triggers compensatory proliferation and is required for regeneration to proceed, *Scientific Reports* **3**, 2084 (2013).

[1867] M. Thauvin, R. Matias de Sousa, M. Alves, M. Volovitch, S. Vriz and C. Rampon, An early Shh-H_2O_2 reciprocal regulatory interaction controls the regenerative program during zebrafish fin regeneration, *Journal of Cell Science* **135**, jcs259664 (2022).

[1868] V. Miskolci, J. Squirrell, J. Rindy, W. Vincent, J. D. Sauer, A. Gibson, K. W. Eliceiri and A. Huttenlocher, Distinct inflammatory and wound healing responses to complex caudal fin injuries of larval zebrafish, *eLife* **8**, e45976 (2019).

[1869] B. Carbonell-M, J. Zapata Cardona and J. P. Delgado, Post-amputation reactive oxygen species production is necessary for axolotls limb regeneration, *Frontiers in Cell and Developmental Biology* **10**, 921520 (2022).

[1870] F. Ferreira, G. Luxardi, B. Reid and M. Zhao, Early bioelectric activities mediate redox-modulated regeneration, *Development* **143**, 45824594 (2016).

[1871] C. Aztekin, T. Hiscock, J. Marioni, J. Gurdon, B. Simons and J. Jullien, Identification of a regeneration-organizing cell in the xenopus tail, *Science* **364**, 653–658 (2019).

[1872] C. L. Bigarella, R. Liang and S. Ghaffari, Stem cells and the impact of ROS signaling, *Development* **141**, 4206–4218 (2014).

[1873] D. Q. Tan and T. Suda, Reactive oxygen species and mitochondrial homeostasis as regulators of stem cell fate and function, *Antioxidants & Redox Signaling* **29**, 149–168 (2018).

[1874] C. M. Madl, S. C. Heilshorn and H. M. Blau, Bioengineering strategies to accelerate stem cell therapeutics, *Nature* **557**, 335–342 (2018).

[1875] N. Khuu, S. Kheiri and E. Kumacheva, Structurally anisotropic hydrogels for tissue engineering, *Trends in Chemistry* **3**, 1002–1026 (2021).

[1876] I. Khan and P. R. Arany, Photobiomodulation therapy promotes expansion of epithelial colony forming units, *Photomedicine and Laser Surgery* **34**, 550–555 (2016).

[1877] A. Pardo, M. Gómez-Florit, S. Barbosa, P. Taboada, R. M. Domingues and M. E. Gomes, Magnetic nanocomposite hydrogels for tissue engineering: Design concepts and remote actuation strategies to control cell fate, *ACS Nano* **15**, 175–209 (2021).

[1878] S. Fekrazad, M. Sohrabi and R. Fekrazad, Angiogenetic and anti-inflammatory effects of photobiomodulation on bone regeneration in rat: A histopathological, immunohistochemical, and molecular analysis, *Journal of Photochemistry and Photobiology B: Biology* **238**, 112599 (2023).

[1879] X. Yuan, W. Zhu, Z. Yang, N. He, F. Chen, X. Han and K. Zhou, Recent advances in 3D printing of smart scaffolds for bone tissue engineering and regeneration, *Advanced Materials* **36**, 2403641 (2024).

[1880] A. Kakkar, A. Singh, S. K. Saraswat, S. Srivastava, N. Khatri, R. K. Nagar, M. Kumar, P. Meena, R. Datt and S. Pandey, Cartilage repair using stem cells & biomaterials: Advancement from bench to bedside, *Molecular Biology Reports* **47**, 8007–8021 (2020).

[1881] S. Trujillo, C. Gonzalez-Garcia, P. Rico, A. Reid, J. Windmill, M. J. Dalby and M. Salmeron-Sanchez, Engineered 3D hydrogels with fulllength fibronectin that sequester and present growth factors, *Biomaterials* **252**, 120104 (2020).

[1882] A. R. Del Bakhshayesh, N. Asadi, A. Alihemmati, H. Tayefi Nasrabadi, A. Montaseri, S. Davaran, S. Saghati, A. Akbarzadeh and A. Abedelahi, An overview of advanced biocompatible and biomimetic materials for creation of replacement structures in the musculoskeletal systems: Focusing on cartilage tissue engineering, *Journal of Biological Engineering* **13**, 85 (2019).

[1883] N. Putthanbut, P. A. B. Su, J.-Y. Lee and C. V. Borlongan, Circadian rhythms in stem cells and their therapeutic potential, *Stem Cell Research & Therapy* **16**, 85 (2025).

[1884] W. Gao, R. Li, M. Ye, L. Zhang, J. Zheng, Y. Yang, X. Wei and Q. Zhao, The circadian clock has roles in mesenchymal stem cell fate decision, *Stem Cell Research & Therapy* **13**, 200 (2022).

[1885] W. Zhang, T. J. Craddock, Y. Li, M. Swartzlander, R. R. Alfano and L. Shi, The circadian clock mediates daily bursts of cell differentiation by periodically restricting cell-differentiation commitment, *Proceedings of the National Academy of Sciences* **119**, e2204470119 (2022).

[1886] H. Elkhenany, A. AlOkda, A. El-Badawy and N. El-Badri, Tissue regeneration: Impact of sleep on stem cell regenerative capacity, *Life Sciences* **214**, 51–61 (2018).

[1887] F. Ramzan, A. Salim and I. Khan, Osteochondral tissue engineering dilemma: Scaffolding trends in regenerative medicine, *Stem Cell Reviews and Reports* **19**, 1615–1634 (2023).

[1888] E. Vardar, E. Samara and P.-Y. Zambelli, A mini-review of current methods and emerging technologies in articular cartilage repair, *SN Comprehensive Clinical Medicine* **3**, 2278–2284 (2021).

[1889] C. Vyas, H. Mishbak, G. Cooper, C. Peach, R. F. Pereira and P. Bartolo, Biological perspectives and current biofabrication strategies in osteochondral tissue engineering, *Biomanufacturing Reviews* **5**, 2 (2020).

[1890] M. Ansari and M. Eshghanmalek, Biomaterials for repair and regeneration of the cartilage tissue, *Bio-Design and Manufacturing* **2**, 41–49 (2019).

[1891] A. Nazempour, C. R. Quisenberry, B. J. Van Wie and N. I. Abu-Lail, Nanomechanics of engineered articular cartilage: Synergistic influences of transforming growth factor-β3 and oscillating pressure, *Journal of Nanoscience and Nanotechnology* **16**, 3136–3145 (2016).

[1892] M. T. Berninger, G. Wexel, E. J. Rummeny, A. B. Imhoff, M. Anton, T. D. Henning and S. Vogt, Treatment of osteochondral defects in the rabbit's knee joint by implantation of allogeneic mesenchymal stem cells in fibrin clots, *JoVE (Journal of Visualized Experiments)*, 75, e4423 (2013).

[1893] L. Jia, P. Zhang, Z. Ci, X. Hao, B. Bai, W. Zhang, H. Jiang and G. Zhou, Acellular cartilage matrix biomimetic scaffold with immediate enrichment of autologous bone marrow mononuclear cells to repair articular cartilage defects, *Materials Today Bio* **15**, 100310 (2022).

[1894] R. Ossendorff, S. G. Walter, F. A. Schildberg, J. Spang, S. Obudzinski, S. Preiss, S. Schneider and G. M. Salzmann, Biologic principles of minced cartilage implantation: A narrative review, *Archives of Orthopaedic and Trauma Surgery* **143**, 3259–3269 (2023).

[1895] X. Hu, J. Xu, W. Li, L. Li, R. Parungao, Y. Wang, S. Zheng, Y. Nie, T. Liu and K. Song, Therapeutic "tool" in reconstruction and regeneration of tissue engineering for osteochondral repair, *Applied Biochemistry and Biotechnology* **191**, 785–809 (2020).

[1896] S. Mallick, Z. Beyene, D. K. Suman, A. Madhual, B. N. Singh and P. Srivastava, Strategies towards orthopaedic tissue engineered graft generation: Current scenario and application, *Biotechnology and Bioprocess Engineering* **24**, 854–869 (2019).

[1897] L. Roseti and B. Grigolo, Current concepts and perspectives for articular cartilage regeneration, *Journal of Experimental Orthopaedics* **9**, 61 (2022).

[1898] S. G. Walter, R. Ossendorff and F. A. Schildberg, Articular cartilage regeneration and tissue engineering models: A systematic review, *Archives of Orthopaedic and Trauma Surgery* **139**, 305–316 (2019).

[1899] J. B. Costa, J. Silva-Correia, R. L. Reis and J. M. Oliveira, Current advances in solid free-form techniques for osteochondral tissue engineering, *Bio-Design and Manufacturing* **1**, 171–181 (2018).

[1900] T.-H. Yoon, M. Jung, C.-H. Choi, H.-S. Kim, Y.-H. Lee, Y.-S. Choi, S.-J. Kim and S.-H. Kim, Arthroscopic gel-type autologous chondrocyte implantation presents histologic evidence of regenerating hyaline-like cartilage in the knee with articular cartilage defect, *Knee Surgery, Sports Traumatology, Arthroscopy* **28**, 941–951 (2020).

[1901] X. Jia, X. Fan, C. Chen, Q. Lu, H. Zhou, Y. Zhao, X. Wang, S. Han, L. Ouyang, H. Yan *et al.*, Chemical and structural engineering of gelatinbased delivery systems for therapeutic applications: A review, *Biomacromolecules* **25**, 564–589 (2024).

[1902] L. Taghiyar, H. Asadi and M. Baghaban Eslaminejad, A bioscaffold of decellularized whole osteochondral sheet improves proliferation and differentiation of loaded mesenchymal stem cells in a rabbit model, *Cell and Tissue Banking* **24**, 711–724 (2023).

[1903] S. Saghati, H. T. Nasrabadi, A. B. Khoshfetrat, K. Moharamzadeh, A. Hassani, S. M. Mohammadi, R. Rahbarghazi and S. Fathi Karkan, Tissue engineering strategies to increase osteochondral regeneration of stem cells; a close look at different modalities, *Stem Cell Reviews and Reports* **17**, 1294–1311 (2021).

[1904] J. Zhou and Y. Shi, Mesenchymal stem/stromal cells (MSCs): Origin, immune regulation, and clinical applications, *Cellular & Molecular Immunology* **20**, 555–557 (2023).

[1905] F. Melo-Fonseca, G. Miranda, H. S. Domingues, I. M. Pinto, M. Gasik and F. S. Silva, Reengineering bone-implant interfaces for improved mechanotransduction and clinical outcomes, *Stem Cell Reviews and Reports* **16**, 1121–1138 (2020).

[1906] H. Gao, Q. Pan, W. Dong and Y. Yao, Progress in osteochondral regeneration with engineering strategies, *Annals of Biomedical Engineering* **50**, 1232–1242 (2022).

[1907] N. L. T. Hang, A. E. Aviña, C.-J. Chang and T.-S. Yang, Photobiomodulation in promoting cartilage regeneration, *International Journal of Molecular Sciences* **26**, 5580 (2025).
[1908] R. F. Lizarelli, A. S. I. Salgado, F. dos Santos Cardoso, K. J. de Oliveira Souza, L. T. Bello, L. Assis, R. Cordon and V. S. Bagnato, Photobiomodulation for degenerative diseases: An overview and perspectives, *Photobiomodulation, Photomedicine, and Laser Surgery* **43**, 383–393 (2025).
[1909] J. W. Weisel and R. I. Litvinov, Fibrin formation, structure and properties, *Fibrous Proteins: Structures and Mechanisms*, 405–456 (2017).
[1910] E. Di Cera, Thrombin, *Molecular Aspects of Medicine* **29**, 203–254 (2008).
[1911] F. J. Vernerey, S. Lalitha Sridhar, A. Muralidharan and S. J. Bryant, Mechanics of 3D cell-hydrogel interactions: Experiments, models, and mechanisms, *Chemical Reviews* **121**, 11085–11148 (2021).
[1912] P. Bikmulina, N. Kosheleva, A. Shpichka, V. Yusupov, V. Gogvadze, Y. Rochev and P. Timashev, Photobiomodulation in 3D tissue engineering, *Journal of Biomedical Optics* **27**, 090901–090901 (2022).
[1913] R. G. Rando, D. V. Buchaim, P. C. Cola and R. L. Buchaim, Effects of photobiomodulation using low-level laser therapy on alveolar bone repair, *Photonics* **10**, 734 (2023).
[1914] M. F. R. Vigliar, L. F. Marega, M. A. H. Duarte, M. P. Alcalde, M. P. d. O. Rosso, R. S. Ferreira Junior, B. Barraviera, C. H. B. Reis, D. V. Buchaim and R. L. Buchaim, Photobiomodulation therapy improves repair of bone defects filled by inorganic bone matrix and fibrin heterologous biopolymer, *Bioengineering* **11**, 78 (2024).
[1915] K. Woo and C. Abueva, Photobiomodulation improves the therapeutic efficacy of mesenchymal stem cells in regenerative medicine, *Medical Lasers; Engineering, Basic Research, and Clinical Application* **11**, 134–142 (2022).
[1916] K. Khorsandi, R. Hosseinzadeh, H. Abrahamse and R. Fekrazad, Biological responses of stem cells to photobiomodulation therapy, *Current Stem Cell Research & Therapy* **15**, 400–413 (2020).
[1917] M. Bozhokin, D. Vcherashnii, S. Yastrebov, L. Beilinson, J. V. Zherebtsova and M. Khotin, Low-intensity photobiomodulation at 632.8 nm increases tgβ3, col2a1, and sox9 gene expression in rat bone marrow mesenchymal stem cells in vitro, *Lasers in Medical Science* **37**, 435–441 (2021).
[1918] B. Ahrabi, M. R. Tavirani, M. S. Khoramgah, M. Noroozian, S. Darabi, S. Khoshsirat and H. A. Abbaszadeh, The effect of photobiomodulation therapy on the differentiation, proliferation,

and migration of the mesenchymal stem cell: A review, *Journal of Lasers in Medical Sciences* **10**, S96 (2019).

[1919] A. Amaroli, C. Pasquale, A. Zekiy, S. Benedicenti, A. Marchegiani, M. G. Sabbieti and D. Agas, Steering the multipotent mesenchymal cells towards an anti-inflammatory and osteogenic bias via photobiomodulation therapy: How to kill two birds with one stone, *Journal of Tissue Engineering* **13**, 20417314221110192 (2022).

[1920] G. Bölükbaşi Ateş, A. Ak, B. Garipcan and M. Gülsoy, Photobiomodulation effects on osteogenic differentiation of adipose-derived stem cells, *Cytotechnology* **72**, 247–258 (2020).

[1921] L. Gholami, S. S. Hendi, M. Saidijam, R. Mahmoudi, R. Tarzemany, A. Arkian, S. Afshar and R. Fekrazad, Near-infrared 940-nm diode laser photobiomodulation of inflamed periodontal ligament stem cells, *Lasers in Medical Science* **37**, 449–459 (2022).

[1922] L. Wang, F. Wu, C. Liu, Y. Song, J. Guo, Y. Yang and Y. Qiu, Low-level laser irradiation modulates the proliferation and the osteogenic differentiation of bone marrow mesenchymal stem cells under healthy and inflammatory condition, *Lasers in Medical Science* **34**, 169–178 (2019).

[1923] G. A. Yen-Patton, W. F. Patton, D. M. Beer and B. S. Jacobson, Endothelial cell response to pulsed electromagnetic fields: Stimulation of growth rate and angiogenesis in vitro, *Journal of Cellular Physiology* **134**, 37–46 (1988).

[1924] E. M. Goodman, B. Greenebaum and M. T. Marron, Effects of electromagnetic fields on molecules and cells, *International Review of Cytology* **158**, 279–338 (1995).

[1925] C. E. Meacham, A. W. DeVilbiss and S. J. Morrison, Metabolic regulation of somatic stem cells in vivo, *Nature Reviews Molecular Cell Biology* **23**, 428–443 (2022).

[1926] P.-F. Lenne and V. Trivedi, Sculpting tissues by phase transitions, *Nature Communications* **13**, 664 (2022).

[1927] P. Romani, L. Valcarcel-Jimenez, C. Frezza and S. Dupont, Crosstalk between mechanotransduction and metabolism, *Nature Reviews Molecular Cell Biology* **22**, 22–38 (2021).

[1928] A. Elosegui-Artola, The extracellular matrix viscoelasticity as a regulator of cell and tissue dynamics, *Current Opinion in Cell Biology* **72**, 1018 (2021).

[1929] J.-M. Choi, A. S. Holehouse and R. V. Pappu, Physical principles underlying the complex biology of intracellular phase transitions, *Annual Review of Biophysics* **49**, 107–133 (2020).

[1930] M. Uehara, K. K. Sakane, H. S. Maciel and W. I. Urruchi, Physics and biology: Bio-plasma physics, *American Journal of Physics* **68**, 450–455 (2000).

[1931] M. Santhamoorthy, D. Thirumalai and K. Thirupathi, Magnetic and colorimetric point-of-care biosensors for infectious disease diagnosis, in *Biosensors for Emerging and Re-Emerging Infectious Diseases.* Elsevier, 253–264 (2022).

[1932] D. Le Sage, K. Arai, D. R. Glenn, S. J. DeVience, L. M. Pham, L. RahnLee, M. D. Lukin, A. Yacoby, A. Komeili and R. L. Walsworth, Optical magnetic imaging of living cells, *Nature* **496**, 486–489 (2013).

[1933] B. B. Lake, R. Menon, S. Winfree, Q. Hu, R. M. Ferreira, K. Kalhor, D. Barwinska, E. A. Otto, M. Ferkowicz, D. Diep *et al.*, An atlas of healthy and injured cell states and niches in the human kidney, *Nature* **619**, 585–594 (2023).

[1934] L. Pio-Lopez and M. Levin, Morphoceuticals: perspectives for discovery of drugs targeting anatomical control mechanisms in regenerative medicine, cancer and aging, *Drug Discovery Today* **28**, 103585 (2023).

[1935] S. Roy and S. Majumdar, *Noise and Randomness in Living System.* Springer (2022).

[1936] C. Blomberg, Fluctuations for good and bad: The role of noise in living systems, *Physics of Life Reviews* **3**, 133–161 (2006).

[1937] G. Courcoubetis, S. Ali, S. V. Nuzhdin, P. Marjoram and S. Haas, Threshold response to stochasticity in morphogenesis, *PLoS One* **14**, e0210088 (2019).

[1938] B. Lehner, Selection to minimise noise in living systems and its implications for the evolution of gene expression, *Molecular Systems Biology* **4**, 170 (2008).

[1939] L. Blanchoin, R. Boujemaa-Paterski, C. Sykes and J. Plastino, Actin dynamics, architecture, and mechanics in cell motility, *Physiological Reviews* **94**, 235–263 (2014).

[1940] A. A. Hyman and E. Karsenti, Morphogenetic properties of microtubules and mitotic spindle assembly, *Cell* **84**, 401–410 (1996).

[1941] A. Thawani, H. A. Stone, J. W. Shaevitz and S. Petry, Spatiotemporal organization of branched microtubule networks, *eLife* **8**, e43890 (2019).

[1942] J. Tabony, N. Glade, J. Demongeot and C. Papaseit, Biological selforganization by way of microtubule reaction-diffusion processes, *Langmuir* **18**, 7196–7207 (2002).

[1943] J. Tabony, Microtubules viewed as molecular ant colonies, *Biology of the Cell* **98**, 603–617 (2006).

[1944] J. Tabony, Self-organization and higher level emergent phenomena in a population of microtubules, *NeuroQuantology* **5**, 100–144 (2007).

[1945] N. Glade, On the nature and shape of tubulin trails: Implications on microtubule self-organization, *Acta Biotheoretica* **60**, 55–82 (2012).

[1946] A. M. Turing, The chemical basis of morphogenesis, *Philosophical Transactions of the Royal Society B* **237**, 37–72 (1952).

[1947] A. R. Houk, A. Jilkine, C. O. Mejean, R. Boltyanskiy, E. R. Dufresne, S. B. Angenent, S. J. Altschuler, L. F. Wu and O. D. Weiner, Membrane tension maintains cell polarity by confining signals to the leading edge during neutrophil migration, *Cell* **148**, 175–188 (2012).

[1948] S. A. Baker and J. Rutter, Metabolites as signalling molecules, *Nature Reviews Molecular Cell Biology* **24**, 355–374 (2023).

[1949] X. Di, X. Gao, L. Peng, J. Ai, X. Jin, S. Qi, H. Li, K. Wang and D. Luo, Cellular mechanotransduction in health and diseases: From molecular mechanism to therapeutic targets, *Signal Transduction and Targeted Therapy* **8**, 282 (2023).

[1950] L. He, M. Ahmad and N. Perrimon, Mechanosensitive channels and their functions in stem cell differentiation, *Experimental Cell Research* **374**, 259–265 (2019).

[1951] F. Martino, A. R. Perestrelo, V. Vinarský, S. Pagliari and G. Forte, Cellular mechanotransduction: From tension to function, *Frontiers in Physiology* **9**, 824 (2018).

[1952] H. Lv, L. Li, M. Sun, Y. Zhang, L. Chen, Y. Rong and Y. Li, Mechanism of regulation of stem cell differentiation by matrix stiffness, *Stem Cell Research & Therapy* **6**, 103 (2015).

[1953] O. Chaudhuri, J. Cooper-White, P. A. Janmey, D. J. Mooney and V. B. Shenoy, Effects of extracellular matrix viscoelasticity on cellular behaviour, *Nature* **584**, 535–546 (2020).

[1954] C. L. Marchant, A. N. Malmi-Kakkada, J. A. Espina and E. H. Barriga, Cell clusters softening triggers collective cell migration in vivo, *Nature Materials* **21**, 1314–1323 (2022).

[1955] K. M. Yamada and M. Sixt, Mechanisms of 3D cell migration, *Nature Reviews Molecular Cell Biology* **20**, 738–752 (2019).

[1956] Y. Sun, C. S. Chen and J. Fu, Forcing stem cells to behave: A biophysical perspective of the cellular microenvironment, *Annual Review of Biophysics* **41**, 519–542 (2012).

[1957] T. E. Angelini, E. Hannezo, X. Trepat, M. Marquez, J. J. Fredberg and D. A. Weitz, Glass-like dynamics of collective cell migration, *Proceedings of the National Academy of Sciences USA* **108**, 4714–4719 (2011).

[1958] H. De Belly, E. K. Paluch and K. J. Chalut, Interplay between mechanics and signalling in regulating cell fate, *Nature Reviews Molecular Cell Biology* **23**, 465–480 (2022).

[1959] C. Argentati, F. Morena, I. Tortorella, M. Bazzucchi, S. Porcellati, C. Emiliani and S. Martino, Insight into mechanobiology: How stem cells feel mechanical forces and orchestrate biological functions, *International Journal of Molecular Sciences* **20**, 5337 (2019).

[1960] A. J. Engler, S. Sen, H. L. Sweeney and D. E. Discher, Matrix elasticity directs stem cell lineage specification, *Cell* **126**, 677–689 (2006).

[1961] F. Decker, D. Oriola, B. Dalton and J. Brugues, Autocatalytic microtubule nucleation determines the size and mass of xenopus laevis egg extract spindles, *eLife* **7**, e31149 (2018).

[1962] M. Bernardello, M. Marsal, E. J. Gualda and P. Loza-Alvarez, Light-sheet fluorescence microscopy for the in vivo study of microtubule dynamics in the zebrafish embryo, *Biomedical Optics Express* **12**, 6237–6254 (2021).

[1963] A. J. Matamoros and P. W. Baas, Microtubules in health and degenerative disease of the nervous system, *Brain Research Bulletin* **126**, 217–225 (2016).

[1964] B. D. Hoffman, C. Grashoff and M. A. Schwartz, Dynamic molecular processes mediate cellular mechanotransduction, *Nature* **475**, 316–323 (2011).

[1965] J. D. Humphrey, E. R. Dufresne and M. A. Schwartz, Mechanotransduction and extracellular matrix homeostasis, *Nature Reviews Molecular Cell Biology* **15**, 802–812 (2014).

[1966] C. Bonnans, J. Chou and Z. Werb, Remodelling the extracellular matrix in development and disease, *Nature Reviews Molecular Cell Biology* **15**, 786–801 (2014).

[1967] R. O. Hynes, The extracellular matrix: Not just pretty fibrils, *Science* **326**, 1216–1219 (2009).

[1968] G. Huang, F. Li, X. Zhao, Y. Ma, Y. Li, M. Lin, G. Jin, T. J. Lu, G. M. Genin and F. Xu, Functional and biomimetic materials for engineering of the three-dimensional cell microenvironment, *Chemical Reviews* **117**, 12764–12850 (2017).

[1969] D. A. C. Walma and K. M. Yamada, The extracellular matrix in development, *Development* **147**, dev175596 (2020).

[1970] S. Seetharaman and S. Etienne-Manneville, Cytoskeletal crosstalk in cell migration, *Trends in Cell Biology* **30**, 720–735 (2020).

[1971] M. Pourjafar and V. K. Tiwari, Plasticity in cell migration modes across development, physiology, and disease, *Frontiers in Cell and Developmental Biology* **12**, 1363361 (2024).

[1972] J. Aseervatham, Cytoskeletal remodeling in cancer, *Biology* **9**, 385 (2020).

[1973] F. Huber, A. Boire, M. P. López and G. H. Koenderink, Cytoskeletal crosstalk: When three different personalities team up, *Current Opinion in Cell Biology* **32**, 39–47 (2015).

[1974] J. Tabony, Self-organization and other emergent properties in a simple biological system of microtubules, *Complexus* **3**, 200–210 (2006).

[1975] L. Farhadi, S. N. Ricketts, M. J. Rust, M. Das, R. M. Robert-sonAnderson and J. L. Ross, Actin and microtubule crosslinkers tune mobility and control co-localization in a composite cytoskeletal network, *Soft Matter* **16**, 7191–7201 (2020).

[1976] K. Nakos, M. N. A. Alam, M. R. Radler, I. A. Kesisova, C. Yang, J. Okletey, M. R. Tomasso, S. B. Padrick, T. M. Svitkina and E. T. Spiliotis, Septins mediate a microtubule-actin crosstalk that enables actin growth on microtubules, *Proceedings of the National Academy of Sciences USA* **119**, e2202803119 (2022).

[1977] S. P. Martínez, J. Greaney and J. Zenker, Beyond the centrosome: The mystery of microtubule organising centres across mammalian preimplantation embryos, *Current Opinion in Cell Biology* **77**, 102114 (2022).

[1978] S. E. Shirmovsky and D. Shulga, Quantum relaxation effects in microtubules, *Physica A: Statistical Mechanics and its Applications* **582**, 126254 (2021).

[1979] S. E. Shirmovsky and D. Shulga, Quantum relaxation processes in microtubule tryptophan system, *Physica A: Statistical Mechanics and its Applications* **617**, 128687 (2023).

[1980] A. Shellard and R. Mayor, All roads lead to directional cell migration, *Trends in Cell Biology* **30**, 852–868 (2020).

[1981] A. Saraswathibhatla, D. Indana and O. Chaudhuri, Cell-extracellular matrix mechanotransduction in 3D, *Nature Reviews Molecular Cell Biology* **24**, 495–516 (2023).

[1982] S. SenGupta, C. A. Parent and J. E. Bear, The principles of directed cell migration, *Nature Reviews Molecular Cell Biology* **22**, 529–547 (2021).

[1983] G. H. Wadhams and J. P. Armitage, Making sense of it all: Bacterial chemotaxis, *Nature Reviews Molecular Cell Biology* **5**, 1024–1037 (2004).

[1984] P. Roca-Cusachs, R. Sunyer and X. Trepat, Mechanical guidance of cell migration: Lessons from chemotaxis, *Current Opinion in Cell Biology* **25**, 543–549 (2013).

[1985] A. Shellard and R. Mayor, Durotaxis: The hard path from in vitro to in vivo, *Developmental Cell* **56**, 227–239 (2021).

[1986] J. Park, D.-H. Kim and A. Levchenko, Topotaxis: A new mechanism of directed cell migration in topographic ECM gradients, *Biophysical Journal* **114**, 1257–1263 (2018).

[1987] S. Van Helvert, C. Storm and P. Friedl, Mechanoreciprocity in cell migration, *Nature Cell Biology* **20**, 8–20 (2018).

[1988] X. Wu, M.-P. Cao, Y.-Y. Shen, K.-P. Chu, W.-B. Tao, W.-T. Song, L.-P. Liu, X.-H. Wang, Y.-F. Zheng, S.-D. Chen *et al.*, Weak power frequency magnetic field acting similarly to EGF stimulation, induces acute activations of the EGFR sensitive actin cytoskeleton motility in human amniotic cells, *PloS One* **9**, e87626 (2014).

[1989] V. Graziani, I. Rodriguez-Hernandez, O. Maiques and V. Sanz-Moreno, The amoeboid state as part of the epithelial-to-mesenchymal transition programme, *Trends in Cell Biology* **32**, 228–242 (2022).

[1990] T. R. Cox, The matrix in cancer, *Nature Reviews Cancer* **21**, 217–238 (2021).

[1991] H. K. Matthews, C. Bertoli and R. A. de Bruin, Cell cycle control in cancer, *Nature Reviews Molecular Cell Biology* **23**, 74–88 (2022).

[1992] S. Taefehshokr, A. Parhizkar, S. Hayati, M. Mousapour, A. Mahmoudpour, L. Eleid, D. Rahmanpour, S. Fattahi, H. Shabani and N. Taefehshokr, Cancer immunotherapy: Challenges and limitations, *Pathology-Research and Practice* **229**, 153723 (2022).

[1993] C. Liu, M. Yang, D. Zhang, M. Chen and D. Zhu, Clinical cancer immunotherapy: Current progress and prospects, *Frontiers in Immunology* **13**, 961805 (2022).

[1994] M. Kciuk, E. B. Yahya, M. Mohamed Ibrahim Mohamed, S. Rashid, M. O. Iqbal, R. Kontek, M. A. Abdulsamad and A. A. Allaq, Recent advances in molecular mechanisms of cancer immunotherapy, *Cancers* **15**, 2721 (2023).

[1995] H. Du, J. M. Bartleson, S. Butenko, V. Alonso, W. F. Liu, D. A. Winer and M. J. Butte, Tuning immunity through tissue mechanotransduction, *Nature Reviews Immunology* **23**, 174–188 (2022).

[1996] G. Crainiciuc, M. Palomino-Segura, M. Molina-Moreno, J. Sicilia, D. G. Aragones, J. L. Y. Li, R. Madurga, J. M. Adrover, A. Aroca-Crevillén, S. Martin-Salamanca *et al.*, Behavioural immune landscapes of inflammation, *Nature* **601**, 415–421 (2022).

[1997] F. D. Brown and S. J. Turley, Fibroblastic reticular cells: Organization and regulation of the T lymphocyte life cycle, *The Journal of Immunology* **194**, 1389–1394 (2015).

[1998] J. Textor, J. N. Mandl and R. J. de Boer, The reticular cell network: A robust backbone for immune responses, *PLoS Biology* **14**, e2000827 (2016).

[1999] O. Cenaj, D. H. Allison, R. Imam, B. Zeck, L. M. Drohan, L. Chiriboga, J. Llewellyn, C. Z. Liu, Y. N. Park, R. G. Wells *et al.*, Evidence for continuity of interstitial spaces across tissue and organ boundaries in humans, *Communications Biology* **4**, 436 (2021).

[2000] M. Buchanan, *Nexus: Small Worlds and the Groundbreaking Science of Networks.* WW Norton (2002).

[2001] M. Novkovic, L. Onder, J. Cupovic, J. Abe, D. Bomze, V. Cremasco, E. Scandella, J. V. Stein, G. Bocharov, S. J. Turley *et al.*, Topological small-world organization of the fibroblastic reticular cell network determines lymph node functionality, *PLoS Biology* **14**, e1002515 (2016).

[2002] M. Novkovic, L. Onder, G. Bocharov and B. Ludewig, Topological structure and robustness of the lymph node conduit system, *Cell Reports* **30**, 893–904 (2020).

[2003] S. Brito, A. Canabarro, D. Cavalcanti and R. Chaves, Satellite-based photonic quantum networks are small-world, *PRX Quantum* **2**, 010304 (2021).

[2004] Z.-W. Wei, B.-H. Wang and X.-P. Han, Renormalization and small-world model of fractal quantum repeater networks, *Scientific Reports* **3**, 1222 (2013).

[2005] C.-P. Chou, Network robustness: Detecting topological quantum phases, *Scientific Reports* **4**, 7526 (2014).

[2006] M. H. Chin, B. Reid, V. Lachina, S. E. Acton and M.-O. Coppens, Bioinspired 3D microprinted cell scaffolds: Integration of graph theory to recapitulate complex network wiring in lymph nodes, *Biotechnology Journal* **19**, 2300359 (2024).

[2007] T. K. Rajab, T. J. O'Malley and V. Tchantchaleishvili, Decellularized scaffolds for tissue engineering: Current status and future perspective, *Artificial Organs* **44**, 1031–1043 (2020).

[2008] M. Lütge, L. Kurz, Y. Stanossek, S. Meili, H.-W. Cheng, A. De Martin, J. Brandstadter, I. Maillard, M. D. Robinson, S. J. Stoeckli *et al.*, Fibroblastic reticular cells form reactive myeloid cell niches in human lymph nodes, *Science Immunology* **10**, 107 (2025).

[2009] A. L. Fletcher, S. E. Acton and K. Knoblich, Lymph node fibroblastic reticular cells in health and disease, *Nature Reviews Immunology* **15**, 350–361 (2015).

[2010] R. Solé, N. Conde-Pueyo, J. Pla-Mauri, J. Garcia-Ojalvo, N. Montserrat and M. Levin, Open problems in synthetic multicellularity, *npj Systems Biology and Applications* **10**, 151 (2024).

[2011] Q. Zhou, Building the future of synthetic organ manufacturing and healthcare, *Cell Proliferation* **56**, 5 (2023).

[2012] D. Song, Y. Xu, S. Liu, L. Wen and X. Wang, Progress of 3D bioprinting in organ manufacturing, *Polymers* **13**, 3178 (2021).
[2013] X. Wang, Bioartificial organ manufacturing technologies, *Cell Transplantation* **28**, 5–17 (2018).
[2014] D. E. Ingber, Human organs-on-chips for disease modelling, drug development and personalized medicine, *Nature Reviews Genetics* **23**, 467–491 (2022).
[2015] M. G. Andrews and A. R. Kriegstein, Challenges of organoid research, *Annual Review of Neuroscience* **45**, 23–39 (2022).
[2016] M. Eisenstein, Artificial organs: Honey, I shrunk the lungs, *Nature* **519**, 7544 (2015).
[2017] X. Fan, K. Hou, G. Liu, R. Shi, W. Wang and G. Liang, Strategies to overcome the limitations of current organoid technology - engineered organoids, *Journal of Tissue Engineering* **16** (2025).
[2018] M. Ratner, A brief history of molecular electronics, *Nature Nanotechnology* **8**, 378–381 (2013).
[2019] E. Lörtscher, Wiring molecules into circuits, *Nature Nanotechnology* **8**, 381–384 (2013).
[2020] A. H. Flood, J. F. Stoddart, D. W. Steuerman and J. R. Heath, Whence molecular electronics? *Science* **306**, 2055–2056 (2004).
[2021] A. Nitzan, Electron transmission through molecules and molecular interfaces, *Annual Review of Physical Chemistry* **52**, 681–750 (2001).
[2022] A. Sarmah, P. Hobza, A. K. Chandra, S. Mitra and T. Nakajima, Many-body effects on electronic transport in molecular junctions: A quantum perspective, *ChemPhysChem* **25**, e202300938 (2024).
[2023] V. N. Uversky, Intrinsic disorder here, there, and everywhere, and nowhere to escape from it, *Cellular and Molecular Life Sciences* **74**, 3065–3067 (2017).
[2024] N. P. Mauranyapin, A. Terrasson and W. P. Bowen, Quantum biotechnology, *Advanced Quantum Technologies* **5**, 2100139 (2022).
[2025] P. W. Jensen, L. B. Kristensen, C. Lavigne and A. Aspuru-Guzik, Toward quantum computing with molecular electronics, *Journal of Chemical Theory and Computation* **18**, 3318–3326 (2022).
[2026] R. Riera Aroche, Y. Ortiz García, M. Martínez Arellano and A. Riera Leal, DNA as a perfect quantum computer based on the quantum physics principles, *Scientific Reports* **14**, 11636 (2024).
[2027] D. Xiang, X. Wang, C. Jia, T. Lee and X. Guo, Molecular-scale electronics: From concept to function, *Chemical Reviews* **116**, 4318–4440 (2016).

[2028] X. Guo, D. Xiang and Y. Li, Integrating molecular functionalities into electrical circuits, in *Molecular-Scale Electronics: Concept, Fabrication and Applications*, chap. 9. Wiley (2020).

[2029] H. Chen and J. Fraser Stoddart, From molecular to supramolecular electronics, *Nature Reviews Materials* **6**, 804–828 (2021).

[2030] M. Levin, Life, death, and self: Fundamental questions of primitive cognition viewed through the lens of body plasticity and synthetic organisms, *Biochemical and Biophysical Research Communications* **564**, 114–133 (2021).

[2031] M. F. Hagan and G. M. Grason, Equilibrium mechanisms of self-limiting assembly, *Reviews of Modern Physics* **93**, 025008 (2021).

[2032] B. R. Parry, I. V. Surovtsev, M. T. Cabeen, C. S. O'Hern, E. R. Dufresne and C. Jacobs-Wagner, The bacterial cytoplasm has glass-like properties and is fluidized by metabolic activity, *Cell* **156**, 183–194 (2014).

[2033] M. Giacomello, A. Pyakurel, C. Glytsou and L. Scorrano, The cell biology of mitochondrial membrane dynamics, *Nature Reviews Molecular Cell Biology* **21**, 204–224 (2020).

[2034] H. M. Cho and W. Sun, Molecular cross talk among the components of the regulatory machinery of mitochondrial structure and quality control, *Experimental & Molecular Medicine* **52**, 730–737 (2020).

[2035] A. V. Nunn, G. W. Guy and J. D. Bell, The quantum mitochondrion and optimal health, *Biochemical Society Transactions* **44**, 1101–1110 (2016).

[2036] A. V. Nunn, G. W. Guy and J. D. Bell, Bioelectric fields at the beginnings of life, *Bioelectricity* **4**, 237–247 (2022).

[2037] Y. Zhang, Q. Zhao, J. Zhang, S. Wei, F. Tao and P. Yang, Bio-inspired adaptive and responsive protein-based materials, *ChemPlusChem* **89**, e202400309 (2024).

[2038] D. Shah, B. Yang, S. Kriegman, M. Levin, J. Bongard and R. Kramer-Bottiglio, Shape changing robots: Bioinspiration, simulation, and physical realization, *Advanced Materials* **33**, 2002882 (2021).

[2039] S. Kriegman, D. Blackiston, M. Levin and J. Bongard, A scalable pipeline for designing reconfigurable organisms, *Proceedings of the National Academy of Sciences USA* **117**, 1853–1859 (2020).

[2040] J. Bongard and M. Levin, There's plenty of room right here: Biological systems as evolved, overloaded, multi-scale machines, *Biomimetics* **8**, 110 (2023).

[2041] M. Levin, The multiscale wisdom of the body: Collective intelligence as a tractable interface for next-generation biomedicine, *BioEssays*, e202400196 (2024).

[2042] J. Davies and M. Levin, Synthetic morphology with agential materials, *Nature Reviews Bioengineering* **1**, 46–59 (2023).

[2043] A. Wagner, A. Upcher, R. Maria, T. Magnesen, E. Zelinger, G. Raposo and B. A. Palmer, Macromolecular sheets direct the morphology and orientation of plate-like biogenic guanine crystals, *Nature Communications* **14**, 589 (2023).

[2044] V. Saranathan, C. O. Osuji, S. G. Mochrie, H. Noh, S. Narayanan, A. Sandy, E. R. Dufresne and R. O. Prum, Structure, function, and self-assembly of single network gyroid ($I4_132$) photonic crystals in butterfly wing scales, *Proceedings of the National Academy of Sciences USA* **107**, 11676–11681 (2010).

[2045] L. Addadi, L. Kronik, L. Leiserowitz, D. Oron and S. Weiner, Organic crystals and optical functions in biology: Knowns and unknowns, *Advanced Materials* **36**, 38 (2024).

[2046] A. Fiorito, D. R. Sheffield, H. Liu, E. Nasirzadeh Orang and N. J. Dawson, Spectral fingerprint of laser emission from rhodamine 6 g infused male indian peafowl tail feathers, *Scientific Reports* **15**, 20938 (2025).

[2047] J.-Q. Chen, Y. Wu and F.-X. Xiao, Single-atom photocatalysis: A new frontier toward solar energy conversion, *Molecular Catalysis* **575**, 114892 (2025).

[2048] B. Xia, Y. Zhang, J. Ran, M. Jaroniec and S.-Z. Qiao, Single-atom photocatalysts for emerging reactions, *ACS Central Science* **7**, 39–54 (2021).

[2049] F. E. Dayan and E. A. Dayan, Porphyrins: One ring in the colors of life: A class of pigment molecules binds King George III, vampires and herbicides, *American Scientist* **99**, 236–243 (2011).

[2050] I. Morita and T. R. Ward, Recent advances in the design and optimization of artificial metalloenzymes, *Current Opinion in Chemical Biology* **81** 102508 (2024).

[2051] I. Rimoldi, G. Coffetti, R. Gandolfi and G. Facchetti, Hybrid metal catalysts as valuable tools in organic synthesis: An overview of the recent advances in asymmetric C-C bond formation reactions, *Molecules* **29**, 5090 (2024).

[2052] A. Das, T. P. Mohammed and M. Sankaralingam, Biological activity of copper porphyrins, *Coordination Chemistry Reviews* **506**, 215661 (2024).

[2053] U. A. Dar, M. Shahnawaz, P. Taneja and M. A. Dar, Recent advances in main group coordination driven porphyrins: A comprehensive review, *ChemistrySelect* **9**, e202304817 (2024).

[2054] V. Nikolaou, E. Nikoloudakis, K. Ladomenou, G. Charalambidis and A. G. Coutsolelos, Porphyrins-valuable pigments of life, *Frontiers in Chemical Biology* **2**, 1346465 (2024).

[2055] P. A. Scattergood, M. Delor, I. V. Sazanovich, O. V. Bouganov, S. A. Tikhomirov, A. S. Stasheuski, A. W. Parker, G. M. Greetham, M. Towrie, E. S. Davies *et al.*, Electron transfer dynamics and excited state branching in a charge-transfer platinum(II) donor-bridge-acceptor assembly, *Dalton Transactions* **43**, 17677–17693 (2014).

[2056] R. H. Holm, P. Kennepohl and E. I. Solomon, Structural and functional aspects of metal sites in biology, *Chemical Reviews* **96**, 2239–2314 (1996).

[2057] J. D. Fischer, G. L. Holliday and J. M. Thornton, The cofactor database: organic cofactors in enzyme catalysis, *Bioinformatics* **26**, 24962497 (2010).

[2058] L. Ma, T. Zhang, M. Li, X. Zhang, L. Li, Y. Shi, R. Cai, X. Yang and C. Duan, Electron transport chain-inspired coordination polymers for macroscopic spatiotemporal scales of charge separation and transport in photocatalysis, *Chemical Science* **15**, 17150–17160 (2024).

[2059] T. K. Baikie, L. T. Wey, J. M. Lawrence, H. Medipally, E. Reisner, M. M. Nowaczyk, R. H. Friend, C. J. Howe, C. Schnedermann, A. Rao *et al.*, Photosynthesis re-wired on the pico-second timescale, *Nature* **615**, 836–840 (2023).

[2060] H. Yong, N. Zotev, J. M. Ruddock, B. Stankus, M. Simmermacher, A. M. Carrascosa, W. Du, N. Goff, Y. Chang, D. Bellshaw *et al.*, Observation of the molecular response to light upon photoexcitation, *Nature Communications* **11**, 2157 (2020).

[2061] E. M. Zhao, N. Suek, M. Z. Wilson, E. Dine, N. L. Pannucci, Z. Gitai, J. L. Avalos and J. E. Toettcher, Light-based control of metabolic flux through assembly of synthetic organelles, *Nature Chemical Biology* **15**, 589–597 (2019).

[2062] J. M. Smith, D. Hartmann and M. J. Booth, Engineering cellular communication between light-activated synthetic cells and bacteria, *Nature Chemical Biology* **19**, 1138–1146 (2023).

[2063] J. Jang, M. Jeong and J. Rho, Quantum nanophotonics, *Nanophotonics* **12**, 335–337 (2023).

[2064] Y. Cho, S. H. Park, J.-H. Huh, A. Gopinath and S. Lee, Dna as grabbers and steerers of quantum emitters, *Nanophotonics* (2022).

[2065] Y. Kim, A. Barulin, S. Kim, L. P. Lee and I. Kim, Recent advances in quantum nanophotonics: Plexcitonic and vibro-polaritonic strong coupling and its biomedical and chemical applications, *Nanophotonics* **12**, 413–439 (2022).

[2066] D. Zoepfl, M. Juan, N. Diaz-Naufal, C. Schneider, L. Deeg, A. Sharafiev, A. Metelmann and G. Kirchmair, Kerr enhanced backaction cooling in magnetomechanics, *Physical Review Letters* **130**, 033601 (2023).

[2067] P. Robin and L. Bocquet, Nanofluidics at the crossroads, *The Journal of Chemical Physics* **158**, 160901 (2023).

[2068] S. N. Gupta, O. Bitton, T. Neuman, R. Esteban, L. Chuntonov, J. Aizpurua and G. Haran, Complex plasmon-exciton dynamics revealed through quantum dot light emission in a nanocavity, *Nature Communications* **12**, 1310 (2021).

[2069] A. Kurt, M. A. Rossi and J. Piilo, Efficient quantum transport in a multi-site system combining classical noise and quantum baths, *New Journal of Physics* **22**, 013028 (2020).

[2070] F. Li, Y. Liu, Y. Dong, Y. Chu, N. Song and D. Yang, Dynamic assembly of DNA nanostructures in living cells for mitochondrial interference, *Journal of the American Chemical Society* **144**, 4667–4677 (2022).

[2071] Y. Wang, Z. Li, F. Mo, T.-J. Chen-Mayfield, A. Saini, A. M. LaMere and Q. Hu, Chemically engineering cells for precision medicine, *Chemical Society Reviews* **52**, 1068–1102 (2023).

[2072] T.-C. Tang, B. An, Y. Huang, S. Vasikaran, Y. Wang, X. Jiang, T. K. Lu and C. Zhong, Materials design by synthetic biology, *Nature Reviews Materials* **6**, 332–350 (2021).

[2073] L. E. Contreras-Llano, C. Meyer, Y. Liu, M. Sarker, S. Lim, M. L. Longo and C. Tan, Holistic engineering of cell-free systems through proteomereprogramming synthetic circuits, *Nature Communications* **11**, 3138 (2020).

[2074] K. Jahnke, N. Ritzmann, J. Fichtler, A. Nitschke, Y. Dreher, T. Abele, G. Hofhaus, I. Platzman, R. R. Schröder, D. J. Müller *et al.*, Proton gradients from light-harvesting E. coli control DNA assemblies for synthetic cells, *Nature Communications* **12**, 3967 (2021).

[2075] Y. Liu, J. Feng, H. Pan, X. Zhang and Y. Zhang, Genetically engineered bacterium: Principles, practices, and prospects, *Frontiers in Microbiology* **13**, 997587 (2022).

[2076] S. Rasmussen, L. Chen, D. Deamer, D. C. Krakauer, N. H. Packard, P. F. Stadler and M. A. Bedau, Transitions from nonliving to living matter, *Science* **303**, 963–965 (2004).

[2077] K. L. Garner, Principles of synthetic biology, *Essays in Biochemistry* **65**, 791–811 (2021).

[2078] D. Chowdhury, Stochastic mechano-chemical kinetics of molecular motors: A multidisciplinary enterprise from a physicist's perspective, *Physics Reports* **529**, 1–197 (2013).

[2079] W.-A. Wang, M. G.-J. Navarro and Z. Gueroui, Self-assembly of magnetically-functionalized molecular motors and microtubules into active gels, *Soft Matter* **15**, 9111–9119 (2019).

[2080] J. J. Keya, A. M. R. Kabir and A. Kakugo, Synchronous operation of biomolecular engines, *Biophysical Reviews* **12**, 401–409 (2020).

[2081] A. M. R. Kabir, D. Inoue and A. Kakugo, Molecular swarm robots: Recent progress and future challenges, *Science and Technology of Advanced Materials* **21**, 323–332 (2020).

[2082] H. Hess and J. L. Ross, Non-equilibrium assembly of microtubules: From molecules to autonomous chemical robots, *Chemical Society Reviews* **46**, 5570–5587 (2017).

[2083] B. Brouwer, F. Della-Felice, J. H. Illies, E. Iglesias-Moncayo, G. Roelfes and I. Drienovská, Noncanonical amino acids: Bringing new-to-nature functionalities to biocatalysis, *Chemical Reviews* **124**, 1087710923 (2024).

[2084] O. Adir, M. R. Albalak, R. Abel, L. E. Weiss, G. Chen, A. Gruber, O. Staufer, Y. Kurman, I. Kaminer, J. Shklover *et al.*, Synthetic cells with self-activating optogenetic proteins communicate with natural cells, *Nature Communications* **13**, 2328 (2022).

[2085] A. Zambrano, G. Fracasso, M. Gao, M. Ugrinic, D. Wang, D. Appelhans, A. deMello and T. D. Tang, Programmable synthetic cell networks regulated by tuneable reaction rates, *Nature Communications* **13**, 3885 (2022).

[2086] D. Gutiérrez-Armayor, S. Ferrara, M. Nieddu, A. Zieleniewska and R. D. Costa, Fluorescent protein pegylation for stable photon manipulation in deep-red light-emitting devices, *Advanced Functional Materials* **35**, 2412921 (2025).

[2087] R. Solé, C. P. Kempes, B. Corominas-Murtra, M. De Domenico, A. Kolchinsky, M. Lachmann, E. Libby, S. Saavedra, E. Smith and D. Wolpert, Fundamental constraints to the logic of living systems, *Interface Focus* **14**, 20240010 (2024).

[2088] J. Li, X. Zhao and J. Liu, Biosimilar liquid-metal living matter, *Matter* **7**, 2033–2065 (2024).

[2089] D. G. Luchinsky and P. V. McClintock, Introduction to the physics of ionic conduction in narrow biological and artificial channels, *Entropy* **23**, 644 (2021).

[2090] K. Li, H. Chen, D. Li, C. Yang, H. Zhang and Z. Zhu, Empowering DNA-based information processing: Computation and data storage, *ACS Applied Materials & Interfaces* **16**, 68749–68771 (2024).

[2091] K. E. Dunn, M. A. Trefzer, S. Johnson and A. M. Tyrrell, Towards a bioelectronic computer: A theoretical study of a multi-layer biomolecular computing system that can process electronic inputs, *International Journal of Molecular Sciences* **19**, 2620 (2018).

[2092] Z. Ezziane, Dna computing: Applications and challenges, *Nanotechnology* **17**, R27 (2005).
[2093] S. Tagore, S. Bhattacharya, M. Islam and M. L. Islam, Dna computation: application and perspectives, *J. Proteomics Bioinform* **3**, 07 (2010).
[2094] L. M. Adleman, Molecular computation of solutions to combinatorial problems, *Science* **266**, 1021–1024 (1994).
[2095] L. M. Adleman, Computing with DNA, *Scientific American* **279**, 54–61 (1998).
[2096] S. Yang, B. W. Bögels, F. Wang, C. Xu, H. Dou, S. Mann, C. Fan and T. F. de Greef, DNA as a universal chemical substrate for computing and data storage, *Nature Reviews Chemistry* **8**, 179–194 (2024).
[2097] L. P. Lee and R. Szema, Inspirations from biological optics for advanced photonic systems, *Science* **310**, 1148–1150 (2005).
[2098] K.-H. Jeong, J. Kim and L. P. Lee, Biologically inspired artificial compound eyes, *Science* **312**, 557–561 (2006).
[2099] K. Liu and L. Jiang, Bio-inspired design of multiscale structures for function integration, *Nano Today* **6**, 155–175 (2011).
[2100] L. Shang, W. Zhang, K. Xu and Y. Zhao, Bio-inspired intelligent structural color materials, *Materials Horizons* **6**, 945–958 (2019).
[2101] M. Li, Q. Lyu, B. Peng, X. Chen, L. Zhang and J. Zhu, Bioinspired colloidal photonic composites: Fabrications and emerging applications, *Advanced Materials* **34**, 2110488 (2022).
[2102] Y. Yao, Y. Zheng, S. Cui, Y. Hou, Q. Zhang and X. Wei, DNA sequence design model for multi-scene fusion, *Neural Computing and Applications* **37**, 5499–5520 (2025).
[2103] C. Zhang, Y. Zhao, X. Xu, R. Xu, H. Li, X. Teng, Y. Du, Y. Miao, H.c. Lin and D. Han, Cancer diagnosis with DNA molecular computation, *Nature Nanotechnology* **15**, 709–715 (2020).
[2104] M. Cao, X. Xiong, L. Li and H. Pei, Fast DNA-based molecular classifier for cancer diagnosis using freeze-thaw cycling, *Chinese Journal of Chemistry* **43**, 1035–1041 (2025).
[2105] D. Bar-Lev, O. Sabary and E. Yaakobi, The zettabyte era is in our DNA, *Nature Computational Science* **4**, 813–817 (2024).
[2106] X. Huang, Z. Hou, W. Qiang, H. Wang, X. Wang, X. Chen, X. Hu, J. Dai, L. Li and G. Zhao, Towards next-generation DNA encryption via an expanded genetic system, *National Science Review* **12**, nwae469 (2025).
[2107] Y. Zhou, K. Bi, Q. Xu, Q. Liu, X. Zhao, Q. Ge and Z. Lu, Ultrafast and accurate DNA storage and reading integrated system via microfluidic magnetic beads polymerase chain reaction, *ACS Nano* **19**, 7306–7316 (2025).

[2108] K. Bharti, A. Cervera-Lierta, T. H. Kyaw, T. Haug, S. Alperin-Lea, A. Anand, M. Degroote, H. Heimonen, J. S. Kottmann, T. Menke, W.K. Mok, S. Sim, L.-C. Kwek and A. Aspuru-Guzik, Noisy intermediatescale quantum algorithms, *Reviews of Modern Physics* **94**, 015004 (2022).

[2109] G. Ding, H. Li, J. Zhao, K. Zhou, Y. Zhai, Z. Lv, M. Zhang, Y. Yan, S.-T. Han and Y. Zhou, Nanomaterials for flexible neuromorphics, *Chemical Reviews* **124**, 12738–12843 (2024).

[2110] C. Yan, C. Fang, J. Gan, J. Wang, X. Zhao, X. Wang, J. Li, Y. Zhang, H. Liu, X. Li *et al.*, From molecular electronics to molecular intelligence, *ACS Nano* **18**, 28531–28556 (2024).

[2111] F. Verstraete, M. M. Wolf and J. Ignacio Cirac, Quantum computation and quantum-state engineering driven by dissipation, *Nature Physics* **5**, 633–636 (2009).

[2112] X. Mi *et al.*, Stable quantum-correlated many-body states through engineered dissipation, *Science* **383**, 1332–1337 (2024).

[2113] A. Currin, K. Korovin, M. Ababi, K. Roper, D. B. Kell, P. J. Day and R. D. King, Computing exponentially faster: implementing a nondeterministic universal turing machine using DNA, *Journal of The Royal Society Interface* **14**, 20160990 (2017).

[2114] D. Valente, Self-replication of a quantum artificial organism driven by single-photon pulses, *Scientific Reports* **11**, 16433 (2021).

[2115] E. Lagasse and M. Levin, Future medicine: from molecular pathways to the collective intelligence of the body, *Trends in Molecular Medicine* **29**, 687–710 (2023).

[2116] N. Kumar and A. Dixit, Role of nanotechnology in futuristic warfare, in *Nanotechnology for Defence Applications*, 301–329 (2019).

www.ingramcontent.com/pod-product-compliance
Lightning Source LLC
LaVergne TN
LVHW051053210926
848486LV00002B/100

* 9 7 8 9 8 1 9 8 3 1 4 1 8 *